Cell Death in Biology and Diseases

Series Editors

Xiao-Ming Yin
Zheng Dong

More information about this series at http://www.springer.com/series/8908

Wen-Xing Ding • Xiao-Ming Yin
Editors

Cellular Injury in Liver Diseases

Springer

Editors
Wen-Xing Ding
Department of Pharmacology, Toxicology and Therapeutics
The University of Kansas Medical Center
Kansas City, KS, USA

Xiao-Ming Yin
Department of Pathology and Laboratory Medicine
Indiana University School of Medicine
Indianapolis, IN, USA

Cell Death in Biology and Diseases
ISBN 978-3-319-53773-3 ISBN 978-3-319-53774-0 (eBook)
DOI 10.1007/978-3-319-53774-0

Library of Congress Control Number: 2017936480

Printed on acid-free paper

This Springer imprint is published by Springer Nature
The registered company is Springer International Publishing AG
The registered company address is: Gewerbestrasse 11, 6330 Cham, Switzerland

Series Preface

Cell death, or conversely cell survival, is a major biological phenomenon. Just as with cell proliferation and cell differentiation, cell death is a choice that a cell has to make, sometimes voluntarily, other times accidentally. As such, cell death serves a purpose in the biology of a multicellular organism. The machinery of cell death and that of cell protection are evolutionarily conserved and their elements can even be found in single-celled organism. The disruption of cell death mechanisms can often cause developmental abnormalities. Factors that can trigger cell death are diverse and the cell death process is intricately connected with other biological processes. Cell death directly contributes to the pathogenesis of many diseases, including cancer, neurodegenerative diseases, and tissue injury in organ failure.

The study of cell death and cell survival has become a multidisciplinary subject, which requires expertise from all fields of the modern biology. Exploring the role of cell death in disease development and the modulation of cell death for the prevention and treatment of devastating disease demands constant updating of our knowledge through the broadest interactions among all investigators, basic and clinical. The rapid expansion of our knowledge in this field has gone beyond what could be summarized in a single book. Thus, this timely series *Cell Death in Biology and Diseases* summarizes new developments in different areas of cell death research in an elaborate and systemic way. Each volume of this series addresses a particular topic of cell death that either has a broad impact on the field or has an in-depth development in a unique direction. As a whole, this series provides a current and encyclopedic view of cell death.

We would like to sincerely thank the editors of each volume in the series and the authors of each chapter in these volumes for their strong commitment and great effort towards making this mission possible. We are also grateful to our team of professional Springer editors. They have worked with us diligently and creatively from the initiation and are continuing this on the development and production of each volume of the series. Finally we hope the readers will enjoy the reading, find the content helpful to their work, and consider this series an invaluable resource.

Indianapolis, IN, USA — Xiao-Ming Yin
Augusta, GA, USA — Zheng Dong

Preface

Cell death is a fundamental biological phenomenon. It is evolutionarily conserved but can assume different forms under different conditions. While apoptosis, necrosis, and necroptosis are perhaps best studied, less known forms of cell death such as pyroptosis, ferroptosis, parthanatos, and entosis are increasingly found in various conditions or organisms. In multicellular organisms, cell death is important for development, shaping organ size, reforming tissue architecture, promoting functional differentiation, and determining mitochondrial inheritance. In the post-development stage, cell death determines the severity in tissue injury and the degree of subsequent response in inflammation, fibrosis, repair, regeneration, and tumorigenesis. The pathological changes in a complex organ, such as the liver, can be greatly affected by the death program in its cellular components.

As the major organ for metabolism and detoxification, the liver is constantly under challenges from both internal and external sources. Viral infection and xenobiotics are the two major environmental stimuli that can cause significant hepatocyte death. Metabolic disturbance (such as in autophagy function) and special food components (such as lipids, ethanol, cholesterol, sugars) are the major internal stress that can also take a significant token on the hepatocytes that leads to cell death. Furthermore, the liver may experience traumatic injury as in cholestasis or ischemia, which often leads to tissue injury to various degrees. The liver is composed of hepatocytes (the majority), cholangiocytes, stellate cells, fibroblasts, macrophages/Kupffer cells, sinusoidal cells, and many more. They respond to these insults with different sensitivities and contribute to the overall liver pathology in various ways. Thus different signaling pathways may be involved in a cell type-specific and/or in a stimulus-specific way, which triggers subcellular organelle (mitochondria and endoplasmic reticulum as well as lysosomes) stress to induce cell death. Significant progresses have been made in the past decades to characterize these cell death pathways and the cells that are involved in liver injury.

We are thus preparing two books devoted to the mechanisms of cell death in the liver and liver diseases. While the first volume mainly discusses liver injury and cell death caused by the various external and internal stimuli, the second volume discusses in great detail the type of cells and intracellular organelles involved in cell death as well as several major signaling pathways involved. We have paid particular attention to the interaction of different cells and aimed to provide a global view of the liver pathology that connects injury/cell death to other pathological changes, such as inflammation, fibrosis, and tumor development. Notably we also discuss the interaction between the liver and the intestine where microbiota can both influence and be influenced by the liver pathology. Authors who are invited to contribute to the two volumes are respected experts in the subject, which should greatly enhance the authority of the chapter and the book. The volume editors are thus indebted to the authors for their outstanding contributions that make these two books so informative and thought-provoking.

Kansas City, KS, USA Wen-Xing Ding
Indianapolis, IN, USA Xiao-Ming Yin

Contents

Contributors

Xiaojuan Chao Department of Pharmacology, Toxicology and Therapeutics, University of Kansas Medical Center, Kansas City, KS, USA

John Y.L. Chiang Department of Integrative Medical Sciences, Northeast Ohio Medical University, Rootstown, OH, USA

Lily Dara Division of GI/Liver, Department of Medicine, Research Center for Liver Disease, Keck School of Medicine, University of Southern California, Los Angeles, CA, USA

Wen-Xing Ding Department of Pharmacology, Toxicology and Therapeutics, University of Kansas Medical Center, Kansas City, KS, USA

Jia Fan Department of Liver Surgery, Liver Cancer Institute, Zhongshan Hospital, Fudan University, Shanghai, China

Key Laboratory of Carcinogenesis and Cancer Invasion of Ministry of Education, Shanghai, China

Gregory J. Gores Division of Gastroenterology and Hepatology, Mayo Clinic, Rochester, MN, USA

Julie Goswami Department of Surgery, University of Pittsburgh Medical Center, Pittsburgh, PA, USA

Michael Hinton Department of Microbiology and Immunology, Medical College of Virginia Campus, Virginia Commonwealth University, Richmond, VA, USA

McGuire VA Medical Center, Richmond, VA, USA

Petra Hirsova Division of Gastroenterology and Hepatology, Mayo Clinic, Rochester, MN, USA

I-Fang Hsin Division of Gastroenterology, Department of Medicine, Cedars-Sinai Medical Center, Los Angeles, CA, USA

Hai Huang Department of Surgery, University of Pittsburgh Medical Center, Pittsburgh, PA, USA

Nazmul Huda Department of Pathology and Laboratory Medicine, Indiana University School of Medicine, Indianapolis, IN, USA

Samar H. Ibrahim Division of Gastroenterology and Hepatology, Mayo Clinic, Rochester, MN, USA

Hartmut Jaeschke Department of Pharmacology, Toxicology and Therapeutics, University of Kansas Medical Center, Kansas City, Kansas, USA

Neil Kaplowitz Division of GI/Liver, Department of Medicine, Research Center for Liver Disease, Keck School of Medicine, University of Southern California, Los Angeles, CA, USA

Bilon Khambu Department of Pathology and Laboratory Medicine, Indiana University School of Medicine, Indianapolis, IN, USA

Eric Kwong Department of Microbiology and Immunology, Medical College of Virginia Campus, Virginia Commonwealth University, Richmond, VA, USA

McGuire VA Medical Center, Richmond, VA, USA

Jiyoung Lee Department of Molecular Microbiology and Immunology, University of Southern California, Keck School of Medicine, Los Angeles, CA, USA

Yunzhou Li Department of Microbiology and Immunology, Medical College of Virginia Campus, Virginia Commonwealth University, Richmond, VA, USA

McGuire VA Medical Center, Richmond, VA, USA

Tiangang Li Department of Pharmacology, Toxicology and Therapeutics, Kansas University Medical Center, Kansas City, KS, USA

Harmeet Malhi Division of Gastroenterology and Hepatology, Mayo Clinic, Rochester, MN, USA

Jing-hsiung James Ou Department of Molecular Microbiology and Immunology, Keck School of Medicine, University of Southern California, Los Angeles, CA, USA

Yuan-Fei Peng Department of Liver Surgery, Liver Cancer Institute, Zhongshan Hospital, Fudan University, Shanghai, China

Key Laboratory of Carcinogenesis and Cancer Invasion of Ministry of Education, Shanghai, China

Ekihiro Seki Division of Gastroenterology, Department of Medicine, Cedars-Sinai Medical Center, Los Angeles, CA, USA

Ying-Hong Shi Department of Liver Surgery, Liver Cancer Institute, Zhongshan Hospital, Fudan University, Shanghai, China

Key Laboratory of Carcinogenesis and Cancer Invasion of Ministry of Education, Shanghai, China

Meng-Xin Tian Department of Liver Surgery, Liver Cancer Institute, Zhongshan Hospital, Fudan University, Shanghai, China

Key Laboratory of Carcinogenesis and Cancer Invasion of Ministry of Education, Shanghai, China

Allan Tsung Department of Surgery, University of Pittsburgh Medical Center, Pittsburgh, PA, USA

Division of Hepatobiliary and Pancreatic Surgery, University of Pittsburgh, Pittsburgh, PA, USA

Han Wang Department of Liver Surgery, Liver Cancer Institute, Zhongshan Hospital, Fudan University, Shanghai, China

Key Laboratory of Carcinogenesis and Cancer Invasion of Ministry of Education, Shanghai, China

Shaogui Wang Department of Pharmacology, Toxicology and Therapeutics, University of Kansas Medical Center, Kansas City, KS, USA

Heather Waring Department of Surgery, University of Pittsburgh Medical Center, Pittsburgh, PA, USA

Benjamin L. Woolbright Department of Pharmacology, Toxicology and Therapeutics, University of Kansas Medical CenterKansas City, Kansas, USA

Shengmin Yan Department of Pathology and Laboratory Medicine, Indiana University School of Medicine, Indianapolis, IN, USA

Xiao-Ming Yin Department of Pathology and Laboratory Medicine, Indiana University School of Medicine, Indianapolis, IN, USA

Huiping Zhou Department of Microbiology and Immunology, Medical College of Virginia Campus, Virginia Commonwealth University, Richmond, VA, USA

McGuire VA Medical Center, Richmond, VA, USA

Jun Zhou Department of Pathology and Laboratory Medicine, Indiana University School of Medicine, Indianapolis, IN, USA

Center of Minimally Invasive Surgery, Xiangya Second Hospital, Central South University, Changsha, Hunan, China

Chapter 1
Cell Death in Drug-Induced Liver Injury

Lily Dara and Neil Kaplowitz

Abbreviations

ALF	Acute liver failure
APAP	Acetaminophen
APAF1	Apoptotic peptidase-activating factor-1
ASK1	Apoptosis signal-regulating kinase 1
ASMase	Acid sphingomyelinase
ATF6	Activating transcription factor 6
ATG	Autophagy-related genes
Cbl-b1	Casitas B lineage lymphoma-B1
CCL4	Carbon tetrachloride
CHOP	CCAAT-enhancer-binding protein homologous protein
cIAP	Cellular inhibitor of apoptosis 1 and 2
Cr	Chromium
CTLA4	Cytotoxic T-lymphocyte-associated protein 4
DILI	Drug-induced liver injury
DR	Death receptor
ER Stress	Endoplasmic reticulum stress
ETC	Electron transport chain
FADD	Fas-associated protein with death domain
GRP78	Glucose regulatory peptide 78
GSH	Glutathione
GSK3b	Glycogen synthase kinase 3 beta
GWAS	Genome-wide association study

L. Dara • N. Kaplowitz (✉)
Division of GI/Liver, Department of Medicine, Research Center for Liver Disease,
Keck School of Medicine, University of Southern California,
2011 Zonal Ave, HMR 101, Los Angeles, CA 90033, USA
e-mail: kaplowit@usc.edu

W.-X. Ding, X.-M. Yin (eds.), *Cellular Injury in Liver Diseases*, Cell Death in Biology and Diseases, DOI 10.1007/978-3-319-53774-0_1

X.-M. Yin and Z. Dong (Series eds.), *Cell Death in Biology and Diseases*

INH	Isonizaide
JNK	c-Jun N terminal kinase
K18	Keratin 18
KCC	King's college criteria
KO	Knockout
HLA	Human leukocyte antigen
HMGB1	High-mobility group box 1
IRE1α	Inositol-requiring enzyme 1α
IDILI	Idiosyncratic drug-induced liver injury
LPS	Lipopolysaccharides
LSEC	Liver sinusoidal endothelial cells
MAPK	Mitogen-activated protein kinase
MDSC	Myeloid-derived suppressor cells
MEF	Mouse embryonic fibroblasts
MELD	Model for end-stage liver disease
MHC	Major histocompatibility complex
miR	Micro RNA
MDB	Mallory-Denk bodies
MKK4	Mitogen-activated protein kinase kinase 4
MLKL	Mixed lineage kinase domain-like
MOMP	Mitochondrial outer membrane permeabilization
MPT	Mitochondrial permeability transition pore
mTOR	Mammalian target of rapamycin
NAFLD	Non-alcoholic fatty liver disease
NASH	Non-alcoholic steatohepatitis
NFκB	Nuclear factor κB
NK	Natural killer
PD-1	Programmed cell death protein-1
PERK	Protein kinase RNA-like ER kinase
PKC	Protein kinase C
POLG	Polymerase γ
PTPN6	Phospho-tyrosine phosphatase, non-receptor type 6
RIPK1	Receptor interacting protein kinase 1
RIPK3	Receptor interacting protein kinase 3
ROS	Reactive oxygen species
Sab	SH3BP5, SH3 domain-binding protein that preferentially associates with Bruton's tyrosine kinase
SMAC	Second mitochondria-derived activator of caspases
TAB2/3	TAK1-binding protein 2/3
TAK1	Transforming growth factor-β-activated kinase 1
TRAF2	TNF receptor-associated factor 2
TRAIL	TNF-related apoptosis-inducing ligand
TNF	Tumor necrosis factor
TNFR	Tumor necrosis factor receptor
TRADD	TNFR1-associated via death domain

UPR	Unfolded protein response
VPA	Valproic acid
WT	Wild type
XIAP	X-chromosome-linked inhibitor of apoptosis protein

1.1 Introduction

Drug-induced liver injury (DILI) presents as a spectrum of clinical presentations such as acute hepatocellular liver injury resulting in acute hepatitis, cholestasis and jaundice, nodular regenerative hyperplasia, sinusoidal obstruction syndrome, or subclinical injury which is detected during routine serum chemistries (Yuan and Kaplowitz 2013). Acute hepatocellular DILI can lead to acute liver failure (ALF), the most feared complication that triggers regulatory responses ranging from warnings, withdrawal, disapproval to recommendation for close monitoring. DILI can present sub-acutely with elevated liver enzymes or gradual jaundice or even manifest as a chronic progressive disease, the latter being less defined and somewhat controversial (Stine and Chalasani 2015). In all these instances, the ongoing liver damage from the insulting drug often leads to inflammation and direct and indirect liver cell death. Although it is unusual for DILI to persist and become chronic, in certain contexts such as in diabetics on long-term methotrexate, low grade injury can become chronic and tissue repair responses can be activated resulting in fibrosis and ultimately cirrhosis. In addition, some drugs can increase weight gain and insulin resistance aggravating underlying NAFLD and thus indirectly promoting NASH. DILI can be further classified into dose-dependent predictable (e.g., acetaminophen) and idiosyncratic injury. In either case, the major concern is hepatocellular death with the fear of massive parenchymal extinction leading to ALF.

Cell death due to DILI and hepatotoxicity can be apoptotic or necrotic depending on the inciting drug and which signaling pathways are activated. Apoptosis, the predominant cell death routine in parenchymal liver cells, hepatocytes, and cholangiocytes, can be activated through intrinsic and extrinsic pathways. Intracellular stress responses, such as mitochondrial toxicity or ER stress from drugs or toxic metabolites, can activate the intrinsic pathway of apoptosis, whereas the extrinsic pathway of apoptosis is activated by the engagement of surface death receptors with their respective ligands, usually via the immune system. Necrosis of hepatocytes to a large part also involves activation of cellular signaling pathways. While this is the subject of much of the current research in the field, the exact mechanistic pathways are not as clearly elucidated as in the case of apoptosis. It is important to point out that hepatotoxicity can occur due to toxic effects of drugs on other cells in the liver such as the liver sinusoidal endothelial cells (LSECs) or changiocytes resulting in their death. In this chapter, we focus on cell death in hepatocytes, the main parenchymal liver cells.

A few drugs such as acetaminophen (APAP), aspirin, and chemotherapeutic agents can cause predictable, dose-dependent, and direct hepatocyte-intrinsic hepatotoxicity.

However, most instances of liver injury from drugs are idiosyncratic. Idiosyncratic Drug-Induced Liver Injury (IDILI) is by definition an unpredictable injury due to a drug, usually occurring after a relatively long latency and only in a small proportion of exposed individuals. The pathogenesis if IDILI is multifactorial and complex but in most instances involves the adaptive immune system as multiple genome-wide association studies (GWAS) have identified strong correlations between certain HLA polymorphisms and IDILI occurrence. A seminal example is the GWAS study by Daly et al. in which patients with flucloxacillin hepatotoxicity were more likely to harbor the HLA-B*57:01 allele compared to exposed individuals with no DILI (Hazard Ratio of 100) (Daly et al. 2009). Additionally, the peripheral blood mononuclear cells of patients with flucloxacillin DILI showed immune activation on re-exposure to flucloxacillin. Flucloxacillin was also able to activate CD8+ T cells in drug naive individuals with the HLA-B*57:01 allele indicating a strong association of the allele with the resultant injurious phenotype (Monshi et al. 2013; Urban et al. 2014; Wuillemin et al. 2013). Other examples of strong associations between HLA polymorphisms and IDILI are the increased risk of amoxicillin-clavulanate toxicity (the single most common form of IDILI) in patients with DRB1*15:01, DQB1*06:02, and HLA-A*02:01 haplotype and ximelagatran in patients with the HLA-DRB1*07:01 polymorphism, among others (Table 1.1) (Kindmark et al. 2008; Mallal et al. 2008; Hautekeete et al. 1999; Lucena et al. 2011; O'Donohue et al. 2000; Chen et al. 2015; Sharma et al. 2002; Daly et al. 2009; Hirata et al. 2008; Kurosaki et al. 2000; Nicoletti et al. 2016; Phillips et al. 2013; Singer et al. 2010; Spraggs et al. 2011; Xu et al. 2015; Monshi et al. 2013; Urban et al. 2014; Wuillemin et al. 2013). Although little mechanistic evidence exists, it is currently hypothesized that once a drug is converted into a "reactive metabolite" or "hapten" it can activate the adaptive immune system through neoantigen formation or molecular mimicry in those individuals with the susceptible HLA haplotype. In addition to the hapten hypothesis two other hypotheses exist for why HLA polymorphisms result in aberrant immune activation and IDILI. These include the pharmacological interaction hypothesis (p-i hypothesis) and the altered peptide repertoire hypothesis. Interactions between antigen-binding grooves and proteins (covalent and noncovalent) are considered to be important in the activation of the adaptive immune system. The "pharmacological interaction hypothesis" postulates that drugs can directly contact MHC molecules on T cells via noncovalent interactions resulting in T cell activation without the presence of an antigenic peptide (Pichler 2002). An example of the possible clinical relevance of the pi hypothesis is in ximelegatran DILI since the drug does not covalently bind proteins to form neoantigens. However, in vitro studies have shown a direct inhibition by ximelegatran of peptide binding to HLA DRB1*0701, supporting direct interaction of the drug to this HLA-binding site, thus pointing to the p-i hypothesis as a possible mechanism for immune activation and DILI (Grove and Aithal 2015). Another potential mechanism is drug-induced mistargeting of endogenous peptides to the wrong HLA leading to autoimmunity, as described for abacavir rash and Stevens Johnson syndrome from carbamazepine (Ostrov et al. 2012; Wei et al. 2012). In experimental IDILI models it has been suggested that subsequent to these immune activation pathways, either CD4+-dependent antibody-mediated cytotoxicity (Chakraborty et al. 2015), or CD8+ T-cells-mediated cytotoxicity (Mak and Uetrecht 2015) results in targeting of hepato-

Table 1.1 DILI drugs with known HLA associations

Drug	Associated HLA haplotypes	Phenotype	Reference
Abacavir	B* 57:01	Hepatocellular and skin rash	Mallal et al. 2008
Amoxicillin-clavulanate	A*02:01, B*18:01, DRB1*1501, DGB1 *0602, DRB1*07	Spectrum	Kindmark et al. 2008, Hautekeete et al. 1999, Lucena et al. 2011
Anti-tuberculous drugs	DQB1*02:01, DQB1 *05:02, DQA1*01:02‡	Hepatocellular and mixed	Chen et al. 2015, Sharma et al. 2002
Fenofibrate	A*33:01	Cholestatic	(Aithal, Liver meeting 2015 Abstract 225)
Flucloxacillin	B* 57:01, DRB1*07:01, DQB1*03:03, DRB1*15‡	Cholestatic	Daly et al. 2009
Flupirtine	DRB1*16:01-DQB1*05:02	Hepatocellular	Nicoletti et al. 2016
Lapatinib	DQA1*02:01, DQB1*02:02, DRB1*07:01	Hepatocellular and mixed (delayed)	Spraggs et al. 2011
Lumiracoxib	DRB1*15:01, DQB1*06:02, DRB5*01:01, DQA1*01:02	Hepatocellular (delayed)	Singer et al. 2010
Minocycline	B*35:02	Hepatocellular	(Urban, Liver meeting 2015 Abstract 1930)
Nevirapine	DRB*01:02	Hepatocellular	Phillips et al. 2013
Pazopanib	B*57:01	Hepatocellular	Xu et al. 2015
Ticlopidine	A*33:03, A*33:01	Cholestatic	Hirata et al. 2008
Tiopronin	A*33	Cholestatic	Kurosaki et al. 2000
Terbinafine	A*33:01	Mixed	(Aithal, Liver meeting 2015 Abstract 225)
Ximelagatran	DRB1*0701, DQA1*0201	Hepatocellular	Kindmark et al. 2008

‡Haplotype decreases risk of toxicity

cytes and death via DR-initiated apoptosis (the de facto cell death pathway in hepatocytes), which will be discussed in detail below. However, the susceptible HLA polymorphism resulting in immune activation does not fully explain hepatotoxicity on its own as most identified HLA haplotypes associated with toxicity are quite common in the general population and IDILI is a rare event. Thus, not all individuals with the susceptible HLA exhibit clinical hepatotoxicity when exposed to the drug. This is likely due to the liver's extraordinary capacity for immune tolerance, a physiologic constant dampening of immune response in the liver which leads to "adaptation." When immune tolerance is inhibited in the liver, for example, by genetic knockout of immune check point inhibitors (such as Cbl-b1−/− mice or PD1−/− mice) or when CTLA4 inhibitors are used, drugs that normally would not result in liver injury can cause T cell activation and DILI (Chakraborty et al. 2015; Metushi et al. 2015; Mak and Uetrecht 2015). Two groups have recently explored this phenomenon using mice deficient in one or more check point inhibitor components. Chakaborty et al. injected two doses of halothane (a drug known to cause allergic DILI), 14 days apart, to mice depleted of myeloid-derived suppressor cells (MDSC) using anti-Gr1 antibody, and reported more severe injury in the MDSC-deficient mice that lack the ability for adaptation (Chakraborty et al. 2015). The anti-Gr-1-treated mice displayed severe inflammation, necrosis, and eosinophilia (Chakraborty et al. 2015). Treatment with anti-CD4 antibody was able to abrogate this allergic response (Chakraborty et al. 2015).

Metushi et al. examined the toxicity of amodiaquine in Cbl-b1 and PD-1-deficient mice (both molecules are immune check point inhibitors). Mice knockout for either gene exhibited more severe DILI than wild-type (WT) mice; however, this resolved due to adaptation and the mice recovered (Metushi et al. 2015). Blocking an additional immune check point pathway using CTLA4 antibody in PD1−/− mice resulted in more severe and sustained injury from amodiaquine (Metushi et al. 2015). In contrast to the halothane immunoallergic model of IDILI, in this model, the resultant liver injury was dependent on the action of cytotoxic CD8+ T lymphocytes (Mak and Uetrecht 2015). These studies suggest that under normal conditions check point immune modulators have a pro-adaptation dampening effect on the activation on T cells in the liver, preventing constant inflammatory response and IDILI (Dara et al. 2015b). Thus, a failure of adaptation is likely an additional or simultaneous requirement for the development of IDILI but the details of this interesting phenomena are beyond the scope of this chapter and is discussed elsewhere (Dara et al. 2015b) (Fig. 1.1).

Morphologically and mechanistically cell death is broadly classified into apoptosis, necrosis, and autophagy (which has close ties to apoptosis). Necroptosis, a form of regulated necrosis, is also an important cell death subroutine that has been studied in the context of DILI. Below we will examine the current evidence and the role of each of these cell death modes in drug hepatotoxicity. It should be pointed out that when extensive injury occurs in humans, it may be very difficult to distinguish which form of cell death predominates. This is less of a problem in animal models where mechanistic biochemical studies can be deployed. Nevertheless, animal models of IDILI are only beginning to be identified (by modulating immune tolerance) but are thus far not very robust, probably because of redundancies in the mechanisms of immune tolerance and the restricted HLA associations specific to humans.

1.2 Apoptosis in DILI

Apoptosis is a programmed mode of cell death that occurs both physiologically and, in many forms of liver disease, as a pathologic process. Apoptosis occurs in two steps, an activation phase where a stimulus or death signal activates a series of biochemical events that lead to the activation of specific proteases termed initiator caspases. Subsequently, these initiator caspases lead to the execution phase (second step), which is carried out by the effector caspases. Apoptosis can be activated via two major and convergent routes: the extrinsic (death receptor) and the intrinsic

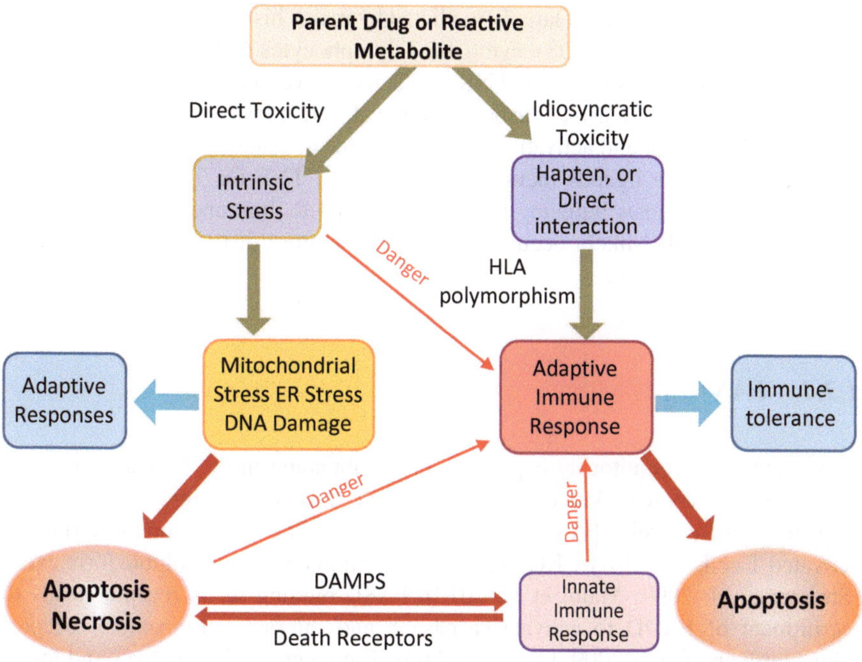

Fig. 1.1 Cell death in DILI. Drugs can be directly toxic to hepatocytes causing "intrinsic stress" or they can activate the adaptive immune system in certain individuals with susceptible HLA polymorphisms causing idiosyncratic, dose-independent, and unpredictable toxicity. Direct toxicity results in intrinsic stress (mitochondrial, ER stress, or DNA damage) that can lead to apoptosis and necrosis when a certain threshold is passed and adaptive responses cannot compensate. Apoptosis and necrosis subsequently lead to the release of DAMPs and a secondary innate immune response in the injured liver. Idiosyncratic DILI occurs via the adaptive immune system but is usually dampened by the process of immune tolerance leading to adaptation. However, if these adaptive responses are overcome then DILI ensues, resulting in cell death (apoptosis). (*ROS* reactive oxygen species, *ER* endoplasmic reticulum, *HLA* human leukocyte antigen, *DAMP* danger-associated molecular patterns)

(mitochondrial) pathways. Both lead to the eventual activation of executioner caspases which results in proteolysis, pyknosis (chromatin condensation), and karyorrhexis (nuclear fragmentation), which are the morphologic features of apoptosis, a.k.a., type I cell death (Galluzzi et al. 2012).

The extrinsic pathway of apoptosis is activated when death receptors (DR) such as TNF, FAS, Trail, etc. are engaged by their ligands resulting in caspase 8 activation and ultimately apoptosis and in some cases necroptosis depending on cell type. Death receptors are type 1 transmembrane proteins with intracellular C-terminal portions that contains a span of amino acids termed the death domain (Guicciardi and Gores 2009). All the members of the death receptor family including TNFR, Fas (CD95), TRAIL-R1 (DR4), and TRAIL-R2 (DR5) are highly expressed in the liver. Immune-mediated killing of hepatocytes occurs through T cells and cytokines (free and membrane bound) and involves these death receptors. Immune-mediated destruction of hepatocytes is largely mediated by major histocompatibility complex (MHC) class I-restricted CD8+ cytotoxic T lymphocytes (CTLs). However, other immune cells such as NK cells and NKT cells may have important roles in the liver and Kupffer cells (KC) also contribute (Tian et al. 2013; Protzer et al. 2012). Death receptor engagement and activation by their ligands activates the extrinsic pathway of apoptosis. These ligands include FasL for FAS, TNF for TNFR, and TRAIL (TNF-related apoptosis inducing ligand) for TRAILR. Immune-mediated hepatocyte killing in IDILI likely occurs through one or more of the receptor-mediated pathways discussed here.

1.2.1 FAS

FAS receptor is ubiquitously expressed and is abundant in the liver among many other organs (pancreas, kidney, heart, brain, etc.). It is also highly expressed by activated mature T cells. FAS Ligand (FASL) is mainly expressed on the surface of activated T cells and also by KC to a lesser extent; in addition, a soluble form exists (Strasser et al. 2009; Malhi et al. 2010). FASL binding to its receptor leads to recruitment of FADD to activate caspase 8 and internalization of the receptor (Guicciardi and Gores 2009; Lee et al. 2006). The receptor oligomerizes and forms large lipid rafts and clusters, is internalized, and localizes to the endosomal compartment for death inducing signaling complex (DISC) assembly that includes FADD, caspase-8, and cFLIP (FLICE (i.e., caspase-8) inhibitory protein). It is important to point out in humans in addition to caspase 8, FAS activates another initiator enzyme, caspase-10, which is not present in mice (Strasser et al. 2009). Caspase 8 activation then leads to activation of down-stream effector caspases and apoptosis in most cells (type I cells) such as immune cells. In hepatocytes (type II cells) FAS-mediated apoptosis depends on caspase-8-mediated cleavage of Bid and subsequent mitochondrial apoptosis pathway involvement. FAS stimulation results in massive hepatocyte apoptosis in WT animals. Bid knockout hepatocytes are resistant to Fas-mediated apoptosis whereas thymocytes are not (Yin et al. 1999). Additionally, FAS-mediated apoptosis can be blocked by overexpression of Bcl-2 or Bcl-XL in type II cells (Scaffidi et al. 1999). In type II cells, X-chromosome-linked inhibitor of apoptosis protein (XIAP) associates with caspase 3, thereby preventing propagation of the caspase 8 signal (Jost et al. 2009). This association can be unhinged by second mitochondria-derived activator of caspases (SMAC/Diablo) which is released from the inter membrane space of mitochondria during MOMP (Jost et al. 2009). Therefore, the extrinsic pathway of apoptosis is insufficient to induce cell death and the intrinsic/mitochondrial pathway needs to be engaged for apoptosis to ensue via MOMP (discussed further in the TNF section). Theoretically, FAS stimulation can result in a mode of hepatocyte killing in IDILI. However, another possible mechanism of FAS contribution to hepatotoxicity is through low level (1 μg) FASL depletion of cellular GSH levels which exacerbates toxicity from drugs such as acetaminophen. This could contribute to patients experiencing toxicity at lower levels when suffering from viral infections (Tinel et al. 2004). While FAS itself does not participate in acetaminophen-induced signaling, which is necrotic, FAS-deficient LPR mice have been shown to have a faster GSH recovery and rebound after APAP resulting in attenuated susceptibility to APAP injury (Williams et al. 2013).

1.2.2 TRAIL

TRAIL-R (a.k.a Apo 2) is a member of the TNF receptor super family that is expressed on hepatocytes. Indeed, TRAIL can induce proapoptotic caspase activity in isolated human hepatocytes (Volkmann et al. 2007). The ligand, TRAIL, is expressed by cells of the immune system but in particular, KC, macrophages, and NK and NKT cells (Malhi et al. 2010). In healthy liver explants TRAIL only induces apoptosis when combined with histone deacetylase inhibitors augmenting hepatotoxic effects of these chemotherapeutic agents, while in diseased liver (livers with HCV, steatosis) TRAIL stimulation resulted in massive apoptosis (Volkmann et al. 2007; Gores and Kaufmann 2001). TRAIL was first described in 1995 and thought to exclusively induce apoptosis in cancer lines, sparing healthy tissues and cells (Walczak et al. 1999; French and Tschopp 1999). Unlike the massive cell death and liver apoptosis seen with FASL, TRAIL administration appeared relatively safe in primates studies (Ashkenazi et al. 1999). Of the five known TRAIL death receptors only two transmit the apoptotic signal (DR4, DR5) through activation of caspases. Binding of TRAIL to DR4 or DR5 similar to FAS results in recruitment of proteins to form a signaling complex and assembly of the DISC. FADD and caspase 8 (and in humans, caspase 10) also play an important role in this death receptor pathway that also needs mitochondrial participation in hepatocytes.

Additionally, TRAIL binding can result in c Jun N-terminal Kinase (JNK) activation, through RIPK1-dependent (early phase) and independent (late phase) pathways (Zhang et al. 2015). Sustained and prolonged JNK activation can also contribute to apoptosis and cell death (see more details in the JNK/MAPK section). Furthermore, JNK itself can upregulate DR4 and DR5, increasing sensitivity to TRAIL in certain contexts (Fu et al. 2010; Malhi 2007; Zou et al. 2004, 2008). Bile acids can also induce DR5 expression (while inhibiting cFLIP function), thereby sensitizing hepatocytes to TRAIL-mediated cell death (Higuchi et al. 2004). This also occurs through a JNK-mediated pathway (Higuchi et al. 2004). The livers of cholestatic mice (bile duct ligation model) also upregulate TRAIL expression, and TRAIL deficiency significantly rescues the liver injury from the cholestasis resulting from bile duct ligation (Takeda et al. 2008; Kahraman et al. 2008). In addition, TRAIL has been shown to play a role in FASL-induced apoptosis in liver through a JNK-dependent mechanism involving activation of proapoptotic BIM (Corazza et al. 2006).

1.2.3 Perforin/Granzyme

Perforin or granzyme-induced apoptosis is the main pathway used by CD8+ cytotoxic lymphocytes and NK cells to eliminate virus-infected cells. While no direct evidence for this exists in DILI models, this is a well-described rout of eliminating virus-infected hepatocytes (Knolle and Thimme 2014). CD8 T cells and NK cells target infected cells, bind to them (conjugate), and release cytotoxic granules containing perforin, granzymes, and granulysin (Voskoboinik et al. 2015). Defects in this pathway result in many human diseases such as familial hemophagocytic lymphohistiocytosis and increased susceptibilities to viral infections and hematological malignancies (Voskoboinik et al. 2015). Although no data exists regarding the role of this pathway in IDILI, possible antigen mimicry can trigger cytotoxic T cells to kill hepatocytes expressing drug antigens thereby killing them and causing DILI.

1.2.4 TNF

The cell death pathways following TNFR stimulation by its ligands are among the most extensively studied signaling cascades in molecular biology (Fig. 1.2). When TNF binds its receptor, multiple proteins are recruited to a membrane-bound supramolecular structure termed complex 1. These include TNFR-associated death domain (TRADD), RIPK1, cellular inhibitor of apoptosis 1 and 2 (cIAP1 and 2), TNFR-associated factor 2 (TRAF2), or TRAF5. cIAPs, which are E3 ubiquitin ligases, are recruited to complex 1 via TRAF2 that prevents their polyubiquitylation. cIAPs catalyze the Lys63-linked polyubiquitylation of RIPK1 which serves as a scaffold for NFκB activation through transforming growth factor-β-activated kinase 1 (TAK1) and TAK1-binding proteins 2 and 3 (TAB2 and TAB3). In order for TNF to result in cell death, a second complex, receptor-free complex II, has to assemble in the cytoplasm. This consists of FADD, RIPK1, or TRADD, and caspase-8 to activate caspase-3 and caspase-7. In many cells such as hepatocytes, formation of complex II is blocked by Fas-associating protein with death domain-like interleukin-1 beta-converting enzyme (FLICE) inhibitory protein (c-FLIP) and NFκB target genes such as A20 (Karin and Lin 2002; Micheau and Tschopp 2003; Irmler et al. 1997).

Initiator and executioner caspases are present in a latent or catytically inactive zymogen form in the cytosol. Once the canonical pathway is engaged, the executioner caspases are activated by cleavage into 10 and 20 kD fragments by the initiator caspase. Activation of the executioner caspases, i.e., caspase −3, −6, and −7, which are effectors of both the extrinsic and intrinsic pathways, results in proteolysis and activation of nucleases leading to DNA fragmentation. In type I cells such as lymphocytes, activation of caspase −8 is sufficient to activate caspase-3 and other executioner caspases resulting in apoptosis. However, in type 2 cells such as hepatocytes, extrinsic pathway apoptosis induction first requires caspase −8-mediated

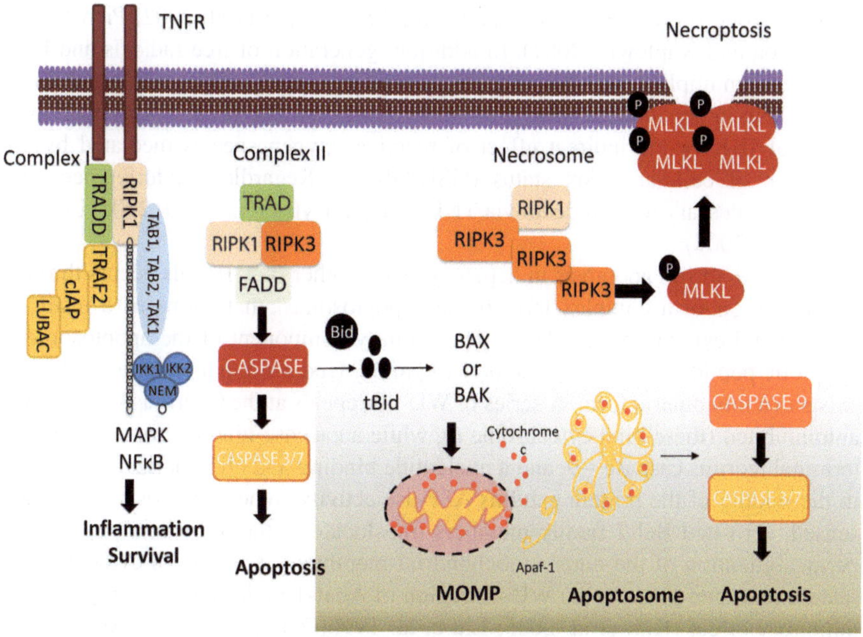

Fig. 1.2 TNF receptor-mediated cell death. TNF binding to TNF receptor (TNFR) results in the formation of complex I that includes TRADD, RIPK1, TRAF2, IAP1, IAP2, and LUBAC. Ubiquitination of RIPK1 results in a platform formation and recruitment of the IKK complex (made up of NEMO, IKK1, and IKK2) and TAB/TAK-1 complexes. This results in the activation of NF-κB and mitogen-activated protein kinase (MAPK). Complex II is made up of TRADD, FADD, and caspase-8 (IIa) or RIPK1/RIPK3, FADD, Caspase 8, and FLIPL (not shown) (IIb) which leads to caspase-8-mediated activation of caspase 3/7 and apoptosis in type I cells. In type II cells caspase 8 mediates cleavage of Bid to tBid resulting in BAX and BAK-mediated MOMP and Cytochrome C release. Cytochrome C binds to APAF-1 to form a wheel-like structure called the apoptosome that activates caspase 9, which in turn activates the effector caspases (caspase 3/7) to induce apoptosis. In certain cells when caspase-8 is inhibited the necrosome is formed, made up of RIPK1, RIPK3, and MLKL which leads to phospho-activation of MLKL, leading to its oligomerization and translocation to cell membrane to induce pore opening and necroptosis

cleavage of Bid, a Bcl-2 family protein, and subsequent mitochondrial participation and mitochondrial outer membrane permeabilization (MOMP) for executioner caspase activation. Proapoptotic Bcl-2 family members that include Bax, which translocates to Bak, which resides in mitochondrial outer membrane are activated by cleaved Bid (tBid) or Bim which leads to MOMP, releasing cytochrome *c* and other intermembrane proteins (Kaufmann et al. 2009; Yin et al. 1999). This Bcl-2 family participation and mitochondrial amplification via activation of the apoptosome is necessary for apoptosis to ensue in liver cells as hepatocytes are extremely resistant to the lethal actions of TNF because formation of the DISC not only activates the apoptosis machinery (and in some cases necroptosis) but also serves to activate NFκB. The resistance to TNF in hepatocytes can be overcome by using translation or transcription inhibitors (such as actinomycin-D or galactosamine), or depletion of glutathione (GSH) (Feng and Kaplowitz 2000; Nagai et al. 2002; Pierce et al. 2000; Lou and Kaplowitz 2007). In addition, generation of free radicals and ROS has also been implicated in hepatocytes to interfere with NFκB signaling and sensitize to TNF in hepatocytes (Han et al. 2006). This could be occurring through inhibition of IKK or the indirect effect of reactive oxygen species mediated by the alteration of cellular redox status (GSH/GSSG). Regardless, interference with NFκB is necessary for cell death via TNF in hepatocytes (Han et al. 2006; Lou and Kaplowitz 2007).

As noted above, apoptosis in hepatocytes and other type II cells occurs through the activation of a complex termed the "apoptosome" that forms as a result of MOMP and cytochrome *c* release. The primary component of the apoptosome is apoptotic peptidase-activating factor-1 (Apaf-1). Under normal conditions, Apaf-1 exists in a monomeric form. A series of WD-40 repeats at the C termus keep Apaf-1 autoinhibited (these bind cytochrome *c*), while a caspase-binding domain at the N terminal recruits caspase −9, and a nucleotide binding and oligomerization domain in the middle of the protein exhibits ATP-ase activity. When apoptosis signals are sensed, activated Bcl-2 family members translocate to the mitochondria, causing permeabilization of the outer mitochondrial membrane resulting in cytochrome *c* release that then binds to the WD-40 region of Apaf-1 molecule releasing its autoinhibitory effect (Kim et al. 2005; Liu et al. 1996; Zou et al. 1997). Apaf-1 then forms a wheel-like structure called the apoptosome that promotes self-activation of caspase 9 which then cleaves executioner caspases (Acehan et al. 2002).

1.3 JNK in DILI

In addition to the Bcl2 family, the mitochondria, and caspases, the mitogen-activated protein kinase (MAPK) family are critical regulators of hepatocyte death. MAPKs are a tiered cascade of kinases and the principal regulators of stress-responsive pathways in hepatocytes (Chang and Karin 2001). MAPKs are governed by their phosphorylation status. C Jun-N-terminal kinase (JNK), an extensively studied member of this family in the context of DILI, plays a critical role in liver homeostasis and disease (Chang and Karin 2001; Sabapathy et al. 2004). Once JNK is activated, the most critical factors dictating cell fate in hepatocytes are the duration of sustained JNK activation and the protein's subcellular localization. Prolonged JNK activation, which likely corresponds to the magnitude and duration of stress, promotes cell demise, while transient JNK activation does not (Bradham et al. 1997; Gunawan et al. 2006; Hanawa et al. 2008; Win et al. 2011). JNK activates the intrinsic mitochondrial pathway of apoptosis by phosphorylation of Bcl2 proteins, Bim and Bax, and mediates phosphorylation and inactivation of antiapoptotic Bcl2 proteins, Bcl-xl and mcl-1 (Lei and Davis 2003; Lei et al. 2002; Tournier et al. 2000). Interestingly, JNK activation and translocation to mitochondria is not only important for apoptosis in hepatocytes but is a critical step in APAP-induced hepatocyte necrosis as well (Win et al. 2011). This delicate balance between JNK activation, mitochondrial ROS generation, and NFκB transcriptional regulation determines hepatocyte death or survival. TNF treatment in hepatocytes causes ROS formation leading to sustained MAP kinase (JNK) activation (Win et al. 2011). ROS formation from mitochondria is required for prolonged JNK activation. JNK can lead to degradation of cFLIP, a pro-survival protein induced by NFκB, and NFκB-responsive genes, such as XIAP, have been implicated in inhibition of JNK (Chang et al. 2006; Liu et al. 2002). Inhibition of NFκB in the presence of TNFR engagement results in sustained JNK activation and promotes TNF-induced hepatocyte death (Chang et al. 2006; Liu et al. 2002). When JNK signals to mitochondria, it binds to SH3BP5 or Sab, which spans the outer mitochondrial membrane (Wiltshire et al. 2002, 2004). Preventing JNK binding to Sab abrogates liver cell death both in apoptotic and in necrotic models of liver injury including DILI from acetaminophen (necrotic model), ER stress apoptosis models, TNF (plus D-galactosamine), and palmitic acid-induced liver cell apoptosis, further supporting the notion that JNK translocation and mitochondrial participation are critical for hepatocyte death in these models (Win et al. 2011, 2014, 2015). The intra-mitochondrial downstream pathways are discussed further in the programmed necrosis section (see below). It should be mentioned that JNK participation in hepatocellular death is not universal; thus in models of direct toxicity of CCL4 and furosemide, JNK is dispensable (Gunawan et al. 2006; McGill et al. 2015).

Some hepatotoxic agents cause injury by activating cellular pathways leading to apoptosis. Actinomycin D and the mushroom toxin α-amanitin induce a transcriptional arrest, thereby inhibiting NFκB and sensitizing to liver cell apoptosis via TNF stimulation and apoptosis (Leist et al. 1994, 1997). Drugs that cause idiosyncratic DILI (IDILI) do so in large part due to aberrant adaptive immunity in susceptible individuals with HLA polymorphisms. This cell death mode in immune-mediated liver injury is death receptor mediated and therefore primarily apoptotic. As an example, diclofenac, a known idiosyncratic hepatotoxin, has been suggested to cause toxicity in large part by inducing MOMP and apoptosis (Siu et al. 2008). In HepG2 cells, diclofenac induced apoptosis through the caspase8/Bid/Apaf1 pathway leading to caspase 3 cleavage. Furthermore, diclofenac induced sustained JNK activation and decreased NFκB transcriptional activity, thereby sensitizing cells to apoptosis through multiple mechanisms (Fredriksson et al. 2011). Interestingly while IDILI toxicity is presumed to be carried out by the adaptive immune system, in vitro drug panels exploring cytotoxicity of IDILI drugs in cultured primary human hepatocytes have demonstrated a direct induction of cytotoxicity by IDILI drugs as measured by decreases in GSH and ATP, increased ROS/ATP ratio, and activation of caspase 3 (Zhang et al. 2016).

1.4 Mitochondria in DILI

Mitochondria are crucial organelles in liver cell fitness and survival and they play an indispensable role in the execution of liver cell apoptosis. Therefore, it is not surprising that many hepatotoxins kill liver cells by targeting mitochondria. It is important to distinguish essential mitochondrial participation downstream of DR signaling, DNA damage, and ER stress from DILI due to direct targeting of drugs to the mitochondria. For example, Reye and colleagues described in the 1960s that children with febrile illness given aspirin developed encephalopathy, profound hypoglycemia, lactic acidosis, and microvesicular steatosis. Mitochondria is a key target of APAP toxicity will be discussed in another chapter. Other drugs such as valproic acid, antiretrovirals, tetracycline, tamoxifen, and amiodarone can also target mitochondrial disrupting oxidative phosphorylation resulting in steatosis and liver injury and even acute liver failure (Lemasters et al. 1998; Silva et al. 1997; Fromenty and Pessayre 1995; Lewis et al. 1989). Drug-induced mitochondrial toxicity can result from interference with mitochondrial DNA synthesis. Fialuridine, an antiviral that was being developed for hepatitis B, exemplifies this type of toxicity. During the clinical trial, patients were noted to have elevated liver enzymes, pancreatitis, lactic acidosis, and even liver failure and death after a long latency. Disruption of mitochondrial function was determined to be the underlying mechanism as evidenced by depletion of mitochondrial DNA and classic clinical features of mitochondrial toxicity (McKenzie et al. 1995). A similar syndrome has been reported with the antiretrovirals zidovudine and didanosine (Bissuel et al. 1994). Due to the serious nature of these toxicities, all antiviral nucleoside and nucleotide analogue drugs have a black-box warning regarding potential mitochondrial toxicity (Dara et al. 2012). Valproic acid (VPA), an anticonvulsant and mood stabilizer, is another drug that kills liver cells by targeting mitochondria in multiple ways. VPA is metabolized by the cytochrome P-450 system to a reactive intermediary, 4-ene VPA, which is conjugated with coA to 4-ene-CoA, which enters mitochondria and then can be converted by β-oxidation to 2,4-diene VPA a highly electrophilic metabolite (Kassahun et al. 1991, 1994). Therefore, these reactive metabolites are generated in larger amounts when patients take other anticonvulsants such as carbamazepine or phenytoin concomitantly (Levy et al. 1990). Also, VPA itself is a fatty acid and can therefore easily cross the mitochondrial membrane to uncouple respiration by interfering with the proton gradient, resulting in microvesicular steatosis and Reye's syndrome (Levy et al. 1990). Valproate also sequesters CoA by forming valproyl-CoA moieties; thus, interfering with natural fatty acyl-CoA-oxidation, as well as pyruvate dehydrogenase, which requires CoA (Ponchaut et al. 1992; Silva et al. 1997). Another important determinant of valproate toxicity is amino acid substitutions that decrease the activity of mt-DNA polymerase γ (POLG) (Naviaux and Nguyen 2004). Alpers-Huttenlocher syndrome, a rare childhood disorder characterized by epilepsy, developmental delay, and liver disease, is associated with severe toxicity from VPA. This VPA toxicity occurs in up to a third of patients with this disorder and is due to the POLG mutation (Naviaux and Nguyen 2004). POLG mutations seem to correlate well with risk of VPA toxicity and interestingly even patients with *heterozygous* mutations in the POLG gene were associated with an increased risk of liver toxicity from valproate in the Drug-Induced Liver Injury Network (DILIN)

(Stewart et al. 2010). The presumption is that the mutation in POLG, by impairing mitochondrial DNA synthesis, sensitizes to the toxicity of VPA although this remains to be proven in model systems. In the general population, VPA-induced

DILI occurs in a small proportion of treated individuals which therefore can be considered IDILI. It is unclear if this represents direct toxicity to mitochondria alone or if an immune mechanism might contribute with sublethal mitochondrial impairment acting as a danger signal costimulating the adaptive immune system. Many IDILI drugs directly impair the function of isolated mitochondria (Porceddu et al. 2012). In addition, antibiotics are the major contributor drugs to IDILI and given mitochondria are descendants of bacteria it is intriguing to speculate that these drugs may cause liver toxicity in part by targeting liver mitochondria.

1.5 ER Stress in DILI

ER stress is another important intracellular stress pathway in hepatocytes that can result in cell death (Dara et al. 2011). ER stress occurs in specialized secretory cells when the amount of protein entering the ER lumen exceeds the organelle's capacity to fold them resulting in the activation of a stress response to cope with and compensate for the resulting gridlock (Hetz et al. 2013; Ron and Walter 2007). Thus, ER stress initiates a series of adaptive mechanisms termed the unfolded protein response or UPR to restore protein folding and homeostasis (by upregulating chaperones, selectively degrading mRNA of secretory proteins to decrease work load, or inhibiting protein synthesis) (Hetz et al. 2013; Ron and Walter 2007). Additionally, the UPR seems to be an essential signaling pathway in many physiologic processes such as lipid and cholesterol metabolism, oxidative stress, inflammatory response pathways, and cancer development (Wang and Kaufman 2012). The three major stress sensors that control the UPR are inositol-requiring enzyme 1α (IRE1α), protein kinase RNA-like ER kinase (PERK), and activating transcription factor 6 (ATF6). These three transmembrane proteins are anchored to the ER and activate downstream signals that travel to the nucleus to activate genes to restore protein folding. The IRE1α branch of the UPR activates X-box-binding protein 1 or XBP1s which is a transcription factor that activates ER-associated protein degradation in the cytosol and protein quality control. When the ER stress is too severe to overcome, IRE1α can activate the JNK-mediated apoptosis pathway (via ASK1 and TRAF2) to induce cell death (Urano et al. 2000). P-JNK then translocates to mitochondria where it induces ROS production; uptake of CA2+ released by the ER amplifies this effect of JNK on mitochondria. Sustained JNK then results in modulation of Bcl2 family leading to MOMP (Win et al. 2014). It has been shown that covalent binding of reactive chemicals activates ER stress, though the precise mechanism is unclear regarding whether covalent binding of client proteins leads to misfolding or direct impairment of UPR mediators (Dara et al. 2011). It is therefore intriguing to hypothesize that certain drugs cause DILI and hepatocyte death via the ER stress pathway. While many proteins are activated through the UPR, one of the main elements of ER stress-mediated apoptosis is CHOP, a transcription repressor that is activated downstream of the PERK and IRE1 pathways of the UPR. CHOP expression in the liver seems to be upregulated with various stressors such as alcohol, LPS, CCL4, APAP although the significance of this induction remains controversial (Ji et al. 2005; Rahman et al. 2007; Campos et al. 2014). Both diclofenac and carbamazepine have been shown to upregulate ER stress genes and CHOP expression in cell culture models of Hep G2 cells and primary human hepatocytes (Fredriksson et al. 2014). The role of ER stress in necrotic models such as APAP is also of interest as there is data that CHOP knockout mice exhibited less liver injury and increased survival after APAP treatment by gavage (Uzi et al. 2013). APAP treatment strongly upregulated CHOP at 12 h and CHOP−/− mice seemed to recover from toxicity faster and exhibited a higher regenerative capacity after injury (Uzi et al. 2013). The role of ER stress in DILI from the combination of protease inhibitors (ritonavir and lopinavir) and alcohol has been investigated (Hu et al. 2015). The drugs and alcohol each mildly induced ER stress markers in primary hepatocytes when used alone, but, the combination of alcohol with ritonavir and lopinavir markedly induced the chaperone glucose regulatory peptide 78 (GRP78) and CHOP (Hu et al. 2015). This was also associated with an increase in cell death markers as well as the autophagy marker LC3 (Hu et al. 2015). Although no conclusive evidence exists, these in vitro studies raise the intriguing possibility that under certain conditions such as alcohol consumption, drugs that cause IDILI can promote cell death via the ER stress pathway. Overall, firm conclusions can be made regarding the role of mitochondrial and ER stress resulting in cell death in DILI from certain drugs and toxins. However, for most IDILI drugs, these stress responses are more likely to contribute to danger signals for adaptive immunity and/or sensitize to immune-mediated killing.

1.6 Regulated Necrosis DILI

Necrosis has been viewed traditionally as an accidental form of cell death, resulting in oncosis or swelling of the cell and intracellular organelles eventually leading to plasma membrane rupture. The accidental nature of this process has been challenged in recent years with the elucidation signaling molecules leading to regulated necrosis pathways. While necroptosis is not the only form of regulated necrosis, it is so far the best described and most studied mode of cell death with necrotic morphology. Necroptosis was first described in cultured cells stimulated with FasL or TNF in the presence of the pancaspase inhibitor Z-VAD-FMK (Holler et al. 2000; Vercammen et al. 1998). Necroptosis can occur with ligation of other death receptors, toll-like receptors, and intracellular signaling (DAI, PKR) (Linkermann and Green 2014). The interesting finding in these initial reports was that the cells continued to die despite caspase inhibition. However, the cell death mode appeared to be morphologically necrosis, often described as "a switch to necroptosis." It was subsequently shown that necroptosis requires the kinase activity of RIPK1 (Holler et al. 2000) and RIPK1 inhibition (by necrostatins) protected against this form of cell death (Degterev et al. 2005). During necroptosis, when caspases are inhibited, RIPK1 recruits RIPK3 that oligomerizes and together they form a complex called the necrosome, which ultimately leads to the activation of MLKL by phosphorylation. P-MLKL then executes necroptosis by translocating to and rupturing the cell membrane, the mechanism of which is still under debate (Chen et al. 2014; Hildebrand et al. 2014; Murphy et al. 2013; Quarato et al. 2016) (Fig. 1.2). In addition to participating in apoptosis and necroptosis, RIPK1 participates in inflammatory pathways and can paradoxically inhibit apoptosis and promote survival under certain conditions (Dannappel et al. 2014; Takahashi et al. 2014; Suda et al. 2016).

The role of necroptosis has been investigated using multiple acute and chronic liver injury models, including models of DILI with conflicting results. Much of the controversy is due to the fact that not all cells undergo necroptosis. Hepatocytes do not robustly express RIPK3 and the induction of RIPK3 after acute DILI, as in the case of APAP, is controversial (Dara et al. 2015a; Ramachandran et al. 2013; Kasof et al. 2000; Sun et al. 1999). RIPK3 mRNA presence in the liver was not conclusive in two reports when the protein was initially described (Kasof et al. 2000; Sun et al. 1999). A limited population of primary cells and cell lines such as L929 fibrosarcoma cells, Jurkat T cells, HT29 cells, MEF, keratinocytes have been shown to switch to an alternate form of cell death with death receptor signaling under conditions where caspases +/−cIAPs are inhibited (Degterev et al. 2014). This has not been the case for all cells; for example, MCF-7, Hela cells and Hek293 cells do not undergo necroptosis due to a lack of RIPK3 expression (Degterev et al. 2014). Absence of RIPK3 in cells predicts whether they can undergo necroptosis (Geserick et al. 2015). While whole liver homogenate is weakly positive for RIPK3 protein using a highly specific monoclonal antibody, this may be due to its presence in the NPC compartment, as freshly isolated hepatocytes do not express RIPK3 (Dara et al. 2015a). Furthermore, most reports citing RIPK3 protein induction have used commercially available nonspecific polyclonal antibodies in injured liver tissue. Interestingly, in immunohistochemical studies, the area of RIPK3 staining corresponds to the area of necrotic or damaged liver which is highly immune-reactive and stains positive even for nonspecific rabbit and mouse IgG (Dara et al. 2015a). In fact even RIPK3−/− liver stains positive for RIPK3 after acetaminophen, providing further evidence that the necrotic area surrounding the central vein is reacting nonspecifically with commercial antisera (Dara et al. 2015a). Another line of evidence against hepatocytes undergoing necroptosis is that blockage of caspase 8 during numerous apoptosis-inducing experimental conditions does not switch the cell death mode from apoptosis to necroptosis, but rather protects (Hatting et al. 2013; Kang et al. 2004, 2015; Kaufmann et al. 2009; Thapaliya et al. 2014; Witek et al. 2009). Additionally, liver-specific ablation of caspase 8 has no effect on the normal liver and is, in fact, protective against Fas and TNF-mediated apoptosis (Kang et al. 2004; Liedtke et al. 2011). APAP-induced hepatic necrosis is perhaps the best studied model of DILI in which necroptosis has been studied. One report by Ramachandran and colleagues has suggested a significant and transient protection in RIPK3−/− mice at 6 h against APAP compared to wild-type (WT) controls and has concluded that in the early phase of toxicity the cell death in hepatocytes can be necroptosis (Ramachandran et al. 2013). Despite the observed protection at 6 h post APAP, at 24 h the RIPK3 WT and knockout (KO) animals had similar liver injury and ALT levels. We have also investigated the role of RIPK3 KO as well as MLKL KO on toxicity from APAP and have found no difference between WT and KO animals at early (6 h) or late time points (24 h) (Dara et al. 2015a). The fact that MLKL KO does not protect mice from liver injury precludes necroptosis as a significant contributor to APAP necrosis in mice. Nevertheless, since MLKL is expressed in hepatocytes it remains possible that RIPK3-independent mechanisms might activate MLKL in certain types of DILI. It is noteworthy that other groups have suggested that necroptosis contributes to APAP toxicity. However, many of these studies were done in vitro and used necrostatin, a RIPK1 kinase inhibitor, and some which used in vivo knockout models did not use strain-matched controls (An et al. 2013; Deutsch et al. 2015; Takemoto et al. 2014; Zhang et al. 2014). Since RIPK1 participates in pathways other than necroptosis we investigated the necroptosis-independent function of RIPK1 in APAP DILI. Knockdown or RIPK1 in WT as well as RIPK3 KO animals protected against APAP. This protection was upstream of JNK activation and dynamin-related protein-1 (DRP-1) translocation to mitochondria (Dara et al. 2015a). Therefore, the effect of necrostatin and RIPK1 knockdown on diminution of APAP toxicity seems to be a form of regulated necrosis and although necroptosis does not occur in the APAP model, a necroptosis-independent role of RIPK1 in APAP toxicity is suggested. The murine model of APAP DILI is a highly reproducible model of liver cell necrosis and closely mimics human pathology. Even though necroptosis is not at play, APAP cell death is a form of regulated necrosis as interference with multiple signaling pathways including the mitogen-activated protein kinases (MAPKs) and in particular JNK and its binding partner Sab, as well as, GSK3b, PKCa, DRP1, cyclophilin D, and MPT have all been shown to abrogate or prevent hepatocyte death (Dara et al. 2015a; Gunawan et al. 2006; Han et al. 2013; Hanawa et al. 2008; LoGuidice and Boelsterli 2011; Win et al. 2011, 2015; Kon et al. 2004; Ramachandran et al. 2011). The JNK pathway plays a particularly important role in this model. Interference with JNK phosphorylation or its binding to mitochondria or the subsequent steps in the pathway all markedly dampen toxicity. Win et al. have identified downstream pathways leading to hepatocyte necrosis after p-JNK binds to Sab. Since p-JNK directly inhibits mitochondrial respiration in the presence of ATP, Win et al. examined the intra-mitochondrial pathway that could explain how Sab, which is an outer membrane protein, signals to the inner membrane. This led to the discovery that p-JNK binding to and phosphorylation of Sab (on the cytoplasmic facing-C terminus of Sab) released a phospho-tyrosine phosphatase, non-receptor type 6 (SHP1, or PTPN6) (Win et al. 2016). SHP1 then translocated to the inner membrane where it dephosphorylated the active form of Src, resulting in its inactivation that required an inner membrane-docking protein called DOK4. This inactivation of mitochondrial Src inhibits the ETC and increases ROS release, sustaining p-JNK activation and promoting cell death and organ injury. Importantly, like interfering with upstream signals of JNK such as ASK1, MKK4, MLK3, RIPK1, blocking of these downstream signals from JNK also abrogates APAP toxicity and protected from cell death. Furthermore, the same protection was seen in TNF-induced hepatic apoptosis (Win et al. 2011). This places the JNK pathway as the key death inducing pathway in hepatocytes in the APAP model as well as apoptosis models such as TNF/galactosamine (Win et al. 2016). In the APAP model, the JNK/Sab/Src/ROS pathway sustains JNK activation, amplifying mitochondrial ROS production and promoting MPT-mediated necrosis whereas in the apoptosis scenarios this pathway sustains high levels of JNK activation which modulate multiple members of the Bcl2 family that promotes MOMP. In the APAP model, cell death is a form of regulated necrosis which is ultimately mediated by MPT. Other forms of regulated cell death such as pyroptosis (a caspase 1-mediated form of inflammatory cell death associated with infections) and ferroptosis (an iron and ROS-dependent form of cell death) have been recently described but their role in DILI is unclear.

1.7 Autophagy in DILI

Autophagy is the lysosomal degradation pathway through which cell content including damaged organelles and protein is recycled (Xie and Klionsky 2007). This occurs through the formation of double-membrane vesicles (autophagosomes) originating from endoplasmic reticulum or cell membrane, which fuse with and deliver their cytoplasmic cargo to lysosomes for degradation by lysosomal hydrolases. The molecular machinery of this pathway was uncovered when the family of autophagy-related genes (ATG) were described (Klionsky et al. 2003). Currently, more than 30 ATG genes implicated in various subtypes of autophagy have been described. Autophagy plays an important role in programmed cell death, especially in close collaboration with apoptosis. Interestingly, it was described even before apoptosis as a mode of cell death (Lockshin and Williams 1965; Lockshin and Zakeri 2004). Autophagic cell death, often referred to as type II cell death with type I being apoptosis and type III necrosis, is morphologically defined as a type of cell death that occurs in the absence of chromatin condensation but is accompanied by massive autophagic vacuolization of the cytoplasm. Cells that die with an autophagic morphology have little or no association with phagocytes (in contrast to apoptotic cells) (Galluzzi et al. 2012). However, autophagy is not solely a cell death mode, it is also a pivotal process in development, immune defense, tumor suppression, and many other cellular processes (Xie and Klionsky 2007). Global Atg7 or Atg 5 knockout mice die shortly after birth. Autophagy is clearly important in liver homeostasis as liver-specific knockout of these genes results in severe liver injury and compensatory Nrf-2 upregulation (Ni et al. 2012b). While many subtypes have been described, for the purpose of DILI we will focus on macroautophagy and mitophagy. Autophagy occurs during homeostasis at low levels in all cells including the liver to recycle proteins. It is rapidly upregulated during starvation via inhibition of the mammalian target of rapamycin (mTOR) pathway (Mizushima et al. 2008). Autophagy is thought to enhance survival as knockout of ATG genes in most instances is associated with acceleration of cell death rather than its inhibition or prevention (Galluzzi et al. 2012). The role of autophagy and, in particular mitophagy, has been studied in APAP-induced hepatotoxicity (Ni et al. 2012a). Since APAP toxicity primarily targets the removal of damaged mitochondria, which generate free radicals and ROS, mitophagy was shown to improve toxicity from APAP (Ni et al. 2012a). Ni and colleagues showed that APAP resulted in the formation of autophagosomes that engulfed mitochondria. Pharmacological inhibition of autophagy by 3-methyladenine or chloroquine exacerbated APAP toxicity, while this was attenuated by rapamycin (Ni et al. 2012a). Importantly, rapamycin was beneficial in attenuating injury even when administered 2 h after APAP (Ni et al. 2012a). Interestingly, in a follow-up study, Ni and colleagues examined the effect of autophagy inhibition on APAP adduct clearance and found that P62 was recruited to APAP adducts and that induction of autophagy cleared adducts and correlated with improved ALT and histology (Ni et al. 2016). The authors concluded that timely removal of adducts via autophagy may, in fact, promote protection from APAP.

Further evidence for the detrimental effects of blocking mitophagy in DILI exists in the context of mitochondrial quality control association with the lysosomal membrane turnover. Acid sphingomyelinase (ASMase) plays a critical role in lysosomal membrane turnover through the hydrolysis of sphingomyelin, which generates ceramide (Baulies et al. 2015). Mitophagy can be impacted by lysosomal cholesterol accumulation that can alter the lysosomal membrane structure and function (Baulies et al. 2015). Acid sphingolyelinase (ASMase) knockout mice exhibited a higher mortality after APAP administration compared to WT littermates and the ASMase knockout hepatocytes display a lower threshold for APAP-induced cell death, defective fusion of mitochondria containing autophagosomes with lysosomes resulting in impaired mitophagy (Baulies et al. 2015). Lysosomal cholesterol was thought to be the linchpin preventing effective mitophagy as cholesterol extraction increased APAP-mediated mitophagy and protected the ASMase knockout mice (Baulies et al. 2015).

Loss of autophagy has also been implicated to worsen TNF-mediated hepatocyte death. Hepatocyte-specific ATG 7 knock-out mice had also exhibited worsening of liver injury, apoptosis, and decreased survival from D-galactosamine/LPS or D-galactosamine/TNF treatment compared to littermate controls (Amir et al. 2013). This increased susceptibility to hepatocyte death was due to increased caspase-8 activation leading to t-Bid cleavage (Amir et al. 2013). Although this is not a DILI model, since idiosyncratic DILI likely mediates cell death via adaptive immunity and through the activation of the DR pathways such as TNFR, this may in fact be relevant to IDILI as well.

Efavirenz, a widely used non-nucleoside reverse transcriptase inhibitor for the treatment of HIV, is known to induce hepatotoxicity. Efavirenz inhibits the ETC by interfering with Complex I resulting in decreased ATP and increased ROS generation. The role of efavirenz toxicity in vitro using human hepatocytes may be due to alterations in mitochondrial morphology and increased induction of autophagy (Apostolova et al. 2011). Inhibition of autophagy using 3-methyladenine has been suggested to worsen efavirenz toxicity and induced apoptosis in these cells (Apostolova et al. 2011). The role of autophagy in a hexavalent chromium (Cr IV) model of liver injury has also been studied. In this study, Cr IV induced mitochondrial toxicity, increased ROS and apoptosis at 24 h. Interestingly, rapamycin helped rescue the cell toxicity while the autophagy inhibitor, 3-methyladenine, worsened toxicity (Xie et al. 2014).

1.8 Cell Death Biomarkers in DILI

DILI has been a diagnosis of exclusion as no specific biomarkers exist to make an accurate diagnosis, except for serum adduct levels in APAP toxicity. It would be useful to identify serum biomarkers that can be used to confirm the liver origin of standard liver blood tests as well as to identify the predominant mode of cell death in humans. In the past few years, a few new biomarkers of liver injury which are relatively specific to cell death subroutines have become available in experimental settings (Watkins 2009). Other biomarkers that pertain specifically to DILI such as anti-liver antibodies, serum adducts, and omics approaches are beyond the scope of this chapter and have been reviewed elsewhere (Watkins 2009). Alanine transaminase (ALT) is the most commonly used biomarker of liver injury both clinically and in the liver research field (Luedde et al. 2014; Kew 2000). While ALT is the best studied and most universally used biomarker of both acute and chronic liver diseases, it does not distinguish between apoptosis and necrosis. Additionally, many new biomarkers are rapidly becoming available that outperform ALT in sensitivity and specificity, these are also cellular proteins that are released from hepatocytes and some are specific to cell death mode. Keeping in mind that this is an evolving and new field, we will discuss the best studied novel biomarkers of DILI: HMGB1, keratin 18, and micro-RNA (miR)-122 (K18).

1.8.1 HMGB1

High-mobility group box 1 protein (HMGB1) is a non-histone DNA-binding protein that is released from necrotic cells (not apoptotic cells) and may be actively secreted in certain conditions from immune cells (Lotze and Tracey 2005; Woolbright and Jaeschke 2015). HMGB1 has signaling properties and functions as a damage associate molecular pattern (DAMP) and acts as a TLR4 agonist. Total and acetylated HMGB1 is increased in serum of patients with APAP toxicity, reflecting necrotic liver injury (Antoine et al. 2009, 2012). In one study using 84 patients with APAP overdose HMGB1 (total and acetylated) and K18 (discussed below) were significantly associated with the degree of injury (ALT). Importantly, there was no significant elevation of these biomarkers in patients with APAP overdose who had normal LFTs, suggesting that these biomarkers are sensitive at identifying liver cell necrosis and injury and are not just induced or secreted during APAP treatment (Antoine et al. 2012). Interestingly, since HMGB1 is a DAMP and is thought to contribute to the post APAP toxicity inflammatory response that occurs in the liver, HMGB1-neutralizing antibodies have been investigated as a possible treatment (Lundback et al. 2016).

1.8.2 Keratin 18 (K18)

Keratins, 7, 8, and 8–20 are intermediary filaments that represent the keratins of single layered and glandular epithelia. They are subdivided into type I (K18-K20) and type II (K7, K8) polypeptides, which are both needed to form obligate noncovalent heteropolymers that include at least one type I and one type II keratin (Strnad et al. 2012; Omary et al. 2009). Hepatocytes only express K8-K18 polymers whereas other simple epithelia tissues such as cholangiocytes also express variable levels of K7/K19, K8/K18, and even K20 (Omary et al. 2009; Strnad et al. 2012). Interestingly, K8 and K18 polymers along with the ubiquitin-binding protein p62 and ubiquitin are the major constituents of Mallory-Denk Bodies (MDB), first described over 100 years ago by Frank Mallory (Omary et al. 2009). In alcoholic liver disease K18 levels strongly correlate with MDB as well as hepatocyte ballooning and fibrosis (Lavallard et al. 2011). The keratins participate in the makeup of cell cytoskeleton and during apoptosis only type I keratins (such as K18) are cleaved by caspases (Ku et al. 2016; Caulin et al. 1997). This has led to the use of serum assays (M30 ELISA) of cleaved K18 as a biomarker for liver death and to identify the cell death subroutine. M30 ELISA detects the cleaved form while M65 detects both caspase-cleaved and un-cleaved (total) K18 (Ku et al. 2016). While theoretically distinguishing between necrotic and apoptotic cell death in DILI may be possible, in practice this proves to be much more complicated. The strongest evidence for K18 predicting clinical outcomes comes from NASH where the cell death mode seems to be apoptosis (Musso et al. 2011). Numerous investigators have examined the levels of M30 and M65 in acute liver injury models such as patients with APAP hepatotoxicity with different results.

Craig and colleagues measured M30, M65, and a few other biomarkers such as HMGB1 and serum nucleosome levels in 39 patients, the majority of which had ALF due to APAP overdose (26 out of 39) (Craig et al. 2011). ALF patients with APAP overdose had significantly higher nucleosome levels than cirrhotics and healthy control subjects, but the levels were similar in APAP vs non-APAP ALF patients. Nucleosome levels correlated with levels of HMGB1, ALT, and M65 but not with M30. None of the cell death markers analyzed improved prognostication in paracetamol patients beyond the King's College criteria (Craig et al. 2011). In another study looking at acute liver failure patients ALF from various causes, caspase activation (increased M30 K18) was associated with spontaneous recovery and better outcomes, indicating that perhaps the M30 assay could be used for prognostic purposes (Volkmann et al. 2008). Interestingly, patients with spontaneous recovery revealed significantly higher caspase activation, accompanied by an elevated expression of the pro-regenerative cytokines interleukin-6 (IL-6) and TNF, than those who did not recover (Volkmann et al. 2008). Patients without spontaneous recovery after ALF showed significantly elevated total K18 using M65 Elisa compared to those who recovered and healthy controls. What is unusual about this work is that the patients APAP toxicity (which is known to be a necrotic form of cell death) exhibited high K18 M30 fragments as well (Volkmann et al. 2008). The K18 assays have also been used in the algorithm for the Model for End-Stage Liver Disease (MELD) score where total K18 replaced bilirubin and the new formula significantly improved the prediction of ALF outcome at the day of hospital admission (Bechmann et al. 2010). The Acute Liver Failure Study Group (ALFSG) has also incorporated the K18 M30 test in their prognostic algorithm for the ALF (Rutherford et al. 2012). The ALFSG study combined hepatic coma grade, serum bilirubin, serum phosphorus, INR, and a log10 value of cleaved cytokeratin-18 (M30) and demonstrated an area under the receiver operator curve (AUC) of 0.822 in a validated cohort of 500 ALF patients. This was superior to the AUCs for both MELD (0.704) and the King's College Criteria (KCC, 0.654) (Rutherford et al. 2012). Of the 250 patients, 75 (30%) had ALF due to APAP overdose and 56 (22%) had other forms of DILI.

Give the current differences in the results, the use of M30/M65 for distinguishing mode of cell death is a work in progress and further studies are also needed for the validation of this test as a prognostic marker in cases of clinically significant DILI and ALF.

1.8.3 miRNA

Micro RNAs are small noncoding RNAs that regulate gene expression and RNA function. Studies of micro RNA in liver injury have focused on miR-122, which is highly abundant and liver specific (Lagos-Quintana et al. 2002). For example, miR-122 is released from hepatocytes after APAP toxicity (Starkey Lewis et al. 2011) and both miR-122 and miR-192 have been detected the liver of APAP-treated mice (Wang et al. 2009). Micro RNAs exhibited dose and exposure duration-dependent changes in the plasma that paralleled serum ALT levels and correlated with histologic injury, and importantly, these changes were detected significantly earlier (Wang et al. 2009). The ratio of miRNA 122/155 has also been shown to be a sensitive biomarker for direct INH-induced toxicity in mice (Song et al. 2016). miR-122/155 expression levels were tracked over a 24-h period and levels of miRNA 122 were elevated as early as 15 mins after INH dosing. miR 122 has been extensively studied as a sensitive and specific biomarker for both acute and chronic liver diseases (Thakral and Ghoshal 2015).

1.9 Conclusion

Cell death is a dominant feature of DILI. There are two main types of cell death, apoptosis, and necrosis and both can be implicated and involved in DILI. It is not clear if other modes of cell death contribute to DILI and autophagy (mitophagy) is mainly protective. The mechanisms of cell death and the subroutine involved largely depend on the drug or toxin, and the magnitude of liver injury, and initiating mechanism of cell death. Most drugs that cause idiosyncratic toxicity or IDILI do so through participation of the adaptive immune system and HLA polymorphisms have been clearly linked to an increased number of drugs potential for hepatotoxicity. In these instances, DILI and the resultant cell death, which is immune mediated, is likely executed via death receptors and is apoptotic. Although this is a logical assumption, it is largely speculative and direct evidence in humans is lacking. The mechanisms of immune-mediated killing of hepatocytes (discussed above) likely resemble what has been shown in other immune-mediated toxicity contexts. More work on defining the possible role of DR-induced necrosis and necroptosis in IDILI is needed. Some drugs exert their toxicity by targeting mitochondria and this can result in apoptosis or necrosis of hepatocytes. APAP-induced DILI, a model of regulated necrotic cell death, is amplified by sustained JNK activation-mediated ROS generation and mitochondrial toxicity ultimately leading to MPT. The role of the JNK-binding partner Sab and its downstream mediator Src on the electron transport chain is also of critical importance. Novel biomarkers of cell death mode can potentially help distinguish between apoptotic and necrotic cell death in DILI and help not only in mechanistic studies but in prognostication in the clinical setting. It is important to better characterize these cell death pathways in various forms of DILI as interfering with cell death signaling is an attractive treatment strategy for patients with DILI, and especially in cases of impending acute liver failure.

Acknowledgment This work was supported by NIH grants K08DK109141(LD), R01DK067215 (NK), P30DK048522 (NK).

References

Acehan D, Jiang X, Morgan DG, Heuser JE, Wang X, Akey CW (2002) Three-dimensional structure of the apoptosome: implications for assembly, procaspase-9 binding, and activation. Mol Cell 9(2):423–432. S1097276502004422 [pii]

Amir M, Zhao E, Fontana L, Rosenberg H, Tanaka K, Gao G, Czaja MJ (2013) Inhibition of hepatocyte autophagy increases tumor necrosis factor-dependent liver injury by promoting caspase-8 activation. Cell Death Differ 20(7):878–887. doi:10.1038/cdd.2013.21

An J, Mehrhof F, Harms C, Lattig-Tunnemann G, Lee SL, Endres M, Li M, Sellge G, Mandic AD, Trautwein C, Donath S (2013) ARC is a novel therapeutic approach against acetaminophen-induced hepatocellular necrosis. J Hepatol 58(2):297–305. doi:10.1016/j.jhep.2012.10.002

Antoine DJ, Williams DP, Kipar A, Jenkins RE, Regan SL, Sathish JG, Kitteringham NR, Park BK (2009) High-mobility group box-1 protein and keratin-18, circulating serum proteins informative of acetaminophen-induced necrosis and apoptosis in vivo. Toxicol Sci 112(2):521–531. doi:10.1093/toxsci/kfp235

Antoine DJ, Jenkins RE, Dear JW, Williams DP, McGill MR, Sharpe MR, Craig DG, Simpson KJ, Jaeschke H, Park BK (2012) Molecular forms of HMGB1 and keratin-18 as mechanistic biomarkers for mode of cell death and prognosis during clinical acetaminophen hepatotoxicity. J Hepatol 56(5):1070–1079. doi:10.1016/j.jhep.2011.12.019

Apostolova N, Gomez-Sucerquia LJ, Gortat A, Blas-Garcia A, Esplugues JV (2011) Autophagy as a rescue mechanism in efavirenz-induced mitochondrial dysfunction: a lesson from hepatic cells. Autophagy 7(11):1402–1404. doi:10.4161/auto.7.11.17653

Ashkenazi A, Pai RC, Fong S, Leung S, Lawrence DA, Marsters SA, Blackie C, Chang L, McMurtrey AE, Hebert A, DeForge L, Koumenis IL, Lewis D, Harris L, Bussiere J, Koeppen H, Shahrokh Z, Schwall RH (1999) Safety and antitumor activity of recombinant soluble Apo2 ligand. J Clin Invest 104(2):155–162. doi:10.1172/JCI6926

Baulies A, Ribas V, Nunez S, Torres S, Alarcon-Vila C, Martinez L, Suda J, Ybanez MD, Kaplowitz N, Garcia-Ruiz C, Fernandez-Checa JC (2015) Lysosomal cholesterol accumulation sensitizes to acetaminophen hepatotoxicity by impairing mitophagy. Sci Rep 5:18017. doi:10.1038/srep18017

Bechmann LP, Jochum C, Kocabayoglu P, Sowa JP, Kassalik M, Gieseler RK, Saner F, Paul A, Trautwein C, Gerken G, Canbay A (2010) Cytokeratin 18-based modification of the MELD score improves prediction of spontaneous survival after acute liver injury. J Hepatol 53(4):639–647. doi:10.1016/j.jhep.2010.04.029

Bissuel F, Bruneel F, Habersetzer F, Chassard D, Cotte L, Chevallier M, Bernuau J, Lucet JC, Trepo C (1994) Fulminant hepatitis with severe lactate acidosis in HIV-infected patients on didanosine therapy. J Intern Med 235(4):367–371

Bradham CA, Stachlewitz RF, Gao W, Qian T, Jayadev S, Jenkins G, Hannun Y, Lemasters JJ, Thurman RG, Brenner DA (1997) Reperfusion after liver transplantation in rats differentially activates the mitogen-activated protein kinases. Hepatology 25(5):1128–1135. doi:10.1002/hep.510250514. S027091399700222X [pii]

Campos G, Schmidt-Heck W, Ghallab A, Rochlitz K, Putter L, Medinas DB, Hetz C, Widera A, Cadenas C, Begher-Tibbe B, Reif R, Gunther G, Sachinidis A, Hengstler JG, Godoy P (2014) The transcription factor CHOP, a central component of the transcriptional regulatory network induced upon CCl4 intoxication in mouse liver, is not a critical mediator of hepatotoxicity. Arch Toxicol 88(6):1267–1280. doi:10.1007/s00204-014-1240-8

Caulin C, Salvesen GS, Oshima RG (1997) Caspase cleavage of keratin 18 and reorganization of intermediate filaments during epithelial cell apoptosis. J Cell Biol 138(6):1379–1394

Chakraborty M, Fullerton AM, Semple K, Chea LS, Proctor WR, Bourdi M, Kleiner DE, Zeng X, Ryan PM, Dagur PK, Berkson JD, Reilly TP, Pohl LR (2015) Drug-induced allergic hepatitis developed in mice when myeloid-derived suppressor cells were depleted prior to halothane treatment. Hepatology. doi:10.1002/hep.27764

Chang L, Karin M (2001) Mammalian MAP kinase signalling cascades. Nature 410(6824):37–40. doi:10.1038/35065000. 35065000 [pii]

Chang L, Kamata H, Solinas G, Luo JL, Maeda S, Venuprasad K, Liu YC, Karin M (2006) The E3 ubiquitin ligase itch couples JNK activation to TNFalpha-induced cell death by inducing c-FLIP(L) turnover. Cell 124(3):601–613. doi:10.1016/j.cell.2006.01.021. S0092-8674(06)00113-9 [pii]

Chen X, Li W, Ren J, Huang D, He WT, Song Y, Yang C, Li W, Zheng X, Chen P, Han J (2014) Translocation of mixed lineage kinase domain-like protein to plasma membrane leads to necrotic cell death. Cell Res 24(1):105–121. doi:10.1038/cr.2013.171

Chen R, Zhang Y, Tang S, Lv X, Wu S, Sun F, Xia Y, Zhan SY (2015) The association between HLA-DQB1 polymorphism and antituberculosis drug-induced liver injury: a case-control study. J Clin Pharm Ther 40(1):110–115. doi:10.1111/jcpt.12211

Corazza N, Jakob S, Schaer C, Frese S, Keogh A, Stroka D, Kassahn D, Torgler R, Mueller C, Schneider P, Brunner T (2006) TRAIL receptor-mediated JNK activation and Bim phosphorylation critically regulate Fas-mediated liver damage and lethality. J Clin Invest 116(9):2493–2499. doi:10.1172/JCI27726

Craig DG, Lee P, Pryde EA, Masterton GS, Hayes PC, Simpson KJ (2011) Circulating apoptotic and necrotic cell death markers in patients with acute liver injury. Liver Int 31(8):1127–1136. doi:10.1111/j.1478-3231.2011.02528.x

Daly AK, Donaldson PT, Bhatnagar P, Shen Y, Pe'er I, Floratos A, Daly MJ, Goldstein DB, John S, Nelson MR, Graham J, Park BK, Dillon JF, Bernal W, Cordell HJ, Pirmohamed M, Aithal GP, Day CP, DILIGEN Study, International SAEC (2009) HLA-B*5701 genotype is a major determinant of drug-induced liver injury due to flucloxacillin. Nat Genet 41(7):816–819. doi:10.1038/ng.379

Dannappel M, Vlantis K, Kumari S, Polykratis A, Kim C, Wachsmuth L, Eftychi C, Lin J, Corona T, Hermance N, Zelic M, Kirsch P, Basic M, Bleich A, Kelliher M, Pasparakis M (2014) RIPK1 maintains epithelial homeostasis by inhibiting apoptosis and necroptosis. Nature 513(7516):90–94. doi:10.1038/nature13608

Dara L, Ji C, Kaplowitz N (2011) The contribution of endoplasmic reticulum stress to liver diseases. Hepatology 53(5):1752–1763. doi:10.1002/hep.24279

Dara L, Han D, Kaplowitz N. (2012) Mechanisms of cell death and relevance to drug toxicity. Drug induced liver disease, 3rd ed. Academic Press is an imprint of ElsevierCambridge, Massachusetts

Dara L, Johnson H, Suda J, Win S, Gaarde W, Han D, Kaplowitz N (2015a) Receptor interacting protein kinase 1 mediates murine acetaminophen toxicity independent of the necrosome and not through necroptosis. Hepatology 62(6):1847–1857. doi:10.1002/hep.27939

Dara L, Liu ZX, Kaplowitz N (2015b) Mechanisms of adaptation and progression in idiosyncratic drug induced liver injury, clinical implications. Liver Int 36(2):158–165. doi:10.1111/liv.12988

Degterev A, Huang Z, Boyce M, Li Y, Jagtap P, Mizushima N, Cuny GD, Mitchison TJ, Moskowitz MA, Yuan J (2005) Chemical inhibitor of nonapoptotic cell death with therapeutic potential for ischemic brain injury. Nat Chem Biol 1(2):112–119. doi:10.1038/nchembio711. nchembio711 [pii]

Degterev A, Zhou W, Maki JL, Yuan J (2014) Assays for necroptosis and activity of RIP kinases. Methods Enzymol 545:1–33. doi:10.1016/B978-0-12-801430-1.00001-9

Deutsch M, Graffeo CS, Rokosh R, Pansari M, Ochi A, Levie EM, Van Heerden E, Tippens DM, Greco S, Barilla R, Tomkotter L, Zambirinis CP, Avanzi N, Gulati R, Pachter HL, Torres-Hernandez A, Eisenthal A, Daley D, Miller G (2015) Divergent effects of RIP1 or RIP3 blockade in murine models of acute liver injury. Cell Death Dis 6:e1759. doi:10.1038/cddis.2015.126

Feng G, Kaplowitz N (2000) Colchicine protects mice from the lethal effect of an agonistic anti-Fas antibody. J Clin Invest 105(3):329–339. doi:10.1172/JCI7398

Fredriksson L, Herpers B, Benedetti G, Matadin Q, Puigvert JC, de Bont H, Dragovic S, Vermeulen NP, Commandeur JN, Danen E, de Graauw M, van de Water B (2011) Diclofenac inhibits tumor necrosis factor-alpha-induced nuclear factor-kappaB activation causing synergistic hepatocyte apoptosis. Hepatology 53(6):2027–2041. doi:10.1002/hep.24314

Fredriksson L, Wink S, Herpers B, Benedetti G, Hadi M, de Bont H, Groothuis G, Luijten M, Danen E, de Graauw M, Meerman J, van de Water B (2014) Drug-induced endoplasmic reticulum and oxidative stress responses independently sensitize toward TNFalpha-mediated hepatotoxicity. Toxicol Sci 140(1):144–159. doi:10.1093/toxsci/kfu072

French LE, Tschopp J (1999) The TRAIL to selective tumor death. Nat Med 5(2):146–147. doi:10.1038/5505

Fromenty B, Pessayre D (1995) Inhibition of mitochondrial beta-oxidation as a mechanism of hepatotoxicity. Pharmacol Ther 67(1):101–154

Fu L, Lin YD, Elrod HA, Yue P, Oh Y, Li B, Tao H, Chen GZ, Shin DM, Khuri FR, Sun SY (2010) c-Jun NH2-terminal kinase-dependent upregulation of DR5 mediates cooperative induction of apoptosis by perifosine and TRAIL. Mol Cancer 9:315. doi:10.1186/1476-4598-9-315

Galluzzi L, Vitale I, Abrams JM, Alnemri ES, Baehrecke EH, Blagosklonny MV, Dawson TM, Dawson VL, El-Deiry WS, Fulda S, Gottlieb E, Green DR, Hengartner MO, Kepp O, Knight RA, Kumar S, Lipton SA, Lu X, Madeo F, Malorni W, Mehlen P, Nunez G, Peter ME, Piacentini M, Rubinsztein DC, Shi Y, Simon HU, Vandenabeele P, White E, Yuan J, Zhivotovsky B, Melino G, Kroemer G (2012) Molecular definitions of cell death subroutines: recommendations of the nomenclature committee on cell death 2012. Cell Death Differ 19(1):107–120. doi:10.1038/cdd.2011.96

Geserick P, Wang J, Schilling R, Horn S, Harris PA, Bertin J, Gough PJ, Feoktistova M, Leverkus M (2015) Absence of RIPK3 predicts necroptosis resistance in malignant melanoma. Cell Death Dis 6:e1884. doi:10.1038/cddis.2015.240

Gores GJ, Kaufmann SH (2001) Is TRAIL hepatotoxic? Hepatology 34(1):3–6. doi:10.1053/jhep.2001.25173

Grove JI, Aithal GP (2015) Human leukocyte antigen genetic risk factors of drug-induced liver toxicology. Expert Opin Drug Metab Toxicol 11(3):395–409. doi:10.1517/17425255.2015.992414

Guicciardi ME, Gores GJ (2009) Life and death by death receptors. FASEB J 23(6):1625–1637. doi:10.1096/fj.08-111005

Gunawan BK, Liu ZX, Han D, Hanawa N, Gaarde WA, Kaplowitz N (2006) c-Jun N-terminal kinase plays a major role in murine acetaminophen hepatotoxicity. Gastroenterology 131(1):165–178. doi:10.1053/j.gastro.2006.03.045. S0016-5085(06)00712-8 [pii]

Han D, Hanawa N, Saberi B, Kaplowitz N (2006) Hydrogen peroxide and redox modulation sensitize primary mouse hepatocytes to TNF-induced apoptosis. Free Radic Biol Med 41(4):627–639. doi:10.1016/j.freeradbiomed.2006.05.002. S0891-5849(06)00294-2 [pii]

Han D, Dara L, Win S, Than TA, Yuan L, Abbasi SQ, Liu ZX, Kaplowitz N (2013) Regulation of drug-induced liver injury by signal transduction pathways: critical role of mitochondria. Trends Pharmacol Sci 34(4):243–253. doi:10.1016/j.tips.2013.01.009

Hanawa N, Shinohara M, Saberi B, Gaarde WA, Han D, Kaplowitz N (2008) Role of JNK translocation to mitochondria leading to inhibition of mitochondria bioenergetics in acetaminophen-induced liver injury. J Biol Chem 283(20):13565–13577. doi:10.1074/jbc.M708916200. M708916200 [pii]

Hatting M, Zhao G, Schumacher F, Sellge G, Al Masaoudi M, Gabetaler N, Boekschoten M, Muller M, Liedtke C, Cubero FJ, Trautwein C (2013) Hepatocyte caspase-8 is an essential modulator of steatohepatitis in rodents. Hepatology 57(6):2189–2201. doi:10.1002/hep.26271

Hautekeete ML, Horsmans Y, Van Waeyenberge C, Demanet C, Henrion J, Verbist L, Brenard R, Sempoux C, Michielsen PP, Yap PS, Rahier J, Geubel AP (1999) HLA association of amoxicillin-clavulanate—induced hepatitis. Gastroenterology 117(5):1181–1186

Hetz C, Chevet E, Harding HP (2013) Targeting the unfolded protein response in disease. Nat Rev Drug Discov 12(9):703–719. doi:10.1038/nrd3976

Higuchi H, Grambihler A, Canbay A, Bronk SF, Gores GJ (2004) Bile acids up-regulate death receptor 5/TRAIL-receptor 2 expression via a c-Jun N-terminal kinase-dependent pathway involving Sp1. J Biol Chem 279(1):51–60. doi:10.1074/jbc.M309476200

Hildebrand JM, Tanzer MC, Lucet IS, Young SN, Spall SK, Sharma P, Pierotti C, Garnier JM, Dobson RC, Webb AI, Tripaydonis A, Babon JJ, Mulcair MD, Scanlon MJ, Alexander WS, Wilks AF, Czabotar PE, Lessene G, Murphy JM, Silke J (2014) Activation of the pseudokinase MLKL unleashes the four-helix bundle domain to induce membrane localization and necroptotic cell death. Proc Natl Acad Sci U S A 111(42):15072–15077. doi:10.1073/pnas.1408987111

Hirata K, Takagi H, Yamamoto M, Matsumoto T, Nishiya T, Mori K, Shimizu S, Masumoto H, Okutani Y (2008) Ticlopidine-induced hepatotoxicity is associated with specific human leukocyte antigen genomic subtypes in Japanese patients: a preliminary case-control study. Pharmacogenomics J 8(1):29–33. doi:10.1038/sj.tpj.6500442

Holler N, Zaru R, Micheau O, Thome M, Attinger A, Valitutti S, Bodmer JL, Schneider P, Seed B, Tschopp J (2000) Fas triggers an alternative, caspase-8-independent cell death pathway using the kinase RIP as effector molecule. Nat Immunol 1(6):489–495. doi:10.1038/82732

Hu J, Han H, Lau MY, Lee H, MacVeigh-Aloni M, Ji C (2015) Effects of combined alcohol and anti-HIV drugs on cellular stress responses in primary hepatocytes and hepatic stellate and kupffer cells. Alcohol Clin Exp Res 39(1):11–20. doi:10.1111/acer.12608

Irmler M, Thome M, Hahne M, Schneider P, Hofmann K, Steiner V, Bodmer JL, Schroter M, Burns K, Mattmann C, Rimoldi D, French LE, Tschopp J (1997) Inhibition of death receptor signals by cellular FLIP. Nature 388(6638):190–195. doi:10.1038/40657

Ji C, Mehrian-Shai R, Chan C, Hsu YH, Kaplowitz N (2005) Role of CHOP in hepatic apoptosis in the murine model of intragastric ethanol feeding. Alcohol Clin Exp Res 29(8):1496–1503. 00000374-200508000-00017 [pii]

Jost PJ, Grabow S, Gray D, McKenzie MD, Nachbur U, Huang DC, Bouillet P, Thomas HE, Borner C, Silke J, Strasser A, Kaufmann T (2009) XIAP discriminates between type I and type II FAS-induced apoptosis. Nature 460(7258):1035–1039. doi:10.1038/nature08229

Kahraman A, Barreyro FJ, Bronk SF, Werneburg NW, Mott JL, Akazawa Y, Masuoka HC, Howe CL, Gores GJ (2008) TRAIL mediates liver injury by the innate immune system in the bile duct-ligated mouse. Hepatology 47(4):1317–1330. doi:10.1002/hep.22136

Kang TB, Ben-Moshe T, Varfolomeev EE, Pewzner-Jung Y, Yogev N, Jurewicz A, Waisman A, Brenner O, Haffner R, Gustafsson E, Ramakrishnan P, Lapidot T, Wallach D (2004) Caspase-8 serves both apoptotic and nonapoptotic roles. J Immunol 173(5):2976–2984

Kang YJ, Bang BR, Han KH, Hong L, Shim EJ, Ma J, Lerner RA, Otsuka M (2015) Regulation of NKT cell-mediated immune responses to tumours and liver inflammation by mitochondrial PGAM5-Drp1 signalling. Nat Commun 6:8371. doi:10.1038/ncomms9371

Karin M, Lin A (2002) NF-kappaB at the crossroads of life and death. Nat Immunol 3(3):221–227. doi:10.1038/ni0302-221. ni0302-221 [pii]

Kasof GM, Prosser JC, Liu D, Lorenzi MV, Gomes BC (2000) The RIP-like kinase, RIP3, induces apoptosis and NF-kappaB nuclear translocation and localizes to mitochondria. FEBS Lett 473(3):285–291

Kassahun K, Farrell K, Abbott F (1991) Identification and characterization of the glutathione and N-acetylcysteine conjugates of (E)-2-propyl-2,4-pentadienoic acid, a toxic metabolite of valproic acid, in rats and humans. Drug Metab Dispos 19(2):525–535

Kassahun K, Hu P, Grillo MP, Davis MR, Jin L, Baillie TA (1994) Metabolic activation of unsaturated derivatives of valproic acid. Identification of novel glutathione adducts formed through coenzyme A-dependent and -independent processes. Chem Biol Interact 90(3):253–275

Kaufmann T, Jost PJ, Pellegrini M, Puthalakath H, Gugasyan R, Gerondakis S, Cretney E, Smyth MJ, Silke J, Hakem R, Bouillet P, Mak TW, Dixit VM, Strasser A (2009) Fatal hepatitis mediated by tumor necrosis factor TNFalpha requires caspase-8 and involves the BH3-only proteins Bid and Bim. Immunity 30(1):56–66. doi:10.1016/j.immuni.2008.10.017. S1074-7613(08)00543-8 [pii]

Kew MC (2000) Serum aminotransferase concentration as evidence of hepatocellular damage. Lancet 355(9204):591–592. doi:10.1016/S0140-6736(99)00219-6

Kim HE, Du F, Fang M, Wang X (2005) Formation of apoptosome is initiated by cytochrome c-induced dATP hydrolysis and subsequent nucleotide exchange on Apaf-1. Proc Natl Acad Sci U S A 102(49):17545–17550. doi:10.1073/pnas.0507900102. 0507900102 [pii]

Kindmark A, Jawaid A, Harbron CG, Barratt BJ, Bengtsson OF, Andersson TB, Carlsson S, Cederbrant KE, Gibson NJ, Armstrong M, Lagerstrom-Fermer ME, Dellsen A, Brown EM, Thornton M, Dukes C, Jenkins SC, Firth MA, Harrod GO, Pinel TH, Billing-Clason SM, Cardon LR, March RE (2008) Genome-wide pharmacogenetic investigation of a hepatic adverse event without clinical signs of immunopathology suggests an underlying immune pathogenesis. Pharmacogenomics J 8(3):186–195. doi:10.1038/sj.tpj.6500458

Klionsky DJ, Cregg JM, Dunn WA Jr, Emr SD, Sakai Y, Sandoval IV, Sibirny A, Subramani S, Thumm M, Veenhuis M, Ohsumi Y (2003) A unified nomenclature for yeast autophagy-related genes. Dev Cell 5(4):539–545

Knolle PA, Thimme R (2014) Hepatic immune regulation and its involvement in viral hepatitis infection. Gastroenterology 146(5):1193–1207. doi:10.1053/j.gastro.2013.12.036

Kon K, Kim JS, Jaeschke H, Lemasters JJ (2004) Mitochondrial permeability transition in acetaminophen-induced necrosis and apoptosis of cultured mouse hepatocytes. Hepatology 40(5):1170–1179. doi:10.1002/hep.20437

Ku NO, Strnad P, Bantel H, Bishr Omary M (2016) Keratins: biomarkers and modulators of apoptotic and necrotic cell death in the liver. Hepatology. doi:10.1002/hep.28493

Kurosaki M, Takagi H, Mori M (2000) HLA-A33/B44/DR6 is highly related to intrahepatic cholestasis induced by tiopronin. Dig Dis Sci 45(6):1103–1108

Lagos-Quintana M, Rauhut R, Yalcin A, Meyer J, Lendeckel W, Tuschl T (2002) Identification of tissue-specific microRNAs from mouse. Curr Biol 12(9):735–739

Lavallard VJ, Bonnafous S, Patouraux S, Saint-Paul MC, Rousseau D, Anty R, Le Marchand-Brustel Y, Tran A, Gual P (2011) Serum markers of hepatocyte death and apoptosis are non invasive biomarkers of severe fibrosis in patients with alcoholic liver disease. PLoS One 6(3):e17599. doi:10.1371/journal.pone.0017599

Lee KH, Feig C, Tchikov V, Schickel R, Hallas C, Schutze S, Peter ME, Chan AC (2006) The role of receptor internalization in CD95 signaling. EMBO J 25(5):1009–1023. doi:10.1038/sj.emboj.7601016

Lei K, Davis RJ (2003) JNK phosphorylation of Bim-related members of the Bcl2 family induces Bax-dependent apoptosis. Proc Natl Acad Sci U S A 100(5):2432–2437. doi:10.1073/pnas.0438011100. 0438011100 [pii]

Lei K, Nimnual A, Zong WX, Kennedy NJ, Flavell RA, Thompson CB, Bar-Sagi D, Davis RJ (2002) The Bax subfamily of Bcl2-related proteins is essential for apoptotic signal transduction by c-Jun NH(2)-terminal kinase. Mol Cell Biol 22(13):4929–4942

Leist M, Gantner F, Bohlinger I, Germann PG, Tiegs G, Wendel A (1994) Murine hepatocyte apoptosis induced in vitro and in vivo by TNF-alpha requires transcriptional arrest. J Immunol 153(4):1778–1788

Leist M, Gantner F, Naumann H, Bluethmann H, Vogt K, Brigelius-Flohe R, Nicotera P, Volk HD, Wendel A (1997) Tumor necrosis factor-induced apoptosis during the poisoning of mice with hepatotoxins. Gastroenterology 112(3):923–934. S0016508597001327 [pii]

Lemasters JJ, Nieminen AL, Qian T, Trost LC, Elmore SP, Nishimura Y, Crowe RA, Cascio WE, Bradham CA, Brenner DA, Herman B (1998) The mitochondrial permeability transition in cell death: a common mechanism in necrosis, apoptosis and autophagy. Biochim Biophys Acta 1366(1–2):177–196

Levy RH, Rettenmeier AW, Anderson GD, Wilensky AJ, Friel PN, Baillie TA, Acheampong A, Tor J, Guyot M, Loiseau P (1990) Effects of polytherapy with phenytoin, carbamazepine, and stiripentol on formation of 4-ene-valproate, a hepatotoxic metabolite of valproic acid. Clin Pharmacol Ther 48(3):225–235

Lewis JH, Ranard RC, Caruso A, Jackson LK, Mullick F, Ishak KG, Seeff LB, Zimmerman HJ (1989) Amiodarone hepatotoxicity: prevalence and clinicopathologic correlations among 104 patients. Hepatology 9(5):679–685

Liedtke C, Bangen JM, Freimuth J, Beraza N, Lambertz D, Cubero FJ, Hatting M, Karlmark KR, Streetz KL, Krombach GA, Tacke F, Gassler N, Riethmacher D, Trautwein C (2011) Loss of caspase-8 protects mice against inflammation-related hepatocarcinogenesis but induces non-apoptotic liver injury. Gastroenterology 141(6):2176–2187. doi:10.1053/j.gastro.2011.08.037. S0016-5085(11)01220-0 [pii]

Linkermann A, Green DR (2014) Necroptosis. N Engl J Med 370(5):455–465. doi:10.1056/NEJMra1310050

Liu X, Kim CN, Yang J, Jemmerson R, Wang X (1996) Induction of apoptotic program in cell-free extracts: requirement for dATP and cytochrome c. Cell 86(1):147–157. S0092-8674(00)80085-9 [pii]

Liu H, Lo CR, Czaja MJ (2002) NF-kappaB inhibition sensitizes hepatocytes to TNF-induced apoptosis through a sustained activation of JNK and c-Jun. Hepatology 35(4):772–778. doi:10.1053/jhep.2002.32534. S0270913902315994 [pii]

Lockshin RA, Williams CM (1965) Programmed cell death-I. Cytology of degeneration in the intersegmental muscles of the pernyi silkmoth. J Insect Physiol 11:123–133

Lockshin RA, Zakeri Z (2004) Apoptosis, autophagy, and more. Int J Biochem Cell Biol 36(12):2405–2419. doi:10.1016/j.biocel.2004.04.011. S1357272504001669 [pii]

LoGuidice A, Boelsterli UA (2011) Acetaminophen overdose-induced liver injury in mice is mediated by peroxynitrite independently of the cyclophilin D-regulated permeability transition. Hepatology 54(3):969–978. doi:10.1002/hep.24464

Lotze MT, Tracey KJ (2005) High-mobility group box 1 protein (HMGB1): nuclear weapon in the immune arsenal. Nat Rev Immunol 5(4):331–342. doi:10.1038/nri1594

Lou H, Kaplowitz N (2007) Glutathione depletion down-regulates tumor necrosis factor alpha-induced NF-kappaB activity via IkappaB kinase-dependent and -independent mechanisms. J Biol Chem 282(40):29470–29481. doi:10.1074/jbc.M706145200

Lucena MI, Molokhia M, Shen Y, Urban TJ, Aithal GP, Andrade RJ, Day CP, Ruiz-Cabello F, Donaldson PT, Stephens C, Pirmohamed M, Romero-Gomez M, Navarro JM, Fontana RJ, Miller M, Groome M, Bondon-Guitton E, Conforti A, Stricker BH, Carvajal A, Ibanez L, Yue QY, Eichelbaum M, Floratos A, Pe'er I, Daly MJ, Goldstein DB, Dillon JF, Nelson MR, Watkins PB, Daly AK, Spanish DR, Eudragene, Dilin, Diligen, International SAEC (2011) Susceptibility to amoxicillin-clavulanate-induced liver injury is influenced by multiple HLA class I and II alleles. Gastroenterology 141(1):338–347. doi:10.1053/j.gastro.2011.04.001

Luedde T, Kaplowitz N, Schwabe RF (2014) Cell death and cell death responses in liver disease: mechanisms and clinical relevance. Gastroenterology 147(4):765–783. doi:10.1053/j.gastro.2014.07.018e764

Lundback P, Lea JD, Sowinska A, Ottosson L, Melin Furst C, Steen J, Aulin C, Clarke JI, Kipar A, Klevenvall L, Yang H, Palmblad K, Park BK, Tracey KJ, Blom AM, Andersson U, Antoine DJ, Erlandsson Harris H (2016) A novel HMGB1 neutralizing chimeric antibody attenuates DILI and post-injury inflammation. Hepatology. doi:10.1002/hep.28736

Mak A, Uetrecht J (2015) The role of CD8 T cells in amodiaquine-induced liver injury in PD1−/− mice cotreated with anti-CTLA-4. Chem Res Toxicol. doi:10.1021/acs.chemrestox.5b00137

Malhi H (2007) TRAILs and tribulation. Hepatology 46(5):1320–1322. doi:10.1002/hep.21913

Malhi H, Guicciardi ME, Gores GJ (2010) Hepatocyte death: a clear and present danger. Physiol Rev 90(3):1165–1194. doi:10.1152/physrev.00061.2009. 90/3/1165 [pii]

Mallal S, Phillips E, Carosi G, Molina JM, Workman C, Tomazic J, Jagel-Guedes E, Rugina S, Kozyrev O, Cid JF, Hay P, Nolan D, Hughes S, Hughes A, Ryan S, Fitch N, Thorborn D, Benbow A, Team P-S (2008) HLA-B*5701 screening for hypersensitivity to abacavir. N Engl J Med 358(6):568–579. doi:10.1056/NEJMoa0706135

McGill MR, Du K, Xie Y, Bajt ML, Ding WX, Jaeschke H (2015) The role of the c-Jun N-terminal kinases 1/2 and receptor-interacting protein kinase 3 in furosemide-induced liver injury. Xenobiotica 45(5):442–449. doi:10.3109/00498254.2014.986250

McKenzie R, Fried MW, Sallie R, Conjeevaram H, Di Bisceglie AM, Park Y, Savarese B, Kleiner D, Tsokos M, Luciano C et al (1995) Hepatic failure and lactic acidosis due to fialuridine (FIAU), an investigational nucleoside analogue for chronic hepatitis B. N Engl J Med 333(17):1099–1105. doi:10.1056/NEJM199510263331702

Metushi IG, Hayes MA, Uetrecht J (2015) Treatment of PD-1(−/−) mice with amodiaquine and anti-CTLA4 leads to liver injury similar to idiosyncratic liver injury in patients. Hepatology 61(4):1332–1342. doi:10.1002/hep.27549

Micheau O, Tschopp J (2003) Induction of TNF receptor I-mediated apoptosis via two sequential signaling complexes. Cell 114(2):181–190. S009286740300521X [pii]

Mizushima N, Levine B, Cuervo AM, Klionsky DJ (2008) Autophagy fights disease through cellular self-digestion. Nature 451(7182):1069–1075. doi:10.1038/nature06639. nature06639 [pii]

Monshi MM, Faulkner L, Gibson A, Jenkins RE, Farrell J, Earnshaw CJ, Alfirevic A, Cederbrant K, Daly AK, French N, Pirmohamed M, Park BK, Naisbitt DJ (2013) Human leukocyte antigen (HLA)-B*57:01-restricted activation of drug-specific T cells provides the immunological basis for flucloxacillin-induced liver injury. Hepatology 57(2):727–739. doi:10.1002/hep.26077

Murphy JM, Czabotar PE, Hildebrand JM, Lucet IS, Zhang JG, Alvarez-Diaz S, Lewis R, Lalaoui N, Metcalf D, Webb AI, Young SN, Varghese LN, Tannahill GM, Hatchell EC, Majewski IJ, Okamoto T, Dobson RC, Hilton DJ, Babon JJ, Nicola NA, Strasser A, Silke J, Alexander WS (2013) The pseudokinase MLKL mediates necroptosis via a molecular switch mechanism. Immunity 39(3):443–453. doi:10.1016/j.immuni.2013.06.018

Musso G, Gambino R, Cassader M, Pagano G (2011) Meta-analysis: natural history of non-alcoholic fatty liver disease (NAFLD) and diagnostic accuracy of non-invasive tests for liver disease severity. Ann Med 43(8):617–649. doi:10.3109/07853890.2010.518623

Nagai H, Matsumaru K, Feng G, Kaplowitz N (2002) Reduced glutathione depletion causes necrosis and sensitization to tumor necrosis factor-alpha-induced apoptosis in cultured mouse hepatocytes. Hepatology 36(1):55–64. doi:10.1053/jhep.2002.33995. S027091390200006X [pii]

Naviaux RK, Nguyen KV (2004) POLG mutations associated with Alpers' syndrome and mitochondrial DNA depletion. Ann Neurol 55(5):706–712. doi:10.1002/ana.20079

Ni HM, Bockus A, Boggess N, Jaeschke H, Ding WX (2012a) Activation of autophagy protects against acetaminophen-induced hepatotoxicity. Hepatology 55(1):222–232. doi:10.1002/hep.24690

Ni HM, Boggess N, McGill MR, Lebofsky M, Borude P, Apte U, Jaeschke H, Ding WX (2012b) Liver-specific loss of Atg5 causes persistent activation of Nrf2 and protects against acetaminophen-induced liver injury. Toxicol Sci 127(2):438–450. doi:10.1093/toxsci/kfs133

Ni HM, McGill MR, Chao X, Du K, Williams JA, Xie Y, Jaeschke H, Ding WX (2016) Removal of acetaminophen protein adducts by autophagy protects against acetaminophen-induced liver injury in mice. J Hepatol 65(2):354–362. doi:10.1016/j.jhep.2016.04.025

Nicoletti P, Werk AN, Sawle A, Shen Y, Urban TJ, Coulthard SA, Bjornsson ES, Cascorbi I, Floratos A, Stammschulte T, Gundert-Remy U, Nelson MR, Aithal GP, Daly AK, International Drug-induced Liver Injury C (2016) HLA-DRB1*16: 01-DQB1*05: 02 is a novel genetic risk factor for flupirtine-induced liver injury. Pharmacogenet Genomics 26(5):218–224. doi:10.1097/FPC.0000000000000209

O'Donohue J, Oien KA, Donaldson P, Underhill J, Clare M, MacSween RN, Mills PR (2000) Co-amoxiclav jaundice: clinical and histological features and HLA class II association. Gut 47(5):717–720

Omary MB, Ku NO, Strnad P, Hanada S (2009) Toward unraveling the complexity of simple epithelial keratins in human disease. J Clin Invest 119(7):1794–1805. doi:10.1172/JCI37762

Ostrov DA, Grant BJ, Pompeu YA, Sidney J, Harndahl M, Southwood S, Oseroff C, Lu S, Jakoncic J, de Oliveira CA, Yang L, Mei H, Shi L, Shabanowitz J, English AM, Wriston A, Lucas A, Phillips E, Mallal S, Grey HM, Sette A, Hunt DF, Buus S, Peters B (2012) Drug hypersensitivity caused by alteration of the MHC-presented self-peptide repertoire. Proc Natl Acad Sci U S A 109(25):9959–9964. doi:10.1073/pnas.1207934109

Phillips E, Bartlett JA, Sanne I, Lederman MM, Hinkle J, Rousseau F, Dunn D, Pavlos R, James I, Mallal SA, Haas DW (2013) Associations between HLA-DRB1*0102, HLA-B*5801, and hepatotoxicity during initiation of nevirapine-containing regimens in South Africa. J Acquir Immune Defic Syndr 62(2):e55–e57. doi:10.1097/QAI.0b013e31827ca50f

Pichler WJ (2002) Pharmacological interaction of drugs with antigen-specific immune receptors: the p-i concept. Curr Opin Allergy Clin Immunol 2(4):301–305

Pierce RH, Campbell JS, Stephenson AB, Franklin CC, Chaisson M, Poot M, Kavanagh TJ, Rabinovitch PS, Fausto N (2000) Disruption of redox homeostasis in tumor necrosis factor-induced apoptosis in a murine hepatocyte cell line. Am J Pathol 157(1):221–236. doi:10.1016/S0002-9440(10)64533-6. S0002-9440(10)64533-6 [pii]

Ponchaut S, van Hoof F, Veitch K (1992) In vitro effects of valproate and valproate metabolites on mitochondrial oxidations. Relevance of CoA sequestration to the observed inhibitions. Biochem Pharmacol 43(11):2435–2442

Porceddu M, Buron N, Roussel C, Labbe G, Fromenty B, Borgne-Sanchez A (2012) Prediction of liver injury induced by chemicals in human with a multiparametric assay on isolated mouse liver mitochondria. Toxicol Sci 129(2):332–345. doi:10.1093/toxsci/kfs197

Protzer U, Maini MK, Knolle PA (2012) Living in the liver: hepatic infections. Nat Rev Immunol 12(3):201–213. doi:10.1038/nri3169

Quarato G, Guy CS, Grace CR, Llambi F, Nourse A, Rodriguez DA, Wakefield R, Frase S, Moldoveanu T, Green DR (2016) Sequential engagement of distinct MLKL phosphatidylinositol-binding sites executes necroptosis. Mol Cell 61(4):589–601. doi:10.1016/j.molcel.2016.01.011

Rahman SM, Schroeder-Gloeckler JM, Janssen RC, Jiang H, Qadri I, Maclean KN, Friedman JE (2007) CCAAT/enhancing binding protein beta deletion in mice attenuates inflammation, endoplasmic reticulum stress, and lipid accumulation in diet-induced nonalcoholic steatohepatitis. Hepatology 45(5):1108–1117. doi:10.1002/hep.21614

Ramachandran A, Lebofsky M, Baines CP, Lemasters JJ, Jaeschke H (2011) Cyclophilin D deficiency protects against acetaminophen-induced oxidant stress and liver injury. Free Radic Res 45(2):156–164. doi:10.3109/10715762.2010.520319

Ramachandran A, McGill MR, Xie Y, Ni HM, Ding WX, Jaeschke H (2013) The receptor interacting protein kinase 3 is a critical early mediator of acetaminophen-induced hepatocyte necrosis in mice. Hepatology 58(6):2099–2108. doi:10.1002/hep.26547

Ron D, Walter P (2007) Signal integration in the endoplasmic reticulum unfolded protein response. Nat Rev Mol Cell Biol 8(7):519–529. doi:10.1038/nrm2199

Rutherford A, King LY, Hynan LS, Vedvyas C, Lin W, Lee WM, Chung RT, Group ALFS (2012) Development of an accurate index for predicting outcomes of patients with acute liver failure. Gastroenterology 143(5):1237–1243. doi:10.1053/j.gastro.2012.07.113

Sabapathy K, Hochedlinger K, Nam SY, Bauer A, Karin M, Wagner EF (2004) Distinct roles for JNK1 and JNK2 in regulating JNK activity and c-Jun-dependent cell proliferation. Mol Cell 15(5):713–725. doi:10.1016/j.molcel.2004.08.028. S1097276504005143 [pii]

Scaffidi C, Schmitz I, Zha J, Korsmeyer SJ, Krammer PH, Peter ME (1999) Differential modulation of apoptosis sensitivity in CD95 type I and type II cells. J Biol Chem 274(32): 22532–22538

Sharma SK, Balamurugan A, Saha PK, Pandey RM, Mehra NK (2002) Evaluation of clinical and immunogenetic risk factors for the development of hepatotoxicity during antituberculosis treatment. Am J Respir Crit Care Med 166(7):916–919. doi:10.1164/rccm.2108091

Silva MF, Ruiter JP, Illst L, Jakobs C, Duran M, de Almeida IT, Wanders RJ (1997) Valproate inhibits the mitochondrial pyruvate-driven oxidative phosphorylation in vitro. J Inherit Metab Dis 20(3):397–400

Singer JB, Lewitzky S, Leroy E, Yang F, Zhao X, Klickstein L, Wright TM, Meyer J, Paulding CA (2010) A genome-wide study identifies HLA alleles associated with lumiracoxib-related liver injury. Nat Genet 42(8):711–714. doi:10.1038/ng.632

Siu WP, Pun PB, Latchoumycandane C, Boelsterli UA (2008) Bax-mediated mitochondrial outer membrane permeabilization (MOMP), distinct from the mitochondrial permeability transition, is a key mechanism in diclofenac-induced hepatocyte injury: multiple protective roles of cyclosporin A. Toxicol Appl Pharmacol 227(3):451–461. doi:10.1016/j.taap.2007.11.030. S0041-008X(07)00523-6 [pii]

Song L, Zhang Z, Zhang J, Zhu X, He L, Shi Z, Gao L, Feng F (2016) Ratio of microRNA-122/155 in isoniazid-induced acute liver injury in mice. Exp Ther Med 12(2):889–894. doi:10.3892/etm.2016.3375

Spraggs CF, Budde LR, Briley LP, Bing N, Cox CJ, King KS, Whittaker JC, Mooser VE, Preston AJ, Stein SH, Cardon LR (2011) HLA-DQA1*02:01 is a major risk factor for lapatinib-induced hepatotoxicity in women with advanced breast cancer. Journal of Clin Oncol 29(6):667–673. doi:10.1200/JCO.2010.31.3197

Starkey Lewis PJ, Dear J, Platt V, Simpson KJ, Craig DG, Antoine DJ, French NS, Dhaun N, Webb DJ, Costello EM, Neoptolemos JP, Moggs J, Goldring CE, Park BK (2011) Circulating microRNAs as potential markers of human drug-induced liver injury. Hepatology 54(5):1767–1776. doi:10.1002/hep.24538

Stewart JD, Horvath R, Baruffini E, Ferrero I, Bulst S, Watkins PB, Fontana RJ, Day CP, Chinnery PF (2010) Polymerase gamma gene POLG determines the risk of sodium valproate-induced liver toxicity. Hepatology 52(5):1791–1796. doi:10.1002/hep.23891

Stine JG, Chalasani N (2015) Chronic liver injury induced by drugs: a systematic review. Liver Int 35(11):2343–2353. doi:10.1111/liv.12958

Strasser A, Jost PJ, Nagata S (2009) The many roles of FAS receptor signaling in the immune system. Immunity 30(2):180–192. doi:10.1016/j.immuni.2009.01.001

Strnad P, Paschke S, Jang KH, Ku NO (2012) Keratins: markers and modulators of liver disease. Curr Opin Gastroenterol 28(3):209–216. doi:10.1097/MOG.0b013e3283525cb8

Suda, J., Dara, L., Yang, L., Aghajan, M., Song, Y., Kaplowitz, N. & Liu, Z. X. 2016. Knockdown of RIPK1 Markedly Exacerbates Murine Immune-Mediated Liver Injury through Massive Apoptosis of Hepatocytes, Independent of Necroptosis and Inhibition of NF-kappaB. J Immunol, 197, 3120–3129

Sun X, Lee J, Navas T, Baldwin DT, Stewart TA, Dixit VM (1999) RIP3, a novel apoptosis-inducing kinase. J Biol Chem 274(24):16871–16875

Takahashi N, Vereecke L, Bertrand MJ, Duprez L, Berger SB, Divert T, Goncalves A, Sze M, Gilbert B, Kourula S, Goossens V, Lefebvre S, Gunther C, Becker C, Bertin J, Gough PJ, Declercq W, van Loo G, Vandenabeele P (2014) RIPK1 ensures intestinal homeostasis by protecting the epithelium against apoptosis. Nature 513(7516):95–99. doi:10.1038/nature13706

Takeda K, Kojima Y, Ikejima K, Harada K, Yamashina S, Okumura K, Aoyama T, Frese S, Ikeda H, Haynes NM, Cretney E, Yagita H, Sueyoshi N, Sato N, Nakanuma Y, Smyth MJ, Okumura K (2008) Death receptor 5 mediated-apoptosis contributes to cholestatic liver disease. Proc Natl Acad Sci U S A 105(31):10895–10900. doi:10.1073/pnas.0802702105

Takemoto K, Hatano E, Iwaisako K, Takeiri M, Noma N, Ohmae S, Toriguchi K, Tanabe K, Tanaka H, Seo S, Taura K, Machida K, Takeda N, Saji S, Uemoto S, Asagiri M (2014) Necrostatin-1 protects against reactive oxygen species (ROS)-induced hepatotoxicity in acetaminophen-induced acute liver failure. FEBS Open Bio 4:777–787. doi:10.1016/j.fob.2014.08.007

Thakral S, Ghoshal K (2015) miR-122 is a unique molecule with great potential in diagnosis, prognosis of liver disease, and therapy both as miRNA mimic and antimir. Curr Gene Ther 15(2):142–150

Thapaliya S, Wree A, Povero D, Inzaugarat ME, Berk M, Dixon L, Papouchado BG, Feldstein AE (2014) Caspase 3 inactivation protects against hepatic cell death and ameliorates fibrogenesis in a diet-induced NASH model. Dig Dis Sci 59(6):1197–1206. doi:10.1007/s10620-014-3167-6

Tian Z, Chen Y, Gao B (2013) Natural killer cells in liver disease. Hepatology 57(4):1654–1662. doi:10.1002/hep.26115

Tinel M, Berson A, Vadrot N, Descatoire V, Grodet A, Feldmann G, Thenot JP, Pessayre D (2004) Subliminal Fas stimulation increases the hepatotoxicity of acetaminophen and bromobenzene in mice. Hepatology 39(3):655–666. doi:10.1002/hep.20094

Tournier C, Hess P, Yang DD, Xu J, Turner TK, Nimnual A, Bar-Sagi D, Jones SN, Flavell RA, Davis RJ (2000) Requirement of JNK for stress-induced activation of the cytochrome c-mediated death pathway. Science 288(5467):870–874. doi:8440 [pii]

Urano F, Wang X, Bertolotti A, Zhang Y, Chung P, Harding HP, Ron D (2000) Coupling of stress in the ER to activation of JNK protein kinases by transmembrane protein kinase IRE1. Science 287(5453):664–666

Urban TJ, Daly AK, Aithal GP (2014) Genetic basis of drug-induced liver injury: present and future. Semin Liver Dis 34(2):123–133. doi:10.1055/s-0034-1375954

Uzi D, Barda L, Scaiewicz V, Mills M, Mueller T, Gonzalez-Rodriguez A, Valverde AM, Iwawaki T, Nahmias Y, Xavier R, Chung RT, Tirosh B, Shibolet O (2013) CHOP is a critical regulator of acetaminophen-induced hepatotoxicity. J Hepatol 59(3):495–503. doi:10.1016/j.jhep.2013.04.024

Vercammen D, Beyaert R, Denecker G, Goossens V, Van Loo G, Declercq W, Grooten J, Fiers W, Vandenabeele P (1998) Inhibition of caspases increases the sensitivity of L929 cells to necrosis mediated by tumor necrosis factor. J Exp Med 187(9):1477–1485

Volkmann X, Fischer U, Bahr MJ, Ott M, Lehner F, Macfarlane M, Cohen GM, Manns MP, Schulze-Osthoff K, Bantel H (2007) Increased hepatotoxicity of tumor necrosis factor-related apoptosis-inducing ligand in diseased human liver. Hepatology 46(5):1498–1508. doi:10.1002/hep.21846

Volkmann X, Anstaett M, Hadem J, Stiefel P, Bahr MJ, Lehner F, Manns MP, Schulze-Osthoff K, Bantel H (2008) Caspase activation is associated with spontaneous recovery from acute liver failure. Hepatology 47(5):1624–1633. doi:10.1002/hep.22237

Voskoboinik I, Whisstock JC, Trapani JA (2015) Perforin and granzymes: function, dysfunction and human pathology. Nat Rev Immunol 15(6):388–400. doi:10.1038/nri3839

Walczak H, Miller RE, Ariail K, Gliniak B, Griffith TS, Kubin M, Chin W, Jones J, Woodward A, Le T, Smith C, Smolak P, Goodwin RG, Rauch CT, Schuh JC, Lynch DH (1999) Tumoricidal activity of tumor necrosis factor-related apoptosis-inducing ligand in vivo. Nat Med 5(2):157–163. doi:10.1038/5517

Wang S, Kaufman RJ (2012) The impact of the unfolded protein response on human disease. J Cell Biol 197(7):857–867. doi:10.1083/jcb.201110131

Wang K, Zhang S, Marzolf B, Troisch P, Brightman A, Hu Z, Hood LE, Galas DJ (2009) Circulating microRNAs, potential biomarkers for drug-induced liver injury. Proc Natl Acad Sci U S A 106(11):4402–4407. doi:10.1073/pnas.0813371106

Watkins PB (2009) Biomarkers for the diagnosis and management of drug-induced liver injury. Semin Liver Dis 29(4):393–399. doi:10.1055/s-0029-1240008

Wei CY, Chung WH, Huang HW, Chen YT, Hung SI (2012) Direct interaction between HLA-B and carbamazepine activates T cells in patients with Stevens-Johnson syndrome. J Allergy Clin Immunol 129(6):1562–1569 . doi:10.1016/j.jaci.2011.12.990e1565

Williams CD, McGill MR, Farhood A, Jaeschke H (2013) Fas receptor-deficient lpr mice are protected against acetaminophen hepatotoxicity due to higher glutathione synthesis and enhanced detoxification of oxidant stress. Food Chem Toxicol 58:228–235. doi:10.1016/j.fct.2013.04.031

Wiltshire C, Matsushita M, Tsukada S, Gillespie DA, May GH (2002) A new c-Jun N-terminal kinase (JNK)-interacting protein, Sab (SH3BP5), associates with mitochondria. Biochem J 367(Pt 3):577–585. doi:10.1042/BJ20020553

Wiltshire C, Gillespie DA, May GH (2004) Sab (SH3BP5), a novel mitochondria-localized JNK-interacting protein. Biochem Soc Trans 32(Pt 6):1075–1077. doi:10.1042/BST0321075

Win S, Than TA, Han D, Petrovic LM, Kaplowitz N (2011) c-Jun N-terminal kinase (JNK)-dependent acute liver injury from acetaminophen or tumor necrosis factor (TNF) requires mitochondrial Sab protein expression in mice. J Biol Chem 286(40):35071–35078. doi:10.1074/jbc.M111.276089. M111.276089 [pii]

Win S, Than TA, Fernandez-Checa JC, Kaplowitz N (2014) JNK interaction with Sab mediates ER stress induced inhibition of mitochondrial respiration and cell death. Cell Death Dis 5:e989. doi:10.1038/cddis.2013.522

Win S, Than TA, Le BH, Garcia-Ruiz C, Fernandez-Checa JC, Kaplowitz N (2015) Sab (Sh3bp5) dependence of JNK mediated inhibition of mitochondrial respiration in palmitic acid induced hepatocyte lipotoxicity. J Hepatol 62(6):1367–1374. doi:10.1016/j.jhep.2015.01.032

Win S, Than TA, Min RW, Aghajan M, Kaplowitz N (2016) c-Jun N-terminal kinase mediates mouse liver injury through a novel Sab (SH3BP5)-dependent pathway leading to inactivation of intramitochondrial Src. Hepatology 63(6):1987–2003. doi:10.1002/hep.28486

Witek RP, Stone WC, Karaca FG, Syn WK, Pereira TA, Agboola KM, Omenetti A, Jung Y, Teaberry V, Choi SS, Guy CD, Pollard J, Charlton P, Diehl AM (2009) Pan-caspase inhibitor VX-166 reduces fibrosis in an animal model of nonalcoholic steatohepatitis. Hepatology 50(5):1421–1430. doi:10.1002/hep.23167

Woolbright BL, Jaeschke H (2015) Sterile inflammation in acute liver injury: myth or mystery? Expert Rev Gastroenterol Hepatol 9(8):1027–1029. doi:10.1586/17474124.2015.1060855

Wuillemin N, Adam J, Fontana S, Krahenbuhl S, Pichler WJ, Yerly D (2013) HLA haplotype determines hapten or p-i T cell reactivity to flucloxacillin. J Immunol 190(10):4956–4964. doi:10.4049/jimmunol.1202949

Xie Z, Klionsky DJ (2007) Autophagosome formation: core machinery and adaptations. Nat Cell Biol 9(10):1102–1109. doi:10.1038/ncb1007-1102

Xie Y, Xiao F, Luo L, Zhong C (2014) Activation of autophagy protects against ROS-mediated mitochondria-dependent apoptosis in L-02 hepatocytes induced by Cr(VI). Cell Physiol Biochem 33(3):705–716. doi:10.1159/000358646

Xu CF, Johnson T, Wang X, Carpenter C, Graves A, Warren L, Xue Z, King KS, Fraser DJ, Stinnett S, Briley LP, Mitrica I, Spraggs CF, Nelson MR, Tada H, du Bois A, Powles T, Kaplowitz N, Pandite LN (2015) HLA-B*57:01 confers susceptibility to pazopanib-associated liver injury in patients with cancer. Clin Cancer Res 22(6):1371–1377. doi:10.1158/1078-0432.CCR-15-2044

Yin XM, Wang K, Gross A, Zhao Y, Zinkel S, Klocke B, Roth KA, Korsmeyer SJ (1999) Bid-deficient mice are resistant to Fas-induced hepatocellular apoptosis. Nature 400(6747):886–891. doi:10.1038/23730

Yuan L, Kaplowitz N (2013) Mechanisms of drug-induced liver injury. Clin Liver Dis 17(4):507–518 . doi:10.1016/j.cld.2013.07.002vii

Zhang YF, He W, Zhang C, Liu XJ, Lu Y, Wang H, Zhang ZH, Chen X, Xu DX (2014) Role of receptor interacting protein (RIP)1 on apoptosis-inducing factor-mediated necroptosis during acetaminophen-evoked acute liver failure in mice. Toxicol Lett 225(3):445–453. doi:10.1016/j.toxlet.2014.01.005

Zhang L, Dittmer MR, Blackwell K, Workman LM, Hostager B, Habelhah H (2015) TRAIL activates JNK and NF-kappaB through RIP1-dependent and -independent pathways. Cell Signal 27(2):306–314. doi:10.1016/j.cellsig.2014.11.014

Zhang J, Doshi U, Suzuki A, Chang CW, Borlak J, Li AP, Tong W (2016) Evaluation of multiple mechanism-based toxicity endpoints in primary cultured human hepatocytes for the identification of drugs with clinical hepatotoxicity: results from 152 marketed drugs with known liver injury profiles. Chem Biol Interact 255:3–11. doi:10.1016/j.cbi.2015.11.008

Zou H, Henzel WJ, Liu X, Lutschg A, Wang X (1997) Apaf-1, a human protein homologous to *C. elegans* CED-4, participates in cytochrome c-dependent activation of caspase-3. Cell 90(3):405–413. S0092-8674(00)80501-2 [pii]

Zou W, Liu X, Yue P, Zhou Z, Sporn MB, Lotan R, Khuri FR, Sun SY (2004) c-Jun NH2-terminal kinase-mediated up-regulation of death receptor 5 contributes to induction of apoptosis by the novel synthetic triterpenoid methyl-2-cyano-3,12-dioxooleana-1, 9-dien-28-oate in human lung cancer cells. Cancer Res 64(20):7570–7578. doi:10.1158/0008-5472.CAN-04-1238

Zou W, Yue P, Khuri FR, Sun SY (2008) Coupling of endoplasmic reticulum stress to CDDO-Me-induced up-regulation of death receptor 5 via a CHOP-dependent mechanism involving JNK activation. Cancer Res 68(18):7484–7492. doi:10.1158/0008-5472.CAN-08-1318

Chapter 2
ER Stress in Drug-Induced Liver Injury

Michael Hinton, Yunzhou Li, Eric Kwong, and Huiping Zhou

Abbreviations

3'UTR	3′-untranslated region
ABCB1	ATP-binding cassette sub-family B member 1
ABCC2	ATP-binding cassette sub-family C member 2
ALT	Alanine aminotransferase
AST	Aspartate aminotransferase
AP	Alkaline phosphatase
ATF	Activating transcription factor
AIDS	Acquired human immunodeficiency syndrome
BiP/GRP78	Binding immunoglobulin protein/78 kDa glucose-regulated protein
CHOP	CCAAT enhancer-binding protein homologous protein
CYP3A	Cytochrome P450 3A4
CYP7A1	Cholesterol 7 alpha-hydroxylase
DILI	Drug-induced liver injury
eIF2α	Eukaryotic initiation factor 2 alpha
ER	Endoplasmic reticulum
ERAD	ER-associated degradation
FDA	The Food and Drug Administration
HAART	Highly active anti-retroviral therapy
HIV-1	Human immunodeficiency virus-1
Insig	The insulin-induced gene
IRE1α	Inositol requiring enzyme 1α
LPS	Lipopolysaccharide

M. Hinton • Y. Li • E. Kwong • H. Zhou (✉)
Department of Microbiology and Immunology, Medical College of Virginia Campus, Virginia Commonwealth University, Richmond, VA 23298, USA

McGuire VA Medical Center, Richmond, VA 23249, USA
e-mail: huiping.zhou@vcuhealth.org

W.-X. Ding, X.-M. Yin (eds.), *Cellular Injury in Liver Diseases*, Cell Death in Biology and Diseases, DOI 10.1007/978-3-319-53774-0_2

X.-M. Yin and Z. Dong (Series eds.), *Cell Death in Biology and Diseases*

MRP2	Multidrug resistance-associated protein 2
PERK	Protein kinase RNA–like endoplasmic reticulum kinase
S1P	Serine protease site-1 protease
S2P	Metalloprotease site-2 protease
SCAP	SREBP cleavage-activating protein
SREBP	Sterol regulatory element-binding proteins
TLR4	Toll-like receptor 4
UPR	Unfolded protein response
XBP1	X-box-binding protein 1
XBP1s	Spliced XBP1
XBP1u	Unspliced XBP1

2.1 Introduction

Drug-induced liver injury (DILI) is a serious public health concern and represents one of the leading causes of liver transplantation and the most frequent single cause for withdrawal of an approved drug from the market or termination of clinical trials of potential drug candidates in the past five decades (Devarbhavi 2012; Giordano et al. 2014; Lee et al. 2016). In the United States, more than 1000 drugs, toxins, and over-the-counter herbal medicines have been reported to cause liver injury (Rangnekar and Fontana 2011; Devarbhavi 2012; Giordano et al. 2014). The number of annual cases of acute liver failure is approximately 2000 and more than half of them are drug-induced (Rangnekar and Fontana 2011; Devarbhavi 2012; Giordano et al. 2014). In addition, drug-induced hepatitis accounts for 10% of all cases of acute hepatitis. The global data of DILI is currently unavailable due to the difficulty in discerning the true incidence of DILI. There is no simple objective test currently available for the diagnosis of DILI and patients are often on multiple medications and dietary supplements (Kim and Naisbitt 2016; Giordano et al. 2014; Devarbhavi 2012).

Most of the hepatic toxins predominantly cause hepatocellular injury (Kim and Naisbitt 2016). Although many drugs also cause cholestasis, this condition is generally reversible after drug discontinuation and rarely leads to liver failure or death. DILI is associated with a variety of risk factors such as race, age, sex, alcohol consumption, preexisting liver disease, genetic factors, virus infection, metabolic syndrome, etc. Multiple pathological mechanisms of DILI have been identified, including apoptosis of hepatocytes, disruption of the hepatic transporters, activation of the immune response, disruption of mitochondrial function, and injury of the bile duct, etc (Gu and Manautou 2012; Lee et al. 2016). Most recently, activation of ER stress has been identified as an important contributor to various liver diseases including DILI (Yao et al. 2016b; Lou et al. 2009; Kaplowitz et al. 2007; Ji and Kaplowitz 2004; Wu et al. 2016; Sharkey et al. 2016; Ashraf and Sheikh 2015; Zhang et al. 2014; Pagliassotti 2012; Kaplowitz and Ji 2006; Mahdi et al. 2016). In this chapter, the current understanding of ER stress and its role in DILI are reviewed with a primary focus on HIV protease inhibitor (PI)-induced hepatic injury.

2.2 ER Stress and the Unfolded Protein Response (UPR)

The ER is an important organelle that plays a critical role in maintaining intracellular homeostasis via regulating intracellular calcium levels, protein synthesis, protein folding and assembly, and lipid and cholesterol synthesis (Jung and Choi 2016). Although the ER has significant adaptive capacity to manage the metabolic demands during feeding and fasting, the duration is limited. It has less flexibility to manage chronic and other escalating challenges. The accumulation of misfolded or unfolded proteins in the ER under a variety of pathological conditions or exposure to certain pharmacological compounds leads to ER stress and activation of an intracellular stress signaling cascade termed the unfolded protein response (UPR) (Schonthal 2012; Dandekar et al. 2015).

The activation of the UPR is a protective mechanism in response to a harmful challenge. The canonical UPR signaling pathways have been well described in eukaryotic cells (Ron and Walter 2007; Todd et al. 2008; Hotamisligil 2010). Three ER-transmembrane proteins (PERK: PKR-like eukaryotic initiation factor 2α kinase; IRE1α: inositol requiring enzyme 1α; and ATF6: activating transcription factor-6) and one chaperone protein (BiP/GRP78: binding immunoglobulin protein/78 kDa glucose-regulated protein) have been identified as crucial regulators of the UPR. As shown in Fig. 2.1, under stress-free conditions, BiP/GRP78 is bound to the intraluminal domains (amino-terminals of IRE1α and PERK and carboxyl-terminal of ATF6) of three UPR transducers to maintain them in an inactivated state (Bertolotti et al. 2000). When ER stress occurs due to accumulation of misfolded or unfolded proteins, BiP/GRP78 is dissociated from these UPR sensors, which results in oligomerization and auto-phosphorylation of PERK and IRE1α. Activation of PERK results in the phosphorylation of eukaryotic initiation factor 2 alpha (eIF2α) and attenuation of global mRNA translation. Activation of PERK also selectively increases the translation of certain mRNAs such as ATF4 (a member of the basic leucine-zipper family of transcription factors) and its downstream targets such as CCAAT enhancer-binding protein homologous protein (CHOP) (Fig. 2.1a). CHOP is a proapoptotic transcription factor that plays a critical role in ER stress-mediated apoptosis (Zhou and Liu 2014). Activated IRE1α has intrinsic endoribonuclease activity, which processes the unspliced X-box-binding protein (XBP1u) mRNA by removing 26-nucleotides to produce the active transcription factor XBP1s. XBP1s is able to promote the expression of genes involved in restoring the ER protein folding capacity or degrading misfolded or unfolded proteins (Fig. 2.1b). Activation of the third UPR pathway requires translocation of ATF6 to the Golgi apparatus where its C-terminal region is cleaved by the serine protease site-1 protease (S1P) and the metalloprotease site-2 protease (S2P) to produce an active soluble transcription factor, ATF6N (Fig. 2.1c) (Hotamisligil 2010). Activation of the canonical UPR response mitigates ER stress by reducing protein synthesis, increasing production of chaperones, and facilitating protein degradation. Sustained ER stress as a result of accumulation of misfolded or unfolded proteins can also activate the ER-associated protein degradation (ERAD) (Schwartz and Ciechanover 2009). If the initial UPR

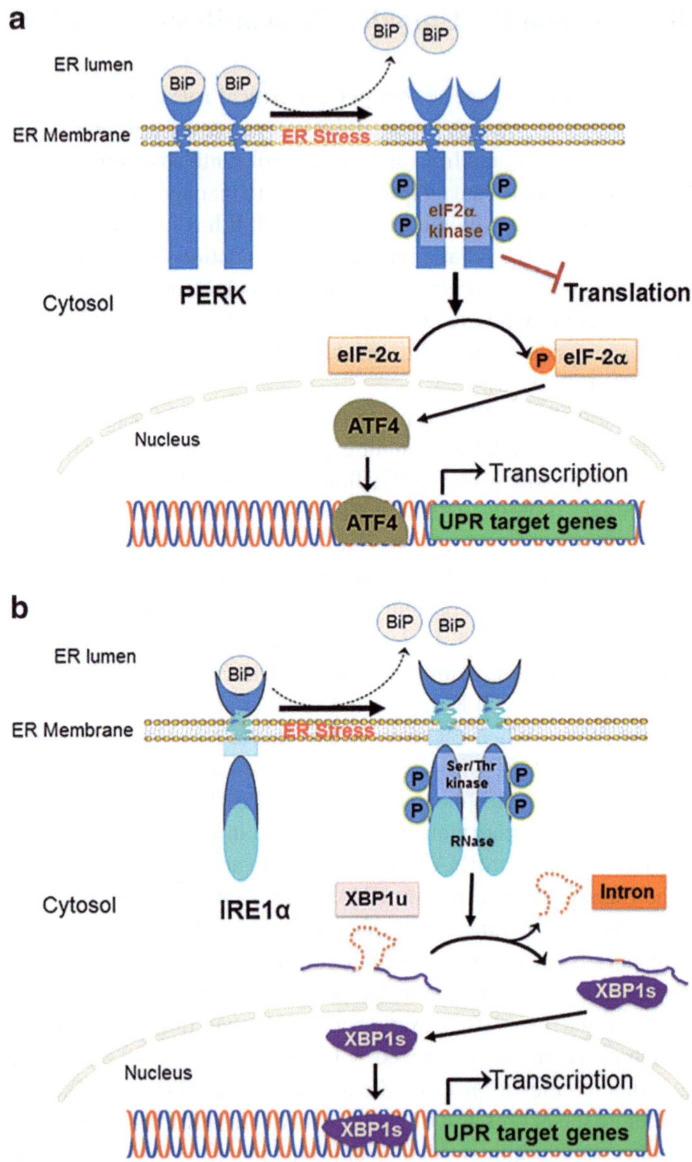

Fig. 2.1 The unfolded protein response (UPR). Three UPR sensors, protein kinase RNA-like endoplasmic reticulum kinase (PERK), inositol-requiring protein 1α (IRE1α) and activating transcript factor (ATF6), and one master regulator Bip/GRP78 have been identified. Under non-stress condition, GRP78 binds to three UPR sensors and prevents their activation. Accumulation of misfolded or unfolded proteins in the ER lumen results in dissociation of GRP78 from the UPR sensors and subsequent activation of PERK, IRE1α, and ATF6. (**a**) The activated PERK phosphorylates the eukaryotic initiation factor 2α (eIF2α), which attenuates the global protein translation and selectively activates ATF4, a key transcription factor regulating the transcription of genes involved in apoptosis, lipid and energy metabolism, autophagy, and stress response. (**b**) The activation of IRE1α results in splicing of the transcription factor X box-binding 1 (XBP1u) and expression of an

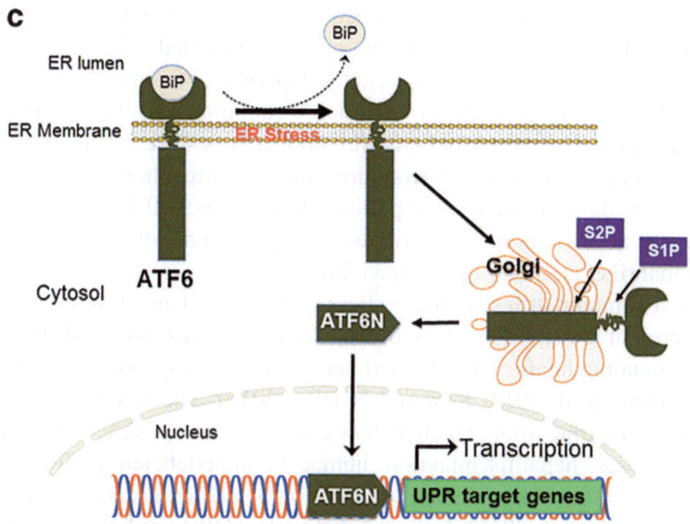

Fig. 2.1 (continued) active transcription factor, spliced XBP1 (XBP1s) by excising a 26-nucleoide intron. XBP1s is involved in regulation of ER-associated degradation (ERAD) and the entry of proteins into the ER and modulation of phospholipid synthesis to accommodate ER membrane expansion under ER stress. (**c**) ATF6 is a member of basic Leu zipper transcription factors. Under ER stress, ATF6 is translocated into Golgi and processed by site 1 protease (S1P) and S2P to generate active ATF6N. ATF6N controls the expression of key genes involved in ERAD and XBP-1. The UPR activation controls multiple physiological processes such as apoptosis, ERAD, protein folding, lipid synthesis, protein secretion, autophagy, etc

fails to resolve ER stress, further activation of the UPR will trigger death responses (Todd et al. 2008; Liu et al. 2016; Schonthal 2012; Jager et al. 2012).

Activation of ER stress by pharmacologic agents or pathophysiologic stimuli results in perturbation of normal cellular function that has been associated with the initiation and progression of numerous human diseases such as cardiovascular diseases, metabolic diseases, inflammatory diseases, immune diseases, neurodegenerative diseases, and various liver diseases including DILI (Chen et al. 2014; Chan et al. 2016; Malhi and Kaufman 2011; Rayavarapu et al. 2012; Back et al. 2012; Cao et al. 2016; Volchuk and Ron 2010; Cui et al. 2016; Cnop et al. 2012; Liu et al. 2016; Go et al. 2016).

2.3 Drug-Induced Liver Injury (DILI)

Most of the drugs that have adverse effects can affect many different organs such as the liver, heart, kidney, lung, skeletal muscle, or central nerve system (Hohenegger 2012; Miltenburg and Boogerd 2014; Marrer and Dieterle 2010; Begriche et al. 2011; Tocchetti et al. 2013; Foufelle and Fromenty 2016). DILI represents the most common indication of adverse drug reaction for the drug withdrawal (Lee et al. 2016). Most of DILIs are unrecognized and underreported. The true incidence is difficult to estimate and varies significantly depending on the setting (Devarbhavi 2012). DILI can be idiosyncratic or intrinsic. Acetaminophen is the leading cause of DILI in the western countries and it is well known to be dose-related or intrinsic in nature. The intrinsic DILI results from drug-induced direct hepatotoxicity over the course of a few days or longer time period. Idiosyncratic DILI is a rare and unpredictable event and occurs in a minority of susceptible individuals with a prolonged latency (Ghabril et al. 2010; Devarbhavi 2012).

The most common drugs leading to liver injury in the United States are antibiotics, nonsteroidal anti-inflammatory drugs, central nervous system drugs, antiviral agents, immunomodulatory agents, and herbal/dietary supplements (Ghabril et al. 2010; Giordano et al. 2014; Kim and Naisbitt 2016). The risk factors associated with DILI are multifactorial, which includes age, sex, race, genetic factors, preexisting liver diseases, hepatitis infection, human immunodeficiency virus-1 (HIV-1) infection, diabetes mellitus, alcohol use, environmental toxins, etc. (Giordano et al. 2014; Kim and Naisbitt 2016). In the United States, DILI has been associated with more than 1000 medications and is the leading cause of acute liver failure in patients referred for liver transplantation (Kim and Naisbitt 2016; Giordano et al. 2014; Rangnekar and Fontana 2011; Devarbhavi 2012). Since the introduction of anti-HIV drugs in the late 1980s, hepatotoxicity has become a major clinical concern for anti-HIV therapy, especially for HIV protease inhibitors (HIV PIs). The incidence of DILI continues to increase with the increasing number of new drugs that have been introduced into clinical use over the past several decades. It has become a significant threat to the public health for an over-medicated society.

2.4 HIV/AIDS and HIV PIs

The acquired human immunodeficiency syndrome (AIDS) epidemic has rapidly expanded since the discovery of HIV as the cause of this disease in 1983. By the end of 2014, the estimated number of people living with HIV reached 36.9 million worldwide and the mortality rate for AIDS reached 1.2 million that year (Bhatti et al. 2016). As HIV biology and events critical to viral replication in the host cell are elucidated, a number of specific pharmacological agents targeting the key steps of HIV life cycle have been developed (Ghosh et al. 2016). The current available anti-HIV medicines include HIV reverse transcriptase inhibitors, HIV PIs, fusion inhibitors, and integrase inhibitors.

The highly active anti-retroviral therapy (HAART) is the most effective treatment currently available for HIV/AIDS, which includes two or three different classes of anti-HIV drugs. HIV PI is the core component of HAART, which specifically inhibits HIV protease activity and prevents the formation of mature HIV virions. The Food and Drug Administration (FDA) approved the first HIV PI, saquinavir, in 1995. Incorporation of saquinavir in HIV therapy greatly improved patient outcomes by reducing viral loads, improving CD4 cell counts, and halting the progression to AIDS (Ghosh et al. 2016). Since then, a total of nine HIV PIs with several different dosages and combinations have been licensed by the FDA: amprenavir (Agenerase), atazanavir (Reyataz), darunavir (Prezista, TMC114), indinavir (Crixivan), fosamprenavir (Lexiva), lopinavir/ritonavir (Kaletra, Aluvia), nelfinavir (Viracept), ritonavir (Norvir), saquinavir (fortovase, soft gel cap), and tipranavir (Aptivus). The incorporation of HIV PIs in HAART has successfully suppressed viral replication in HIV patients, significantly reduced morbidity and mortality, and changed HIV infection from an acute disease with a high morbidity and mortality to a manageable chronic disease (Poorolajal et al. 2016; Krentz et al. 2005). Unfortunately, current anti-HIV drugs are very toxic, especially HIV PIs. The benefits of HAART are compromised by serious side effects (Bhatti et al. 2016). Antiretroviral hepatotoxicity occurs in approximately 10% of patients and is higher in those with underlying liver disease (Lana et al. 2001; Nunez et al. 2001; Merwat and Vierling 2011). Although HIV-infected patients under HAART have a similar lifespan as non HIV-infected population, the quality of life of HIV-infected patients is compromised by the hepatic and metabolic toxicities of anti-HIV drugs, especially with HIV PIs (Guira et al. 2016; Hurwitz et al. 2004; Jain et al. 2001; Parakh et al. 2009). As of June 2015, the United Nations Program on HIV/AIDS (UNAIDS) approximates 15.8 million people living with HIV have access to antiretroviral therapy. Hepatic toxicity and lipodystrophy specifically associated with HIV PIs have become a matter of particular concern in the clinical realm (McGovern 2004; Surgers and Lacombe 2013).

2.5 ER Stress and HIV PI-Induced Hepatotoxicity

Hepatotoxicity is characterized by an abnormal liver function test, which monitors the changes in key liver enzymes, such as serum alanine aminotransferase (ALT), aspartate aminotransferase (AST), and alkaline phosphatase (AP) levels. Preclinical and clinical studies have shown elevated serum ALT and/or AST with HIV PI treatment, particularly with ritonavir administration (Sulkowski 2004; Brunet et al. 2012; Macias et al. 2012). HIV PI-induced liver injury results in the disruption of the normal cellular function of hepatocytes, which are the cells responsible for lipid homeostasis, bile acid synthesis, and gluconeogenesis. Long-term HAART treatment has been associated with metabolic side effects including dyslipidemia, insulin resistance, and cardiovascular complications including atherosclerosis (Sulkowski 2004; Dieterich 2003; Djedaini et al. 2009; Ioannou et al. 2015; Gleason et al. 2016; Parakh et al. 2009). Although the mechanism underlying HIV PI-induced hepatotoxicity remains to be fully identified, an increasing body of evidence suggests that multiple mechanisms may be involved and individual PIs may have different effects on hepatic liver injury (Wu et al. 2014). Studies from our group and others' suggest that activation of the ER stress could be a critical cellular event involved in HIV PI-induced heptic toxicity and metabolic syndrome, in addition to other key mechanisms such as inflammation, oxidative stress, and mitochondrial dysfunction (Bruning 2011; Taura et al. 2013; Zha et al. 2013a; Zhang et al. 2009; Zhou 2011; Zha and Zhou 2012; Foufelle and Fromenty 2016).

2.5.1 HIV PI-Induced Lipogenesis

The liver plays a central role in regulating nutrient absorption, cholesterol metabolism, lipogenesis, and gluconeogenesis (Bechmann et al. 2012; Lee et al. 2012). In addition, it is also important for hormone production, xenobiotic metabolism, and detoxification. Hepatic lipid homeostasis is maintained by numerous nuclear receptors and transcription factors (Zhou and Liu 2014). The sterol regulatory element-binding proteins (SREBPs) are a family of membrane-bound transcription factors responsible for regulating more than two dozen genes involved in the synthesis and uptake of cholesterol, fatty acids, triglycerides, and phospholipids and lipid metabolism (Horton et al. 2002; Foufelle et al. 2007). There are three isoforms of SREBPs including SREBP-1a, SREBP-1c, and SREBP-2. With alternative splicing both SREBP-1a and SREBP-1c are obtained from a single gene. SREBP-1a has the general effect of activating all SREBP-responsive genes, while SREBP-1c and SREBP-2 are specific for enhancing the transcription of genes required for fatty acid and cholesterol synthesis (Horton et al. 2002; Foufelle et al. 2007). The functionality of SREBPs depends on association with the SREBP cleavage-activating protein (SCAP) and the insulin-induced gene (Insig) complex. SREBPs are processed into active forms by the same proteases (S1P and S2P) that process ATF6 in the Golgi. ER stress-mediated activation of PERK and IRE1α-XBP-1 signaling pathways results in the attenuation of protein translation and promotion of protein degradation, which are responsible for the depletion of insig-1 and subsequently the proteolytic activation of SREBPs leading to lipogenesis (Lee and Ye 2004; Kammoun et al. 2009b). Overexpression of GRP78 or insig-1 inhibits SREBP-1c activation and hepatic steatosis (Chen et al. 2011; Kammoun et al. 2008, 2009a; Zhang et al. 2012). The disruption of hepatic cholesterol, bile acid, and lipid homeostasis from an imbalance of biosynthesis and metabolism leads to hepatic lipid accumulation, inflammation, and eventually hepatic injury (Sanyal 2005, 2013, 2015; Puri et al. 2009).

Regulation of hepatic lipogenesis has been linked to ER stress and the activity of the UPR (Zhang and Wang 2016; Lee et al. 2012; Zhou and Liu 2014). Our previous studies have shown that individual HIV PIs had different effects on the activation of ER stress in hepatocytes and macrophages, which correlated to the incidence of dyslipidemia associated with different HIV PIs in the clinic (Zhou et al. 2005, 2006). Initial studies done by Williams, K et al. with indinavir in primary rodent hepatocytes indicated that indinavir specifically upregulated the expression and activation of SREBPs (Williams et al. 2004). The mRNA levels of cholesterol 7 alpha-hydroxylase (CYP7A1) were markedly decreased, while fatty acid synthase mRNA levels were up-regulated (Williams et al. 2004). The ability of HIV PI to activate the UPR is linked to its effect on hepatic lipid accumulation and lipotoxicity (Zhou et al. 2007). Our previous studies also found that ritonavir had the most significant effect on the UPR activation, while amprenavir did not activate the UPR in primary hepatocytes (Zhou et al. 2006). This may explain why ritonavir induces hepatic lipid accumulation and cell apoptosis, but not amprenavir. In addition, ritonavir has the most adverse metabolic effects, including insulin resistance, lipodystrophy, and hyperlipidemia in HIV patients and currently used only as a pharmacoenhancer of other HIV PIs (Mateo et al. 2014; Pere et al. 2008; Putcharoen et al. 2015). HIV PIs are extensively metabolized by cytochrome P450 3A4 (CYP3A) in the liver. The plasma half-life is remarkably short when used alone (Putcharoen et al. 2015). Ritonavir is a strong inhibitor of CYP3A and in combination with other HIV PIs significantly increases their half-life. In addition, ritonavir also inhibits the drug transporters ABCB1 (ATP-binding cassette subfamily B member 1, also called P-glycoprotein) and ABCC2 (ATP-binding cassette subfamily C member 2; also called MRP2: Multidrug resistance-associated protein), which have been shown to pump out the HIV PIs from the intestinal cells and macrophages (Holmstock et al. 2012; Zha et al. 2013b).

Activation of the ER stress and extended upregulation of CHOP have been shown to be involved in various liver injuries, including DILI, lipotoxicity, and cholestasis (Pfaffenbach et al. 2010; Tamaki et al. 2008; Uzi et al. 2013; Willy et al. 2015). Our studies demonstrated that HIV PI-induced hepatic lipotoxicity is closely linked to the upregulation of CHOP in hepatocytes and in liver (Wang et al. 2013; Zhou et al. 2007). Lopinavir and ritonavir significantly induced lipogenesis, hepatic lipid accumulation, and apoptosis in wild-type mice, but not in CHOP knockout mice (Wang et al. 2013). These studies suggest that CHOP is an important molecular link to ER stress and hepatic lipotoxicity and that increased expression of CHOP represents a critical factor underlying events leading to HIV PI-induced hepatic injury (Wang et al. 2013).

2.5.2 HIV PI-Induced Inflammation

Inflammation and ER stress are important adaptive defense responses that help promote cell survival under various stress conditions (Dandekar et al. 2015). However, chronic inflammation and prolonged activation of the ER stress have been identified as important contributors to various liver injuries and metabolic diseases (Adolph et al. 2012; Ashraf and Sheikh 2015; Cao et al. 2016; Dandekar et al. 2015; Duwaerts and Maher 2014; Hasnain et al. 2012; Hotamisligil 2010). Our previous studies showed that HIV PIs induced inflammatory cytokines, TNF-α, and IL-6 in macrophages via ER stress/CHOP-mediated ERK1/2 activation, and as a result, increased the cytosolic translocation of RNA-binding protein HuR and subsequent binding to the 3′UTR (3′-untranslated region) of TNF-α and IL-6 mRNAs in macrophages (Chen et al. 2009). The activation of the innate immune response has recently been shown to play an important role in promoting DILI (Goto et al. 2015). Hepatic macrophages play critical roles in maintaining homeostasis in the liver and in the pathogenesis of various hepatic injuries (Ju and Tacke 2016). However, the heterogeneity of macrophages in the liver is very complex. Macrophages derived from different origins can have distinct effects on hepatic metabolic homeostasis and liver injury. In addition, macrophages can be polarized to different subpopulations, including classically activated and inflammatory M1 and alternatively activated/anti-inflammatory M2, in response to various external signals and insults (Ju and Tacke 2016; Harvey et al. 2015; Zhou et al. 2014).

The Kupffer cell (hepatic resident macrophages)-mediated inflammation is of critical importance to hepatic injuries induced by drugs, toxins, and lipids via secreting various pro- and anti-inflammatory mediators (Goto et al. 2015; Arguello et al. 2015; Ju and Tacke 2016; Li and Diehl 2003; Meli et al. 2014; Ni et al. 2016; Wan et al. 2014; Wenfeng et al. 2014). Recently, it has been reported that ER stress-mediated signaling pathways are involved in regulation of macrophage polarization (Xiu et al. 2015). CHOP and ER stress are implicated in the induction and differentiation of M2 macrophages (Yao et al. 2016a). Moreover, the gut-derived endotoxin lipopolysaccharide (LPS) activates the toll-like receptor 4 (TLR4) and triggers hepatocyte apoptosis (Wenfeng et al. 2014). Our previous studies also reported that HIV PIs induce ER stress in intestinal epithelial cells. HIV PI-mediated upregulation of CHOP in intestine is responsible for dysfunction of intestinal barrier function, microbial translocation, and induction of systemic inflammation (Wu et al. 2010). ER stress and inflammation have also been linked to inhibition of E-cadherin and zonula occludens-1 expression, the key components of intestinal epithelia, and result in a defective epithelial barrier (Fan et al. 2014). Gut-derived microbial products and inflammatory mediators significantly promote the progression of various liver diseases including DILI, fatty liver diseases, and alcoholic liver disease (Bieghs and Trautwein 2014; Chen and Schnabl 2014; Duwaerts and Maher 2014; Schnabl and Brenner 2014).

2.6 Summary

The liver is the largest internal organ of the human body and plays a critical role in metabolizing nutrients, drugs, and environmental toxicants. It is the most frequent site of drug-induced toxicity. Although severe DILI is relatively rare, drug-induced hepatic injury is the most common indication for drug withdrawal and the most frequent cause of acute liver failure. The clinical impact of DILI is substantial because of the number of drugs used and the number of patients treated. The diagnosis and treatment of DILI remain significant challenges due to the complexities of disease pathogenesis and a lack of understanding of the underlying cellular/molecular mechanisms. During the last decade, numerous studies have shown that the activation of ER stress represents a key step in the development of DILI even at therapeutic doses, especially for HIV PIs. HIV-1 infection continues to be a serious global health problem. As the key component of HAART, a life-long treatment for HIV infection, reduction of HIV PI-induced hepatic injury remains a difficult task. Identification of the ER stress as a critical player in HIV PI-induced inflammation,

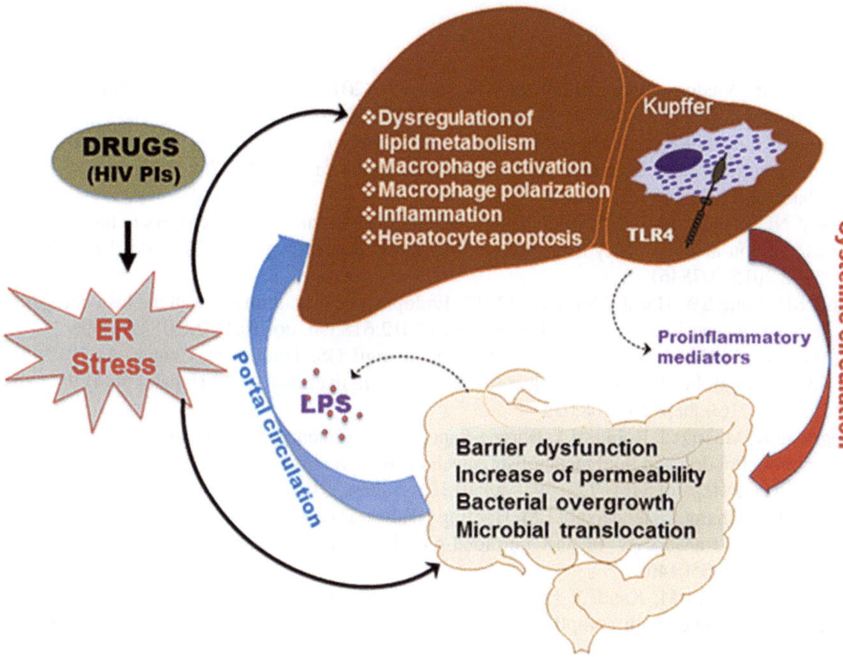

Fig. 2.2 Mechanisms of HIV PI-mediated liver injury. Activation of ER stress in hepatic cells results in dysfunction of hepatocytes, dysregulation of lipid metabolism, activation of Kupffer cells, and production of proinflammatory mediators. Liver-derived inflammatory mediators reach intestine via system circulation. Drug-induced activation of the ER stress in intestinal epithelial cells and systemic inflammatory mediators disrupt intestinal barrier integrity and increase of bacterial overgrowth. Increase of gut permeability results in high circulating endotoxin that reaches the liver via portal circulation. Endotoxin (lipopolysaccharide or LPS) is recognized by the Toll-like receptor (TLR)-4 complex on resident macrophages or Kupffer cells in the liver, leading to production of proinflammatory cytokines and resulting in injury to hepatocytes

disruption of intestinal barrier integrity, and dysregulation of hepatic lipid metabolism opened new direction for the development of preventive and therapeutic strategies for HIV PI-induced liver injury, as well as for other DILIs (Fig. 2.2). In addition, elucidating of the underlying mechanism of DILI not only has high scientific values but also has significant economic impact on improving new drug development by developing new reliable screening systems to eliminate the candidates with hepatotoxicity at an early stage.

Taken together, the rapid progress in understanding ER stress in DILI and other liver diseases gives rise to the expectation that ER stress may be used as a biomarker and therapeutic target for DILI as well as other liver diseases.

Acknowledgment This study was supported by National Natural Science Foundation of China Grants (81070245 and 81270489 to HZ); National Institutes of Health Grants (R21AI068432; R01AT004148; 1R01DK104893; DK-057543), VA Merit Awards (I01BX001390); NIAAA 1F31AA024713; VCU Massey Cancer Pilot grant (A35362 to HZ). We would like to thank Elaine Studer Kennedy for English editing and proof-reading.

References

Adolph TE, Niederreiter L, Blumberg RS, Kaser A (2012) Endoplasmic reticulum stress and inflammation. Dig Dis 30(4):341–346. doi:10.1159/000338121

Arguello G, Balboa E, Arrese M, Zanlungo S (2015) Recent insights on the role of cholesterol in non-alcoholic fatty liver disease. Biochim Biophys Acta 1852(9):1765–1778. doi:10.1016/j.bbadis.2015.05.015

Ashraf NU, Sheikh TA (2015) Endoplasmic reticulum stress and oxidative stress in the pathogenesis of non-alcoholic fatty liver disease. Free Radic Res 49(12):1405–1418. doi:10.3109/10715762.2015.1078461

Back SH, Kang SW, Han J, Chung HT (2012) Endoplasmic reticulum stress in the beta-cell pathogenesis of type 2 diabetes. Exp Diabetes Res 2012:618396. doi:10.1155/2012/618396

Bechmann LP, Hannivoort RA, Gerken G, Hotamisligil GS, Trauner M, Canbay A (2012) The interaction of hepatic lipid and glucose metabolism in liver diseases. J Hepatol 56(4):952–964. doi:10.1016/j.jhep. 2011.08.025

Begriche K, Massart J, Robin MA, Borgne-Sanchez A, Fromenty B (2011) Drug-induced toxicity on mitochondria and lipid metabolism: mechanistic diversity and deleterious consequences for the liver. J Hepatol 54(4):773–794. doi:10.1016/j.jhep.2010.11.006

Bertolotti A, Zhang Y, Hendershot LM, Harding HP, Ron D (2000) Dynamic interaction of BiP and ER stress transducers in the unfolded-protein response. Nat Cell Biol 2(6):326–332. doi:10.1038/35014014

Bhatti AB, Usman M, Kandi V (2016) Current scenario of HIV/AIDS, treatment options, and major challenges with compliance to antiretroviral therapy. Cureus 8(3):e515. doi:10.7759/cureus.515

Bieghs V, Trautwein C (2014) Innate immune signaling and gut-liver interactions in non-alcoholic fatty liver disease. Hepatobiliary Surg Nutr 3(6):377–385. doi:10.3978/j.issn.2304-3881.2014.12.04

Brunet C, Reliquet V, Jovelin T, Venisse N, Winer N, Bui E et al (2012) Effectiveness and safety of saquinavir/ritonavir in HIV-infected pregnant women: INEMA cohort. Med Mal Infect 42(9):421–428. doi:10.1016/j.medmal.2012.07.010

Bruning A (2011) Analysis of nelfinavir-induced endoplasmic reticulum stress. Methods Enzymol 491:127–142. doi:10.1016/b978-0-12-385928-0.00008-0

Cao SS, Luo KL, Shi L (2016) Endoplasmic reticulum stress interacts with inflammation in human diseases. J Cell Physiol 231(2):288–294. doi:10.1002/jcp.25098

Chan P, Stolz J, Kohl S, Chiang WC, Lin JH (2016) Endoplasmic reticulum stress in human photoreceptor diseases. Brain Res 1648(Pt B):538–541. doi:10.1016/j.brainres.2016.04.021

Chen P, Schnabl B (2014) Host-microbiome interactions in alcoholic liver disease. Gut Liver 8(3):237–241. doi:10.5009/gnl.2014.8.3.237

Chen L, Jarujaron S, Wu X, Sun L, Zha W, Liang G et al (2009) HIV protease inhibitor lopinavir-induced TNF-alpha and IL-6 expression is coupled to the unfolded protein response and ERK signaling pathways in macrophages. Biochem Pharmacol 78(1):70–77. doi:10.1016/j.bcp.2009.03.022

Chen K, Jin P, He HH, Xie YH, Xie XY, Mo ZH (2011) Overexpression of Insig-1 protects beta cell against glucolipotoxicity via SREBP-1c. J Biomed Sci 18:57. doi:10.1186/1423-0127-18-57

Chen S, Melchior WB Jr, Guo L (2014) Endoplasmic reticulum stress in drug- and environmental toxicant-induced liver toxicity. J Environ Sci Health C Environ Carcinog Ecotoxicol Rev 32(1):83–104. doi:10.1080/10590501.2014.881648

Cnop M, Foufelle F, Velloso LA (2012) Endoplasmic reticulum stress, obesity and diabetes. Trends Mol Med 18(1):59–68. doi:10.1016/j.molmed.2011.07.010

Cui Y, Zhao D, Barrow PA, Zhou X (2016) The endoplasmic reticulum stress response: a link with tuberculosis? Tuberculosis (Edinb) 97:52–56. doi:10.1016/j.tube.2015.12.009

Dandekar A, Mendez R, Zhang K (2015) Cross talk between ER stress, oxidative stress, and inflammation in health and disease. Methods Mol Biol 1292:205–214. doi:10.1007/978-1-4939-2522-3_15

Devarbhavi H (2012) An update on drug-induced liver injury. J Clin Exp Hepatol 2(3):247–259. doi:10.1016/j.jceh.2012.05.002

Dieterich D (2003) Managing antiretroviral-associated liver disease. J Acquir Immune Defic Syndr 34(Suppl 1):S34–S39

Djedaini M, Peraldi P, Drici MD, Darini C, Saint-Marc P, Dani C et al (2009) Lopinavir co-induces insulin resistance and ER stress in human adipocytes. Biochem Biophys Res Commun 386(1):96–100. doi:10.1016/j.bbrc.2009.05.148

Duwaerts CC, Maher JJ (2014) Mechanisms of liver injury in non-alcoholic steatohepatitis. Curr Hepatol Rep 13(2):119–129. doi:10.1007/s11901-014-0224-8

Fan L, Hu L, Yang B, Fang X, Gao Z, Li W et al (2014) Erlotinib promotes endoplasmic reticulum stress-mediated injury in the intestinal epithelium. Toxicol Appl Pharmacol 278(1):45–52. doi:10.1016/j.taap.2014.04.015

Foufelle F, Fromenty B (2016) Role of endoplasmic reticulum stress in drug-induced toxicity. Pharmacol Res Perspect 4(1):e00211. doi:10.1002/prp2.211

Foufelle F, Hainault I, Kammoun H, Chabanon H, Ferre P (2007) SREBP-1c and the transcriptional control of lipogenesis. Int J Obes (Lond) 31:S3–S3

Ghabril M, Chalasani N, Bjornsson E (2010) Drug-induced liver injury: a clinical update. Curr Opin Gastroenterol 26(3):222–226. doi:10.1097/MOG.0b013e3283383c7c

Ghosh AK, Osswald HL, Prato G (2016) Recent progress in the development of HIV-1 protease inhibitors for the treatment of HIV/AIDS. J Med Chem 59(11):5172–5208. doi:10.1021/acs.jmedchem.5b01697

Giordano C, Rivas J, Zervos X (2014) An update on treatment of drug-induced liver injury. J Clin Transl Hepatol 2(2):74–79. doi:10.14218/JCTH.2014.00005

Gleason RL Jr, Caulk AW, Seifu D, Rosebush JC, Shapiro AM, Schwartz MH et al (2016) Efavirenz and ritonavir-boosted lopinavir use exhibited elevated markers of atherosclerosis across age groups in people living with HIV in Ethiopia. J Biomech 49(13):2584–2592. doi:10.1016/j.jbiomech.2016.05.018

Go BS, Kim J, Yang JH, Choe ES (2016) Psychostimulant-induced endoplasmic reticulum stress and neurodegeneration. Mol Neurobiol. doi:10.1007/s12035-016-9969-0

Goto S, Deguchi J, Nishio N, Nomura N, Funabashi H (2015) Hepatotoxicants induce cytokine imbalance in response to innate immune system. J Toxicol Sci 40(3):389–404. doi:10.2131/jts.40.389

Gu X, Manautou JE (2012) Molecular mechanisms underlying chemical liver injury. Expert Rev Mol Med 14:e4. doi:10.1017/s1462399411002110

Guira O, Tieno H, Diendere AE, Sagna Y, Diallo I, Yameogo B et al (2016) Features of metabolic syndrome and its associated factors during highly active antiretroviral therapy in Ouagadougou (Burkina Faso). J Int Assoc Provid AIDS Care 15(2):159–163. doi:10.1177/2325957415601503

Harvey LD, Yin Y, Attarwala IY, Begum G, Deng J, Yan HQ et al (2015) Administration of DHA reduces endoplasmic reticulum stress-associated inflammation and alters microglial or macrophage activation in traumatic brain injury. ASN Neuro 7(6). doi:10.1177/1759091415618969

Hasnain SZ, Lourie R, Das I, Chen AC, McGuckin MA (2012) The interplay between endoplasmic reticulum stress and inflammation. Immunol Cell Biol 90(3):260–270. doi:10.1038/icb.2011.112

Hohenegger M (2012) Drug induced rhabdomyolysis. Curr Opin Pharmacol 12(3):335–339. doi:10.1016/j.coph.2012.04.002

Holmstock N, Annaert P, Augustijns P (2012) Boosting of HIV protease inhibitors by ritonavir in the intestine: the relative role of cytochrome P450 and P-glycoprotein inhibition based on Caco-2 monolayers versus in situ intestinal perfusion in mice. Drug Metab Dispos 40(8):1473–1477. doi:10.1124/dmd.112.044677

Horton JD, Goldstein JL, Brown MS (2002) SREBPs: activators of the complete program of cholesterol and fatty acid synthesis in the liver. J Clin Invest 109(9):1125–1131. doi:10.1172/JCI15593

Hotamisligil GS (2010) Endoplasmic reticulum stress and the inflammatory basis of metabolic disease. Cell 140(6):900–917. doi:10.1016/j.cell.2010.02.034

Hurwitz BE, Klimas NG, Llabre MM, Maher KJ, Skyler JS, Bilsker MS et al (2004) HIV, metabolic syndrome X, inflammation, oxidative stress, and coronary heart disease risk : role of protease inhibitor exposure. Cardiovasc Toxicol 4(3):303–316

Ioannou GN, Bryson CL, Weiss NS, Boyko EJ (2015) Associations between lipodystrophy or antiretroviral medications and cirrhosis in patients with HIV infection or HIV/HCV coinfection. Eur J Gastroenterol Hepatol 27(5):577–584. doi:10.1097/meg.0000000000000290

Jager R, Bertrand MJ, Gorman AM, Vandenabeele P, Samali A (2012) The unfolded protein response at the crossroads of cellular life and death during endoplasmic reticulum stress. Biol Cell 104(5):259–270. doi:10.1111/boc.201100055

Jain RG, Furfine ES, Pedneault L, White AJ, Lenhard JM (2001) Metabolic complications associated with antiretroviral therapy. Antiviral Res 51(3):151–177

Ji C, Kaplowitz N (2004) Hyperhomocysteinemia, endoplasmic reticulum stress, and alcoholic liver injury. World J Gastroenterol 10(12):1699–1708

Ju C, Tacke F (2016) Hepatic macrophages in homeostasis and liver diseases: from pathogenesis to novel therapeutic strategies. Cell Mol Immunol 13(3):316–327. doi:10.1038/cmi.2015.104

Jung TW, Choi KM (2016) Pharmacological modulators of endoplasmic reticulum stress in metabolic diseases. Int J Mol Sci 17(2). doi:10.3390/ijms17020192

Kammoun H, Chabanon H, Hainault I, Ferre P, Foufelle F (2008) Endoplasmic reticulum stress prevention improves hepatic steatosis of ob/ob mice through an inhibition of SREBP-1c proteolytic cleavage. Diabetologia 51:S315–S316

Kammoun HL, Chabanon H, Hainault I, Luquet S, Magnan C, Koike T et al (2009a) GRP78 expression inhibits insulin and ER stress-induced SREBP-1c activation and reduces hepatic steatosis in mice. J Clin Invest 119(5):1201–1215. doi:10.1172/JCI37007

Kammoun HL, Hainault I, Ferre P, Foufelle F (2009b) Nutritional related liver disease: targeting the endoplasmic reticulum stress. Curr Opin Clin Nutr Metab Care 12(6):575–582. doi:10.1097/MCO.0b013e32833189db

Kaplowitz N, Ji C (2006) Unfolding new mechanisms of alcoholic liver disease in the endoplasmic reticulum. J Gastroenterol Hepatol 21(Suppl 3):S7–S9. doi:10.1111/j.1440-1746.2006.04581.x

Kaplowitz N, Than TA, Shinohara M, Ji C (2007) Endoplasmic reticulum stress and liver injury. Semin Liver Dis 27(4):367–377. doi:10.1055/s-2007-991513

Kim SH, Naisbitt DJ (2016) Update on advances in research on idiosyncratic drug-induced liver injury. Allergy Asthma Immunol Res 8(1):3–11. doi:10.4168/aair.2016.8.1.3

Krentz HB, Kliewer G, Gill MJ (2005) Changing mortality rates and causes of death for HIV-infected individuals living in Southern Alberta, Canada from 1984 to 2003. HIV Med 6(2):99–106. doi:10.1111/j.1468-1293.2005.00271.x

Lana R, Nunez M, Mendoza JL, Soriano V (2001) Rate and risk factors of liver toxicity in patients receiving antiretroviral therapy. Med Clin (Barc) 117(16):607–610

Lee JN, Ye J (2004) Proteolytic activation of sterol regulatory element-binding protein induced by cellular stress through depletion of Insig-1. J Biol Chem 279(43):45257–45265. doi:10.1074/jbc.M408235200

Lee JS, Mendez R, Heng HH, Yang ZQ, Zhang K (2012) Pharmacological ER stress promotes hepatic lipogenesis and lipid droplet formation. Am J Transl Res 4(1):102–113

Lee SJ, Lee YJ, Park KK (2016) The pathogenesis of drug-induced liver injury. Expert Rev Gastroenterol Hepatol:1–11. doi:10.1080/17474124.2016.1196133

Li Z, Diehl AM (2003) Innate immunity in the liver. Curr Opin Gastroenterol 19(6):565–571

Liu MQ, Chen Z, Chen LX (2016) Endoplasmic reticulum stress: a novel mechanism and therapeutic target for cardiovascular diseases. Acta Pharmacol Sin 37(4):425–443. doi:10.1038/aps.2015.145

Lou LX, Geng B, Chen Y, Yu F, Zhao J, Tang CS (2009) Endoplasmic reticulum stress involved in heart and liver injury in iron-loaded rats. Clin Exp Pharmacol Physiol 36(7):612–618. doi:10.1111/j.1440-1681.2008.05114.x

Macias J, Neukam K, Mallolas J, Lopez-Cortes LF, Carton JA, Domingo P et al (2012) Liver toxicity of initial antiretroviral drug regimens including two nucleoside analogs plus one non-nucleoside analog or one ritonavir-boosted protease inhibitor in HIV/HCV-coinfected patients. HIV Clin Trials 13(2):61–69. doi:10.1310/hct1302-61

Mahdi AA, Rizvi SH, Parveen A (2016) Role of endoplasmic reticulum stress and unfolded protein responses in health and diseases. Indian J Clin Biochem 31(2):127–137. doi:10.1007/s12291-015-0502-4
Malhi H, Kaufman RJ (2011) Endoplasmic reticulum stress in liver disease. J Hepatol 54(4):795–809. doi:10.1016/j.jhep.2010.11.005
Marrer E, Dieterle F (2010) Impact of biomarker development on drug safety assessment. Toxicol Appl Pharmacol 243(2):167–179. doi:10.1016/j.taap.2009.12.015
Mateo MG, Gutierrez Mdel M, Vidal F, Domingo P (2014) An update on the pharmacological strategies in the treatment of HIV-1-associated adipose redistribution syndromes. Expert Opin Pharmacother 15(12):1749–1760. doi:10.1517/14656566.2014.928694
McGovern B (2004) Hepatic safety and HAART. J Int Assoc Physicians AIDS Care (Chic) *3*(Suppl 2):S24–S40
Meli R, Mattace Raso G, Calignano A (2014) Role of innate immune response in non-alcoholic fatty liver disease: metabolic complications and therapeutic tools. Front Immunol 5:177. doi:10.3389/fimmu.2014.00177
Merwat SN, Vierling JM (2011) HIV infection and the liver: the importance of HCV-HIV coinfection and drug-induced liver injury. Clin Liver Dis 15(1):131–152. doi:10.1016/j.cld.2010.09.012
Miltenburg NC, Boogerd W (2014) Chemotherapy-induced neuropathy: a comprehensive survey. Cancer Treat Rev 40(7):872–882. doi:10.1016/j.ctrv.2014.04.004
Ni Y, Zhuge F, Nagashimada M, Ota T (2016) Novel action of carotenoids on non-alcoholic fatty liver disease: macrophage polarization and liver homeostasis. Nutrients 8(7). doi:10.3390/nu8070391
Nunez M, Lana R, Mendoza JL, Martin-Carbonero L, Soriano V (2001) Risk factors for severe hepatic injury after introduction of highly active antiretroviral therapy. J Acquir Immune Defic Syndr 27(5):426–431
Pagliassotti MJ (2012) Endoplasmic reticulum stress in nonalcoholic fatty liver disease. Annu Rev Nutr 32:17–33. doi:10.1146/annurev-nutr-071811-150644
Parakh A, Dubey AP, Kumar A, Maheshwari A (2009) Lipodystrophy and metabolic complications of highly active antiretroviral therapy. Indian J Pediatr 76(10):1017–1021. doi:10.1007/s12098-009-0216-9
Pere D, Ignacio SL, Ramon T, Fernando L, Alberto T, Pompeyo V et al (2008) Dyslipidemia and cardiovascular disease risk factor management in HIV-1-infected subjects treated with HAART in the Spanish VACH cohort. Open AIDS J 2:26–38. doi:10.2174/1874613600802010026
Pfaffenbach KT, Gentile CL, Nivala AM, Wang D, Wei Y, Pagliassotti MJ (2010) Linking endoplasmic reticulum stress to cell death in hepatocytes: roles of C/EBP homologous protein and chemical chaperones in palmitate-mediated cell death. Am J Physiol Endocrinol Metab 298(5):E1027–E1035. doi:10.1152/ajpendo.00642.2009
Poorolajal J, Hooshmand E, Mahjub H, Esmailnasab N, Jenabi E (2016) Survival rate of AIDS disease and mortality in HIV-infected patients: a meta-analysis. Public Health. doi:10.1016/j.puhe.2016.05.004
Puri P, Wiest MM, Cheung O, Mirshahi F, Sargeant C, Min HK et al (2009) The plasma lipidomic signature of nonalcoholic steatohepatitis. Hepatology 50(6):1827–1838. doi:10.1002/hep.23229
Putcharoen O, Do T, Avihingsanon A, Ruxrungtham K (2015) Rationale and clinical utility of the darunavir-cobicistat combination in the treatment of HIV/AIDS. Drug Des Devel Ther 9:5763–5769. doi:10.2147/dddt.s63989
Rangnekar AS, Fontana RJ (2011) An update on drug induced liver injury. Minerva Gastroenterol Dietol 57(2):213–229
Rayavarapu S, Coley W, Nagaraju K (2012) Endoplasmic reticulum stress in skeletal muscle homeostasis and disease. Curr Rheumatol Rep 14(3):238–243. doi:10.1007/s11926-012-0247-5
Ron D, Walter P (2007) Signal integration in the endoplasmic reticulum unfolded protein response. Nat Rev Mol Cell Biol 8(7):519–529. doi:10.1038/nrm2199
Sanyal AJ (2005) Mechanisms of disease: pathogenesis of nonalcoholic fatty liver disease. Nat Clin Pract Gastroenterol Hepatol 2(1):46–53. doi:10.1038/ncpgasthep0084

Sanyal AJ (2013) An integrated view of liver injury and disease progression in nonalcoholic steatohepatitis. Hepatol Int 7(Suppl 2):800–805. doi:10.1007/s12072-013-9479-3

Sanyal AJ (2015) Novel therapeutic targets for steatohepatitis. Clin Res Hepatol Gastroenterol 39(Suppl 1):S46–S50. doi:10.1016/j.clinre.2015.05.012

Schnabl B, Brenner DA (2014) Interactions between the intestinal microbiome and liver diseases. Gastroenterology 146(6):1513–1524. doi:10.1053/j.gastro.2014.01.020

Schonthal AH (2012) Endoplasmic reticulum stress: its role in disease and novel prospects for therapy. Scientifica (Cairo) 2012:857516. doi:10.6064/2012/857516

Schwartz AL, Ciechanover A (2009) Targeting proteins for destruction by the ubiquitin system: implications for human pathobiology. Annu Rev Pharmacol Toxicol 49:73–96. doi:10.1146/annurev.pharmtox.051208.165340

Sharkey LM, Davies SE, Kaser A, Woodward JM (2016) Endoplasmic reticulum stress is implicated in intestinal failure-associated liver disease. JPEN J Parenter Enteral Nutr 40(3):431–436. doi:10.1177/0148607115571014

Sulkowski MS (2004) Drug-induced liver injury associated with antiretroviral therapy that includes HIV-1 protease inhibitors. Clin Infect Dis 38(Suppl 2):S90–S97. doi:10.1086/381444

Surgers L, Lacombe K (2013) Hepatoxicity of new antiretrovirals: a systematic review. Clin Res Hepatol Gastroenterol 37(2):126–133. doi:10.1016/j.clinre.2013.02.008

Tamaki N, Hatano E, Taura K, Tada M, Kodama Y, Nitta T et al (2008) CHOP deficiency attenuates cholestasis-induced liver fibrosis by reduction of hepatocyte injury. Am J Physiol Gastrointest Liver Physiol 294(2):G498–G505. doi:10.1152/ajpgi.00482.2007

Taura M, Kariya R, Kudo E, Goto H, Iwawaki T, Amano M et al (2013) Comparative analysis of ER stress response into HIV protease inhibitors: lopinavir but not darunavir induces potent ER stress response via ROS/JNK pathway. Free Radic Biol Med 65:778–788. doi:10.1016/j.freeradbiomed.2013.08.161

Tocchetti CG, Gallucci G, Coppola C, Piscopo G, Cipresso C, Maurea C et al (2013) The emerging issue of cardiac dysfunction induced by antineoplastic angiogenesis inhibitors. Eur J Heart Fail 15(5):482–489. doi:10.1093/eurjhf/hft008

Todd DJ, Lee AH, Glimcher LH (2008) The endoplasmic reticulum stress response in immunity and autoimmunity. Nat Rev Immunol 8(9):663–674. doi:10.1038/nri2359

Uzi D, Barda L, Scaiewicz V, Mills M, Mueller T, Gonzalez-Rodriguez A et al (2013) CHOP is a critical regulator of acetaminophen-induced hepatotoxicity. J Hepatol 59(3):495–503. doi:10.1016/j.jhep.2013.04.024

Volchuk A, Ron D (2010) The endoplasmic reticulum stress response in the pancreatic beta-cell. Diabetes Obes Metab 12(Suppl 2):48–57. doi:10.1111/j.1463-1326.2010.01271.x

Wan J, Benkdane M, Alons E, Lotersztajn S, Pavoine C (2014) M2 kupffer cells promote hepatocyte senescence: an IL-6-dependent protective mechanism against alcoholic liver disease. Am J Pathol 184(6):1763–1772. doi:10.1016/j.ajpath.2014.02.014

Wang Y, Zhang L, Wu X, Gurley EC, Kennedy E, Hylemon PB et al (2013) The role of CCAAT enhancer-binding protein homologous protein in human immunodeficiency virus protease-inhibitor-induced hepatic lipotoxicity in mice. Hepatology 57(3):1005–1016. doi:10.1002/hep.26107

Wenfeng Z, Yakun W, Di M, Jianping G, Chuanxin W, Chun H (2014) Kupffer cells: increasingly significant role in nonalcoholic fatty liver disease. Ann Hepatol 13(5):489–495

Williams K, Rao YP, Natarajan R, Pandak WM, Hylemon PB (2004) Indinavir alters sterol and fatty acid homeostatic mechanisms in primary rat hepatocytes by increasing levels of activated sterol regulatory element-binding proteins and decreasing cholesterol 7alpha-hydroxylase mRNA levels. Biochem Pharmacol 67(2):255–267

Willy JA, Young SK, Stevens JL, Masuoka HC, Wek RC (2015) CHOP links endoplasmic reticulum stress to NF-kappaB activation in the pathogenesis of nonalcoholic steatohepatitis. Mol Biol Cell 26(12):2190–2204. doi:10.1091/mbc.E15-01-0036

Wu X, Sun L, Zha W, Studer E, Gurley E, Chen L et al (2010) HIV protease inhibitors induce endoplasmic reticulum stress and disrupt barrier integrity in intestinal epithelial cells. Gastroenterology 138(1):197–209. doi:10.1053/j.gastro.2009.08.054

Wu X, Li Y, Peng K, Zhou H (2014) HIV protease inhibitors in gut barrier dysfunction and liver injury. Curr Opin Pharmacol 19:61–66. doi:10.1016/j.coph.2014.07.008

Wu FL, Liu WY, Van Poucke S, Braddock M, Jin WM, Xiao J et al (2016) Targeting endoplasmic reticulum stress in liver disease. Expert Rev Gastroenterol Hepatol:1–12. doi:10.1080/17474124.2016.1179575

Xiu F, Catapano M, Diao L, Stanojcic M, Jeschke MG (2015) Prolonged endoplasmic reticulum-stressed hepatocytes drive an alternative macrophage polarization. Shock 44(1):44–51. doi:10.1097/shk.0000000000000373

Yao Y, Wang Y, Zhang Z, He L, Zhu J, Zhang M et al (2016a) Chop deficiency protects mice against Bleomycin-induced pulmonary fibrosis by attenuating M2 macrophage production. Mol Ther 24(5):915–925. doi:10.1038/mt.2016.36

Yao X, Li Y, Cheng X, Li H (2016b) ER stress contributes to alpha-naphthyl isothiocyanate-induced liver injury with cholestasis in mice. Pathol Res Pract 212(6):560–567. doi:10.1016/j.prp.2016.05.001

Zha BS, Zhou H (2012) ER stress and lipid metabolism in adipocytes. Biochem Res Int 2012:312943. doi:10.1155/2012/312943

Zha BS, Wan X, Zhang X, Zha W, Zhou J, Wabitsch M et al (2013a) HIV protease inhibitors disrupt lipid metabolism by activating endoplasmic reticulum stress and inhibiting autophagy activity in adipocytes. PLoS One 8(3):e59514. doi:10.1371/journal.pone.0059514

Zha W, Wang G, Xu W, Liu X, Wang Y, Zha BS et al (2013b) Inhibition of P-glycoprotein by HIV protease inhibitors increases intracellular accumulation of berberine in murine and human macrophages. PLoS One 8(1):e54349. doi:10.1371/journal.pone.0054349

Zhang L, Wang HH (2016) The essential functions of endoplasmic reticulum chaperones in hepatic lipid metabolism. Dig Liver Dis. doi:10.1016/j.dld.2016.03.016

Zhang S, Carper MJ, Lei X, Cade WT, Yarasheski KE, Ramanadham S (2009) Protease inhibitors used in the treatment of HIV+ induce beta-cell apoptosis via the mitochondrial pathway and compromise insulin secretion. Am J Physiol Endocrinol Metab 296(4):E925–E935. doi:10.1152/ajpendo.90445.2008

Zhang C, Chen X, Zhu RM, Zhang Y, Yu T, Wang H et al (2012) Endoplasmic reticulum stress is involved in hepatic SREBP-1c activation and lipid accumulation in fructose-fed mice. Toxicol Lett 212(3):229–240. doi:10.1016/j.toxlet.2012.06.002

Zhang XQ, Xu CF, Yu CH, Chen WX, Li YM (2014) Role of endoplasmic reticulum stress in the pathogenesis of nonalcoholic fatty liver disease. World J Gastroenterol 20(7):1768–1776. doi:10.3748/wjg.v20.i7.1768

Zhou H (2011) HIV protease inhibitors induce endoplasmic reticulum stress and disrupt barrier integrity in intestinal epithelial cells. Methods Enzymol 490:107–119. doi:10.1016/b978-0-12-385114-7.00006-4

Zhou H, Liu R (2014) ER stress and hepatic lipid metabolism. Front Genet 5:112. doi:10.3389/fgene.2014.00112

Zhou H, Pandak WM Jr, Lyall V, Natarajan R, Hylemon PB (2005) HIV protease inhibitors activate the unfolded protein response in macrophages: implication for atherosclerosis and cardiovascular disease. Mol Pharmacol 68(3):690–700. doi:10.1124/mol.105.012898

Zhou H, Gurley EC, Jarujaron S, Ding H, Fang Y, Xu Z et al (2006) HIV protease inhibitors activate the unfolded protein response and disrupt lipid metabolism in primary hepatocytes. Am J Physiol Gastrointest Liver Physiol 291(6):G1071–G1080. doi:10.1152/ajpgi.00182.2006

Zhou H, Jarujaron S, Gurley EC, Chen L, Ding H, Studer E et al (2007) HIV protease inhibitors increase TNF-alpha and IL-6 expression in macrophages: involvement of the RNA-binding protein HuR. Atherosclerosis 195(1):e134–e143. doi:10.1016/j.atherosclerosis.2007.04.008

Zhou D, Huang C, Lin Z, Zhan S, Kong L, Fang C et al (2014) Macrophage polarization and function with emphasis on the evolving roles of coordinated regulation of cellular signaling pathways. Cell Signal 26(2):192–197. doi:10.1016/j.cellsig.2013.11.004

Chapter 3
Mechanisms of Acetaminophen-Induced Liver Injury

Benjamin L. Woolbright and Hartmut Jaeschke

Abbreviations

AIF	Apoptosis-inducing factor
ALF	Acute liver failure
AMAP	N-Acetyl-m-Aminophenol
APAP	Acetaminophen
ASK1	Apoptosis signal-regulating kinase 1
ATG	Autophagy-related genes
Bax	Bcl-2-associated X protein
Bcl-2	B-cell lymphoma-2
Bid	Bh3-interacting-domain death agonist
CCR2	C-C Chemokine receptor 2
CXCR2	CXC Chemokine receptor 2
DAMPs	Damage-associated molecular patterns
ETC	Electron transport chain
GCLC	Glutamate-cysteine ligase catalytic subunit
GSH	Glutathione
JNK	c-Jun N terminal kinase
HMGB1	High mobility group box 1
IL-10	Interleukin-10
iNOS	Inducible nitric oxide synthase
MAPK	Mitogen-activated protein kinase
miR	Micro RNA
MKK4	Mitogen-activated protein kinase kinase 4
MLK3	Mixed lineage kinase 3

B.L. Woolbright • H. Jaeschke (✉)
Department of Pharmacology, Toxicology and Therapeutics, University of Kansas Medical Center, Kansas City, KS, USA
e-mail: hjaeschke@kumc.edu

W.-X. Ding, X.-M. Yin (eds.), *Cellular Injury in Liver Diseases*, Cell Death in Biology and Diseases, DOI 10.1007/978-3-319-53774-0_3

X.-M. Yin and Z. Dong (Series eds.), *Cell Death in Biology and Diseases*

MLKL	Mixed lineage kinase domain-like
MOMP	Mitochondrial outer membrane permeabilization
MPTP	Mitochondrial permeability transition pore
NAC	N-Acetylcysteine
NAPQI	N-Acetyl-p-benzoquinone imine
NQO1	NAD(P)H dehydrogenase [quinone] 1
Nrf2	Nuclear factor (erythroid-derived 2)-like 2
RIP1	Receptor-interacting protein kinase 1
RIP3	Receptor-interacting protein kinase 3
RNS	Reactive nitrogen species
ROS	Reactive oxygen species
Sab	SH3BP5, SH3 domain-binding protein that preferentially associates with Bruton's tyrosine kinase
SMAC	Second mitochondria-derived activator of caspases
SOD	Superoxide dismutase
STAT3	Signal transducer and activator of transcription 3
TLR	Toll-like receptor
TNF-α	Tumor necrosis factor-α

3.1 Introduction

Acute liver failure (ALF) due to acetaminophen- (APAP)-induced liver injury is the number one cause of ALF in the West (Lee 2013). Combination of APAP with a number of other pharmaceutical agents can lead to accidental overdose. For many of these patients, an overdose can be rapidly diagnosed using serum acetaminophen levels and serum biochemistry (Rumack and Matthew 1975). Patients that present quickly after the overdose can be treated with N-acetylcysteine (NAC), and recover in a majority of cases (Bebarta et al. 2010). Patients who are treated with NAC distally from their overdose, i.e., more than 24 h commonly progress to ALF and many have a poor prognosis (Lee 2013). Some of these patients undergo liver transplantation to survive, which is curative, but places a life-time need for anti-rejection medication on the patient. Thus, new therapeutic approaches are needed that can be used to limit injury and prevent ALF when treatment is initiated during the progression of injury. Targeting established mediators in APAP overdose is critical, as is the discovery of novel therapeutic targets. The focus of this review will be on currently understood mechanisms of hepatotoxicity after APAP overdose.

3.2 Metabolism and Initiation of Toxicity

At therapeutic doses, the parent compound APAP is largely nontoxic as it is rapidly excreted after enzymatic conjugation with glucuronic acid or sulfate, and only a small fraction is metabolized by cytochrome P450 enzymes to form a reactive metabolite (McGill and Jaeschke 2013). After an overdose, sulfation is quickly saturated (Clements et al. 1984). In contrast, glucuronidation, which is a high capacity conjugation pathway, can be dramatically increased but does not appear to be saturated (Xie et al. 2015a). Importantly, a substantial amount of the drug is then metabolized by the cytochrome P450 enzymes Cyp2E1 and Cyp1A2 (Zaher et al. 1998) forming enhanced amounts of the electrophile N-acetyl-*p*-benzoquinone imine (NAPQI) (Dahlin et al. 1984). Some of the NAPQI generated can be detoxified by glutathione (GSH) (Rosen et al. 1984) leading to extensive depletion of hepatic GSH levels, which serves as a biochemical marker of metabolic activation of APAP (Jaeschke et al. 2011). However, an increasing fraction of NAPQI covalently binds to cellular proteins (Mitchell et al. 1973). The antidote NAC works in this pathway, but does not directly bind NAPQI. In contrast, NAC stimulates the production of cellular GSH and thus more rapidly detoxifies NAPQI (Corcoran and Wong 1986). Although NAPQI binds GSH and is detoxified, because of the rapid depletion of the hepatic GSH content there is a fraction that adducts cellular proteins, especially mitochondrial proteins, and initiates the characteristic mitochondrial oxidative stress found after APAP overdose (Jaeschke 1990; Knight et al. 2001). While metabolism by CYP2E1 also generates small quantities of ROS during oxidative metabolism (Nieto et al. 2002), the primary oxidant stress is generated by dysfunctional mitochondria (Jaeschke et al. 2012a; Du et al. 2017). Thus, the primary step in the toxicity of APAP is oxidative metabolism of APAP to NAPQI by CYP2E1, and its subsequent binding to cellular proteins.

Much of the recent work on APAP metabolism has focused on protein adducts of APAP, with an emphasis on the presence of APAP adducts in the serum of both mice and human patients (McGill et al. 2013a, b; Davern et al. 2006). Serum adducts were originally thought to be byproducts of cellular necrosis, and were only present when there was considerable ALT release (Pumford et al. 1990). Recent data have shown that serum adducts are present even at therapeutic doses in humans (Heard et al. 2011; James et al. 2013) and in animals (McGill et al. 2012b, 2013). Moreover, supra-therapeutic dosing can generate APAP protein adducts even in the absence of overt hepatotoxicity (Hu et al. 2016b; McGill et al. 2012b, 2013; O'Malley et al. 2015). This would suggest that the presence of adducts alone is not sufficient to initiate toxicity and that there is likely a threshold level of adducts necessary for toxicity (James et al. 2009). This hypothesis has been used to develop a novel point-of-care-based test for detecting the potential for APAP toxicity in overdose patients (Roberts et al. 2017).

For a number of years it has been known that APAP-induced injury does not correlate with the overall protein adduct formation but with the presence of mitochondrial adducts. Sid Nelson's group first demonstrated that APAP and its nontoxic regioisomer *N*-acetyl-*meta*-aminophenol (AMAP) show similar total protein adducts formation but only APAP causes mitochondrial protein adducts, which correlate with toxicity in the mouse (Tirmenstein and Nelson 1989). These findings in mice were confirmed by others (Qiu et al. 2001). However, AMAP appears to be hepatotoxic in human liver slices (Hadi et al. 2013) and in primary human hepatocytes (Xie et al. 2015b). It could be confirmed that the lack of AMAP toxicity in mouse hepatocytes correlates with the absence of mitochondrial adducts; however, the toxicity in human hepatocytes clearly involves mitochondrial AMAP protein adducts (Xie et al. 2015b). In a recent dose-response study, it was shown that mitochondrial protein adducts formation correlates with c-jun *N*-terminal kinase (JNK) activation and mitochondrial dysfunction in the mouse (Hu et al. 2016b). Furthermore, in the rat model, there is no relevant toxicity even after very high overdoses of APAP due to the limited mitochondrial protein adducts and absence of JNK activation (McGill et al. 2012b). Together, these findings strongly suggest a critical role of mitochondrial protein adducts formation as an initiating event in APAP hepatotoxicity.

3.3 The C-JUN-N-Terminal Kinase Pathway in APAP Toxicity

The mitogen-activated protein kinase (MAPK) JNK exists in two isoforms in hepatocytes, termed JNK1 and JNK2. As early as two hours after the initial overdose, the JNK pathway is activated, and a phosphorylated form of JNK translocates to the mitochondria (Hanawa et al. 2008, Xie et al. 2015c). JNK activation requires the activation of upstream MAPK such as apoptosis signal-regulating kinase 1 (ASK1) (Nakagawa et al. 2008; Xie et al. 2015c), mixed lineage kinase-3 (MLK3) (Sharma et al. 2012), and mitogen-activated protein kinase kinase 4 (MKK4) (Win et al. 2011). While the function of JNK in the mitochondria is not well understood, the mitochondrial outer membrane protein SH3 domain-binding protein that preferentially associates with Btk (Sab) serves as the initial molecular target for JNK (Win et al. 2011). Activation of Sab on the mitochondrial outer membrane inactivates proto-oncogene tyrosine-protein kinase Src on the inner mitochondrial membrane, which disrupts the electron transport chain (ETC) and enhances mitochondrial ROS leakage (Win et al. 2016). Consistent with this hypothesis, the JNK inhibitor SP600125 also prevents the accumulation of peroxynitrite in the mitochondria (Saito et al. 2010a), which plays a key role in the mitochondrial pathology (Knight et al. 2002). Peroxynitrite is formed inside the mitochondria by the reaction of nitric oxide with superoxide (Cover et al. 2005). The mitochondrial superoxide production is well defined in APAP-induced liver injury and is the central source of ROS (reviewed in Jaeschke et al. 2012a) and JNK activation results in a self-sustaining feedback loop that dramatically promotes cellular dysfunction through generation of ROS from the mitochondria (Han et al. 2013; Du et al. 2015). This hypothesis is not only supported by direct inhibition of JNK (Hanawa et al. 2008; Saito et al. 2010a) but also by effects on upstream kinases responsible for JNK activation. Pharmacological inhibition of ASK-1 is highly protective (Xie et al. 2015c) as is the genetic depletion of ASK-1 (Nakagawa et al. 2008) or MLK3 (Sharma et al. 2012). Likewise, deletion of protein tyrosine phosphatase 1B (Mobasher et al. 2013), or mitogen-activated protein kinase phosphatase-1 (Wancket et al. 2012), which reduce

JNK activation, aggravated APAP-induced liver injury. In addition, a number of other pharmacological agents that protect against APAP hepatotoxicity are also proposed to work through JNK (reviewed in Du et al. 2015). For example, administration of metformin either before or after APAP overdose reduced JNK activity and protected against the injury (Kim et al. 2015); although, it should be noted that others obtained similar protection in this model with metformin without the noted change in JNK activity (Du et al. 2016b). Leflunomide, an immunosuppressant and pyrimidine synthesis inhibitor, also protects against APAP-induced liver injury through suppression of JNK activity (Latchoumycandane et al. 2007). Importantly, there is evidence for JNK activation and a role for JNK in human samples as well. Human hepatocytes exposed to a JNK inhibitor are partially protected against APAP-induced liver injury (Xie et al. 2014) and samples from patients with APAP overdose have elevated levels of JNK in hepatocytes (Cubero et al. 2016). Despite this substantial evidence in favor of a pathological role for JNK, a recent study has indicated that hepatocyte-specific deletion of JNK may actually predispose mice to increased injury after APAP overdose indicating some activity of JNK might be protective (Cubero et al. 2016). However, one still needs to keep in mind that this is a single experiment that shows the opposite effect compared to a large number of different approaches that all demonstrate a pathophysiological role of JNK activation (Jaeschke 2016). Cubero et al. (2016) also showed that SP600125 is protective even in hepatocyte-specific JNK knockout animals, which raises the concern that the pharmacological actions of this particular JNK inhibitor may involve off-target effects (Cubero et al. 2016). Again, some concerns were raised regarding the solvent use in these experiments; importantly, the fact that SP600125 protects in a JNK knock-out mouse does not preclude the idea that this inhibitor still acts through JNK inhibition in a normal animal (Jaeschke 2016). Thus, the preponderance of data still strongly supports an important role for the JNK signaling pathway in APAP-induced liver injury. *These mechanisms are summarized in Fig. 3.1.*

3.4 The Role of the BCL-2 Protein Family in APAP Hepatotoxicity

Concurrent to activation of the JNK pathway, a number of other signaling pathways are activated, including members of the B-cell lymphoma-2 (Bcl-2) family. The Bcl-2 family is a diverse family of proteins involved in cell turnover, many of which also play a pivotal role in cell death processes in multiple disease states (Gross et al. 1998; Korsmeyer et al. 1993). Bcl-2 family members Bh3-interacting-domain death agonist (Bid) and Bcl-2-associated X protein (Bax) are both activated upon overdose of APAP (El-Hassan et al. 2003; Jaeschke and Bajt 2006). Cleavage of bid in the cytoplasm leads to translocation of the truncated form of bid to the mitochondria where it can form pores leading to mitochondrial outer membrane permeabilization (MOMP) (Yin et al. 1999). Bid translocation occurs rapidly after APAP overdose

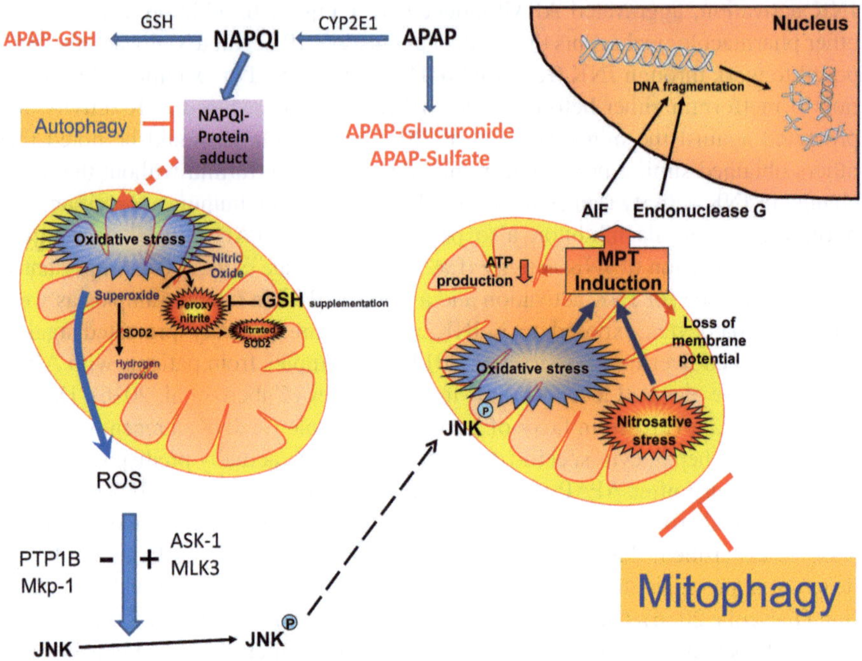

Fig. 3.1 Intracellular mechanisms of acetaminophen hepatotoxicity. Therapeutic doses of acetaminophen are conjugated with glucuronide or sulfate via Phase II metabolism. Toxic doses of acetaminophen are converted to NAPQI by cytochrome P4502E1. NAPQI adducts intracellular proteins, especially mitochondrial proteins, which causes oxidative stress. This can be attenuated by stimulation of the removal of the protein adducts through autophagy. Under normal conditions though, this results in significant mitochondrial oxidative stress and the activation of the mitogen activated protein kinase cascade ultimately causing the activation of JNK and the translocation of the phosphorylated form to the mitochondria where it can further amplify the oxidant stress. Multiple upstream components of the phosphorylation and dephosphorylation of JNK are involved in acetaminophen hepatotoxicity. The combined oxidative stress causes the opening of the mitochondrial permeability transition pore (MPTP) and loss of mitochondrial stability. Permeability of the mitochondrial outer membrane results in release of apoptosis-inducing factor and endonuclease G from the mitochondria that translocate to the nucleus where they cleave DNA, resulting in oncotic necrosis. AIF, apoptosis-inducing factor; APAP, acetaminophen; ASK-1, apoptosis signal-regulating kinase 1; EndoG, endonuclease G; GSH, glutathione; JNK, c-Jun N-terminal kinase; Mkp-1, mitogen-activated protein kinase phosphatase 1; MLK3, mixed lineage kinase 3; MPT, mitochondrial permeability transition; NAPQI, N-acetyl p-benzoquinone imine; PTP1B, protein-tyrosine phosphatase 1B; ROS, reactive oxygen species; SOD2, superoxide dismutase 2

(Jaeschke and Bajt 2006). Similarly, there is a rapid translocation of the cytosolic Bax protein to the mitochondria, which also causes MOMP (Bajt et al. 2008). Bax gene knockout mice show reduced mitochondrial intermembrane protein release, less DNA fragmentation, and less early injury (Bajt et al. 2008). However, Bax deficiency does not affect the mitochondrial oxidant stress, which eventually triggers the MPT resulting in mitochondrial matrix swelling, rupture of the outer membrane, and release of intermembrane proteins followed by DNA fragmentation and cell death (Bajt et al. 2008). While it is not well characterized what cellular stressors activate these proteins in the APAP model, inhibition of JNK reduces Bax activity (Saito et al. 2010a), indicating oxidative stress and JNK activation are both likely candidates. APAP also results in downregulation of cytoprotective Bcl-2 family members such as Bcl-xL and Bcl-2 itself (Latchoumycandane et al. 2007). Mitochondrial protective agents such as leflunomide mitigate this interaction and are also protective against APAP overdose (Latchoumycandane et al. 2007). Surprisingly, mice overexpressing Bcl-2 are actually more susceptible to APAP-induced liver injury, although it is not well understood why this is the case (Adams et al. 2001). It should be noted that Bcl-2 family member-induced cell death is commonly indicated to be through apoptosis in the literature (reviewed in Tait and Green 2013). Despite this, there is no evidence for caspase activation in the standard APAP overdose model (Gujral et al. 2002; Lawson et al. 1999; Jaeschke et al. 2006). Regardless, while Bcl-2 family members are unequivocally activated, and likely play a role in the injury, their promotion of APAP-induced cell death is through promotion of mitochondrial dysfunction, but independent of traditional apoptosis pathways.

3.5 The Role of Receptor-Interacting Protein Kinase Family Members

The recent identification of the receptor-interacting protein kinase 1/3 (RIP1/RIP3) protein family's role in cell death (He et al. 2009; Zhang et al. 2009) has led to a number of studies examining these proteins in liver disease (Ramachandran et al. 2013; Roychowdhury et al. 2013; Vucur et al. 2013; McGill et al. 2015). These proteins were first identified as critical mediators of TNF-α-induced apoptosis (He et al. 2009; Zhang et al. 2009). Under conditions of caspase inhibition, RIP1 and RIP3 can accumulate, dimerize, and initiate a signaling pathway that results in non-apoptotic cell death (He et al. 2009; Zhang et al. 2009). These proteins mediate a molecular mechanism for cell death (termed necroptosis) under conditions of caspase inhibition, which likely serves as a survival mechanism under normal physiological conditions. Critical for this necroptosis pathway is the mixed lineage kinase domain-like (MLKL), which is activated by the RIP1/RIP3 complex, and moves to the cell membrane to form membrane-disrupting pores, causing membrane leakage and necrotic cell death (Wang et al. 2014; Zhang et al. 2016). APAP overdose dramatically upregulates RIP3 in the liver, and genetic knockout of RIP3 is protective, as is depletion of RIP3 with morpholino-based antisense reagents (Ramachandran et al. 2013; Deutsch et al. 2015). Pharmacological treatment with dabrafenib, which inhibits RIP3-mediated Ser358 phosphorylation of MLKL and disrupted the interaction between RIP3 and MLKL, attenuated APAP-induced liver injury in mice and human hepatocytes (Li et al. 2014). Treatment with the RIP1 inhibitor necrostatin is also protective against APAP in vitro and in vivo, indicating there might be interactions between RIP1 and RIP3 (Ramachandran et al. 2013; Zhang et al. 2014; Dara et al. 2015). However, there appears to be evidence that MLKL is not involved in APAP hepatotoxicity (Dara et al. 2015; Gunther et al. 2016). Furthermore, RIP3-deficiency protects only against the early phase of APAP-induced liver injury (Ramachandran et al. 2013) and others have questioned the role of RIP3 all together (Dara et al. 2015). There could be one explanation for this controversy, i.e., that RIP3 may be more important in human than in mouse hepatocytes (Li et al. 2014). However, the specific mechanisms of RIP1 and RIP3 involvement in the pathophysiology of APAP hepatotoxicity remain poorly understood.

3.6 Oxidative and Nitrosative Stress in APAP Hepatotoxicity

Despite being controversially discussed for a very long time, especially in the 1980s, the essential role of a mitochondrial oxidant stress in APAP hepatotoxicity is now firmly established (Jaeschke et al. 2012a; Du et al. 2016a). Mitochondrial dysfunction leads to leakage of electrons from the mitochondrial ETC thereby forming superoxide, which can give rise to further reactive oxygen (ROS) and nitrogen species (RNS). Detoxifying these ROS and RNS through supporting the internal antioxidant systems is highly protective, even when done after the initial overdose, but before the generation of overt mitochondrial dysfunction (Knight et al. 2002; Bajt et al. 2003; James et al. 2003b; Saito et al. 2010b; Du et al. 2016a, 2017). NAC, the primary antidote used clinically, is thought to work primarily as a precursor for hepatic GSH production, which can scavenge NAPQI (Corcoran and Wong 1986; Lauterburg et al. 1983). However, subsequent studies have unequivocally demonstrated that the recovery of mitochondrial GSH after treatment with GSH precursors is critical for the detoxification of ROS and more importantly for peroxynitrite (Knight et al. 2002; Saito et al. 2010b). Superoxide can dismutate either spontaneously or in a SOD-catalyzed mechanism to molecular oxygen and hydrogen peroxide, which can be detoxified by glutathione peroxidase. However, the key effect of mitochondrial MnSOD (SOD2) is to rapidly and effectively remove superoxide to prevent the formation of the potent oxidant and nitrating species peroxynitrite. In fact, peroxynitrite is a critical toxicant in APAP-induced liver injury (Knight et al. 2002). Consequently, even partial deficiency of MnSOD dramatically exacerbates peroxynitrite formation and cytotoxicity after APAP overdose (Fujimoto et al. 2009; Ramachandran et al. 2011). Furthermore, treatment with a mitochondrial-homing SOD mimetic, Mito-TEMPO, effectively prevented peroxynitrite formation and protected against APAP toxicity (Du et al. 2017). This mitochondrial oxidative stress is enhanced by concurrent lysosomal dysfunction, as a number of lysosomal entities are also released during APAP toxicity (Woolbright et al. 2012). This includes release of lysosomal iron, which can translocate to the mitochondria where it supports hydrogen peroxide-induced cytotoxicity through Fenton-like reactions (Kon et al. 2010; Hu et al. 2016a). In support of these data, exposing hepatocytes to high concentrations of oxygen during cell culture results in a significant increase in the susceptibility to APAP toxicity due to increased nitrosative stress (Burke et al. 2010); whereas lowering oxygen tension attenuates APAP toxicity (Yan et al. 2010). This is likely due to increased/decreased oxygen availability that alters mitochondrial electron leakage and ROS production (Yan et al. 2010). Overdose of APAP results in depletion of both mitochondrial and total hepatic GSH (Knight et al. 2002; Saito et al. 2010b), which may be due to the presence of CYP2E1 in the mitochondria as well as the cytoplasm (Knockaert et al. 2011). Female mice are actually resistant to APAP-induced liver as they recover mitochondrial GSH more rapidly, and thus detoxify ROS and RNS from the mitochondria more effectively (Du et al. 2014). This massive oxidative response after APAP overdose causes activation of the nuclear factor (erythroid-derived 2)-like 2 (Nrf2) signaling pathway (Enomoto et al. 2001). This response results in upregulation of a number of antioxidant genes, including genes in the GSH synthesis pathway (Wild et al. 1999). Knockout of Nrf2 results in dramatic sensitization toward APAP underscoring the important role of the antioxidant response (Enomoto et al. 2001). In the cases of toxic overdoses, this antioxidant response is overwhelmed and significant oxidant stress combined with mitochondrial pore formation by Bcl-2 family proteins results in outer mitochondrial membrane disruption and leakage of mitochondrial components to the cytoplasm. This triggers the mitochondria permeability transition pore (MPTP) opening (Kon et al. 2004). While this has previously been thought to be an irreversible commitment to cell death, more recent data indicate that MPTP opening might be a reversible step if JNK-mediated amplification is diminished (Hu et al. 2016b). In any case, MPTP opening results in matrix swelling and rupture of the outer membrane, which causes the release of a number of mitochondrial proteins that begin to execute cell death functions.

3.7 Mitochondria-Mediated DNA Fragmentation in APAP Toxicity

MPTP formation after APAP overdose results in release of a number of mitochondrial proteins into the cytoplasm including cytochrome c, second mitochondria-derived activator of caspases (Smac), endonuclease G, and apoptosis-inducing factor (AIF) (Bajt et al. 2006, 2008). Once released from the mitochondria, only endonuclease G and AIF translocate to the nucleus where they serve as endonucleases that cleave DNA and directly cause cell death (Bajt et al. 2006, 2008; Cover et al. 2005). DNA fragmentation is clearly evident in the area of necrosis after APAP-induced liver injury as only the necrotic zone stains positive with the TUNEL assay (Lawson et al. 1999; Gujral et al. 2002). The TUNEL staining pattern (nucleus and the cytoplasm) of the cells is consistent with necrotic cell death (Lawson et al. 1999; Cover et al. 2005) confirming previous concerns that the TUNEL assay is not specific for apoptosis but actually detects DNA fragmentation independent of the mode of cell death (Grasl-Kraupp et al. 1995). Partial ablation of AIF results in substantial protection against the injury (Bajt et al. 2011) and protection by NAC or GSH results in no nuclear translocation of endonuclease G or AIF (Bajt et al. 2006, 2008; Cover et al. 2005). Because of the lack of caspase activation during APAP toxicity (Lawson et al. 1999; Jaeschke et al. 2006), caspase-activated DNase cannot be responsible for the nuclear DNA fragmentation (Cover et al. 2005). Thus, the mitochondria-derived nuclease activity of endonuclease G and AIF is the executor step in APAP-induced liver injury as this effect results in complete nuclear breakdown, plasma membrane disruption, and cellular rupture. In addition to these critical proteins, cytochrome c and second mitochondria-derived activator of caspase (Smac) are also released into the cytoplasm, although no direct role has been found for them in the mechanism of toxicity. It remains unclear, why despite the mitochondrial release of these pro-apoptotic proteins, there is no apoptotic cell death. In contrast, release of well-defined biomarkers of necrosis from the liver including full-length cytokeratin-18 (M65), mitochondrial glutamate dehydrogenase, HMGB1 protein, microRNA-122, mitochondrial DNA, and nuclear DNA fragments and morphological evidence of necrosis are all present after APAP-induced liver injury (Antoine et al. 2012; McGill et al. 2012a). The reason for this switch to necrosis is not currently well defined, but may be due to a lack of ATP necessary for forming the apoptosome (Williams et al. 2011). While there is evidence for minor, temporary caspase activation under certain conditions in mice (Antoine et al. 2010), and clinically (Antoine et al. 2012), it likely plays a minimal role in the overall disease state. In support of this conclusion, APAP-induced cell death in primary mouse hepatocytes (Bajt et al. 2004; Ramachandran et al. 2013), in the metabolically competent hepatoma cell line HepaRG (McGill et al. 2011) and in primary human hepatocytes (Xie et al. 2014), is consistently necrosis. Thus, mitochondrial dysfunction during APAP causes release of mitochondrial proteins and lysis of nuclear DNA, directly leading to necrotic cell death.

3.8 Autophagy as an Adaptive Mechanism to APAP-Induced Cellular Stress

Autophagy is an increasingly well-studied process wherein cells break down damaged or unneeded molecules or organelles in lysosomes and then reuse that material for critical processes (Takeshige et al. 1992; Yin et al. 2008). Component-specific autophagy exists for both autophagy of fat droplets termed lipophagy (Singh et al. 2009; Liu and Czaja 2013) and autophagy of mitochondria termed mitophagy (Dagda et al. 2009; Kim et al. 2007; Williams and Ding 2015). Recent work indicates that autophagy of protein adducts and mitochondria are critical adaptive processes that limit APAP-induced liver injury (Ni et al. 2012a, 2016). In fact, chemical stimulation of autophagy attenuates APAP hepatotoxicity and inhibition of autophagy has the opposite effect (Ni et al. 2012a, 2016). The effect of autophagy on cell death is most obvious at the edge of the necrotic area where cells struggle to survive

(Ni et al. 2013). Surprisingly, depletion of the critical autophagy protein 5 (Atg5) eliminates autophagy but the mice are highly resistant to APAP toxicity (Ni et al. 2012b). The chronic loss of autophagy causes extensive cellular stress in hepatocytes leading to a cycle of continuous apoptotic cell death and regeneration (Ni et al. 2012b). This cellular stress leads to suppression of Cyp2E1protein expression with reduced reactive metabolite formation and less protein adducts. In addition, Atg5-deficiency also causes upregulation of Nrf2 and related protective genes including NAD(P)H dehydrogenase [quinone] 1(NQO1) and glutamate-cysteine ligase catalytic subunit (gclc), which supports an accelerated recovery of hepatic GSH levels (Ni et al. 2012b). The combination of these adaptive mechanisms to the cellular stress of inhibited autophagy ultimately contributes to the resistance of Atg5-deficient mice to APAP-induced liver injury (Ni et al. 2012b). However, acute activation of autophagy likely protects against APAP hepatotoxicity partially through elimination of damaged mitochondria that generate reactive oxygen (Ni et al. 2012a). Although mitochondrial fission promotes this oxidant stress (Ramachandran et al. 2013), the removal of these damaged mitochondrial fragments is ultimately beneficial (Ni et al. 2012a, 2013). APAP protein adducts can also be removed by autophagy (Ni et al. 2016). The acute elimination of protein adducts, especially in the cytosol, may not be as critical as mitophagy, however, given the fact that even therapeutic doses of APAP result in low levels of adducts in hepatocytes (McGill et al. 2012a, 2013), and some patients take APAP daily for months or even years, the chronic accumulation of adducts needs to be avoided. Thus, adduct removal through autophagy is critical for long-term health of liver cells during APAP exposure (Ni et al. 2016). More work is required in this area to better understand the molecular mechanisms that control both the removal of mitochondria and protein adducts during APAP hepatotoxicity.

3.9 The Sterile Inflammatory Response During APAP Hepatotoxicity

One of the inevitable outcomes of acute APAP-induced necrosis is the development of a sterile inflammatory response. Necrotic cell death causes release of intracellular constituent molecules (damage-associated molecular patterns, DAMPs) such as nuclear DNA fragments, mitochondrial DNA, and proteins like HMGB1 that stimulate toll-like receptors (TLRs) on local inflammatory cells such as Kupffer cells in the liver (Woolbright and Jaeschke 2015; 2017). In response to these DAMPs macrophages produce a number of cytokines and chemokines that activate and recruit a variety of inflammatory cells into the liver (Cover et al. 2006; Dambach et al. 2002; Holt et al. 2008; Lawson et al. 2000). Neutrophils are the first cell to accumulate in the liver after APAP overdose followed by bone-marrow-derived monocytes. The main purpose of these phagocytes is to remove the necrotic cell debris in preparation for regeneration and wound healing. However, these inflammatory cells have

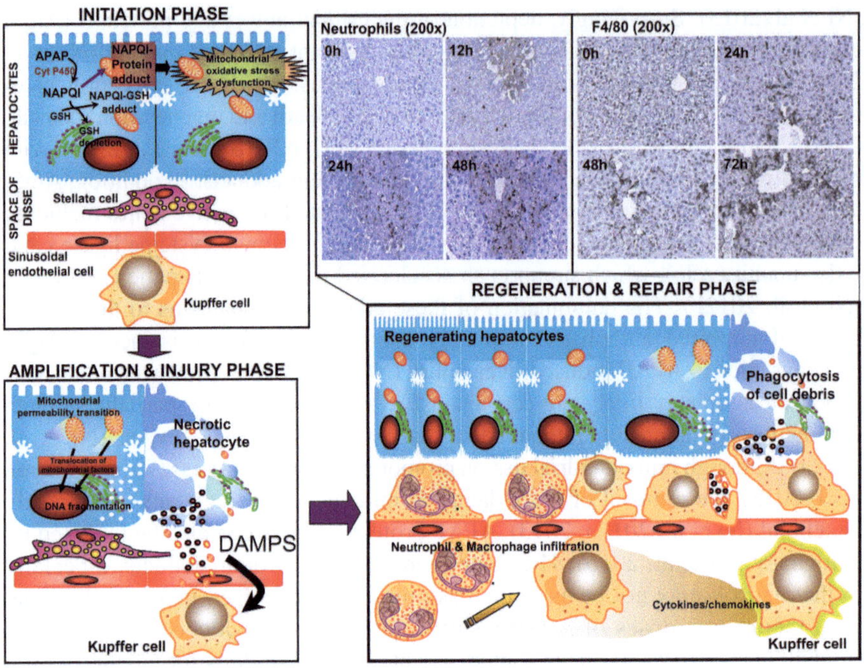

Fig. 3.2 The multiple stages of acetaminophen hepatotoxicity. During an overdose, APAP is converted to NAPQI resulting in mitochondria oxidative stress and eventual necrosis. Necrosis results in the release of DAMPs, in the form of mitochondrial and nuclear DNA, from hepatocytes. DAMPs activate resident macrophages (Kupffer cells) (F4/80 positive) via TLRs. Kupffer cells produce cytokines, which attract neutrophils and monocytes largely to the area of necrosis (*see* above), and stimulate phagocytosis of the cellular debris to create space for regenerating hepatocytes. NAPQI – N-acetyl *p*-benzoquinone imine, GSH – glutathione, DAMPs – damage-associated molecular patterns, TLR – toll-like receptor (not depicted). Figure adapted from Jaeschke et al. 2012a with permission

the capacity to cause additional tissue injury through generation of reactive oxygen species (Jaeschke 2011). Substantial controversy exists in this field, with a number of groups arguing in favor of the inflammatory process directly contributing to the injury, while other groups have argued that the inflammatory process is secondary to the injury (Jaeschke et al. 2012b; Woolbright and Jaeschke 2017). As a full review on the controversy is outside the scope of this chapter, we will only briefly discuss mechanisms of inflammation relevant to the direct injury. *Figure 3.2 depicts the various stages of APAP-induced liver injury associated with inflammation.*

Depletion of neutrophils with a neutralizing antibody 24 h before the administration of APAP is highly protective (Liu et al. 2006). However, this is largely due to the induction of cytoprotective genes in the liver caused by the activated Kupffer cell during the removal of the inactivated neutrophils (Jaeschke and Liu 2007). Furthermore, administration of neutrophil-depleting antibodies at time points close to APAP did not provide protection (Cover et al. 2006). Similarly, although gadolinium chloride-mediated inactivation of Kupffer cells was reported to be protective (Michael et al. 1999), total elimination of Kupffer cells with liposomal clodronate actually aggravated the injury (Ju et al. 2002). In addition, inhibition of monocyte recruitment in CCR2-deficient mice did not protect against APAP hepatotoxicity, as reported by a number of groups (Dambach et al. 2002; Hogaboam et al. 1999; Holt et al. 2008; You et al. 2013). However, recently, the same CCR2-deficient mice showed reduced APAP-induced liver injury (Mossanen et al. 2016). Importantly, there is no protection when neutrophil- and macrophage-derived ROS production is inhibited (James et al. 2003a; Williams et al. 2014), which makes it highly unlikely that these phagocytes contribute to the injury as cytotoxicity of neutrophils is critically dependent on the capacity to generate ROS (Jaeschke 2011). In addition, neutrophils are generally not fully activated until after the injury phase (Williams et al. 2014). Neutrophil elastase has recently been hypothesized to be the direct cytotoxic agent as bone marrow transplantation with elastase-deficient cells was protective (Huebener et al. 2015). However, whole body elastase knockout mice are not protected against APAP overdose (Jaeschke, unpublished). In addition, proteases such as elastase are generally important for neutrophil migration in the extravascular space and are unlikely used as a cytotoxic mediator by these inflammatory cells (Jaeschke 2006). Thus, the preponderance of the experimental evidence does not support a direct cytotoxicity of a specific inflammatory cell during APAP hepatotoxicity.

Despite this, a number of interventions targeting inflammatory mediators are protective, and reasonable explanations for these effects are noted. IL-10 regulates the induction of inducible nitric oxide synthase (iNOS), which can be a source of nitric oxide production in the APAP model (Bourdi et al. 2002). Knockout of IL-10 predisposes mice to APAP-induced liver injury due to enhanced iNOS production, and double knockout of IL-10 and iNOS ameliorates this increase (Bourdi et al. 2002). Similarly, knockout of IL-4 made mice more susceptible to APAP-induced liver injury as these mice had reduced GSH synthesis capacity due to reduced γ-glutamylcysteine ligase (γ-GCL) gene expression (Ryan et al. 2012). C-X-C chemokine receptor 2 (CXCR2) is a primary chemokine receptor and mediates the inflammatory response in a number of murine models such as hepatic ischemia-reperfusion (Kuboki et al. 2008). CXCR2 also may be protective against APAP-induced liver injury as the protective effect of administration of the chemokine CXC ligand 10 is mediated by CXCR2 (Bone-Larson et al. 2001). However, it is currently not well understood how either of these proteins mediates their effect. Administration of IL-22 also protected mice against APAP toxicity through a signal transducer and activator of transcription 3 (STAT3)-mediated mechanism that was independent of inflammation (Feng et al. 2014). Animals treated with neutralizing antibodies against HMGB1 were partially protected (Lundback et al. 2016), as were hepatocyte-specific HMGB1 knockout mice (Huebener et al. 2015), although the mechanisms of protection are currently not well defined. While these serve as examples, a number of other experimental studies have indicated that DAMPs and cytokines or chemokines released during inflammation can later affect APAP toxicity. However, many of these protective mechanisms could be adaptive responses to the loss of a specific protein or mediator. Alternatively, there are increasing concerns about mismatch of background strains between gene knock-out animals and wild-type mice, which could be the reason for contradictory reports in the literature using the same knockout animals (Bourdi et al. 2011; Duan et al. 2016).

While inflammation may not directly kill hepatocytes, mounting evidence indicates inflammation is entirely necessary for recovery. A number of chemokines and cytokines are associated with nonsurvival in overdose patients including IL-10, MCP-1, IL-6, and more (Antoniades et al. 2006; Berry et al. 2010). Many of these cytokines are typically considered "anti-inflammatory" as they are either secreted by anti-inflammatory type 2 macrophages, or are directly signaling to reduce inflammation and resolve injury (Berry et al. 2010; Antoniades et al. 2012, 2014). Resolution of the injury is critical for recovery and liver regeneration is tightly linked to survival (Schmidt and Dalhoff 2005). The lack of recovery of liver function seen in non-survivors may be due to defunct signaling mechanisms or aberrant signaling mechanisms in these patients that fail to turn off normal inflammation. Then in the case of a prolonged inflammatory response, regeneration is inhibited and the patients are at enhanced risk for sepsis and increased mortality. Infection and sepsis is responsible for nearly 10% of all deaths due to ALF (Reuben et al. 2016). Global blockade of inflammation becomes highly risky in this population as significant adverse events have been seen when inflammation is blocked and patients are at risk for sepsis or systemic inflammatory response syndrome (SIRS) and multiple organ failure (Boetticher et al. 2008).

3.10 Conclusions

The mechanisms that control APAP-induced liver injury are largely derived from the formation of a reactive metabolite and protein adducts that generate an initial mitochondrial dysfunction and oxidant stress. This activates MAPK pathways that serve to enhance mitochondrial dysfunction. While a number of cellular defense mechanisms are in place to counteract the progression to cell death, an overdose of APAP can overwhelm these defenses and stress adaptations leading to extensive necrosis. Although the current clinical antidote NAC is highly effective at the early, metabolism stages of the toxicity, there are no therapeutic interventions available that target the later stages of the injury or regeneration. Thus, better understanding of the mechanisms of cell death, the development of acute liver failure, and regeneration is critical for the discovery of novel intervention strategies to improve outcome after APAP-induced liver injury and ALF.

Acknowledgment Work in the authors' laboratory was supported in part by the National Institutes of Health grants R01 DK070195 and R01 AA12916, and by grants from the National Institute of General Medical Sciences (P20 GM103549 and P30GM118247) of the National Institutes of Health. Additional support came from the "Training Program in Environmental Toxicology" T32 ES007079-26A2 (to B.L.W.) from the National Institute of Environmental Health Sciences.

References

Adams ML, Pierce RH, Vail ME, White CC, Tonge RP, Kavanagh TJ, Fausto N, Nelson SD, Bruschi SA (2001) Enhanced acetaminophen hepatotoxicity in transgenic mice overexpressing BCL-2. Mol Pharmacol 60(5):907–915

Antoine DJ, Williams DP, Kipar A, Laverty H, Park BK (2010) Diet restriction inhibits apoptosis and HMGB1 oxidation and promotes inflammatory cell recruitment during acetaminophen hepatotoxicity. Mol Med 16(11–12):479–490

Antoine DJ, Jenkins RE, Dear JW, Williams DP, McGill MR, Sharpe MR, Craig DG, Simpson KJ, Jaeschke H, Park BK (2012) Molecular forms of HMGB1 and keratin-18 as mechanistic biomarkers for mode of cell death and prognosis during clinical acetaminophen hepatotoxicity. J Hepatol 56(5):1070–1079

Antoniades CG, Khamri W, Abeles RD, Taams LS, Triantafyllou E, Possamai LA, Bernsmeier C, Mitry RR, O'Brien A, Gilroy D, Goldin R, Heneghan M, Heaton N, Jassem W, Bernal W, Vergani D, Ma Y, Quaglia A, Wendon J, Thursz M (2014) Secretory leukocyte protease inhibitor: a pivotal mediator of anti-inflammatory responses in acetaminophen-induced acute liver failure. Hepatology 59(4):1564–1576

Antoniades CG, Quaglia A, Taams LS, Mitry RR, Hussain M, Abeles R, Possamai LA, Bruce M, McPhail M, Starling C, Wagner B, Barnardo A, Pomplun S, Auzinger G, Bernal W, Heaton N, Vergani D, Thursz MR, Wendon J (2012) Source and characterization of hepatic macrophages in acetaminophen-induced acute liver failure in humans. Hepatology 56(2):735–746

Antoniades CG, Berry PA, Davies ET, Hussain M, Bernal W, Vergani D, Wendon J (2006) Reduced monocyte HLA-DR expression: a novel biomarker of disease severity and outcome in acetaminophen-induced acute liver failure. Hepatology 44(1):34–43

Bajt ML, Cover C, Lemasters JJ, Jaeschke H (2006) Nuclear translocation of endonuclease G and apoptosis-inducing factor during acetaminophen-induced liver cell injury. Toxicol Sci 94(1):217–225

Bajt ML, Farhood A, Lemasters JJ, Jaeschke H (2008) Mitochondrial bax translocation accelerates DNA fragmentation and cell necrosis in a murine model of acetaminophen hepatotoxicity. J Pharmacol Exp Ther 324(1):8–14

Bajt ML, Knight TR, Farhood A, Jaeschke H (2003) Scavenging peroxynitrite with glutathione promotes regeneration and enhances survival during acetaminophen-induced liver injury in mice. J Pharmacol Exp Ther 307(1):67–73

Bajt ML, Knight TR, Lemasters JJ, Jaeschke H (2004) Acetaminophen-induced oxidant stress and cell injury in cultured mouse hepatocytes: protection by N-acetyl cysteine. Toxicol Sci 80(2):343–349

Bajt ML, Ramachandran A, Yan HM, Lebofsky M, Farhood A, Lemasters JJ, Jaeschke H (2011) Apoptosis-inducing factor modulates mitochondrial oxidant stress in acetaminophen hepatotoxicity. Toxicol Sci 122(2):598–605

Bebarta VS, Kao L, Froberg B, Clark RF, Lavonas E, Qi M, Delgado J, McDonagh J, Arnold T, Odujebe O, O'Malley G, Lares C, Aguilera E, Dart R, Heard K, Stanford C, Kokko J, Bogdan G, Mendoza C, Mlynarchek S, Rhyee S, Hoppe J, Haur W, Tan HH, Tran NN, Varney S, Zosel A, Buchanan J, Al-Helial M (2010) A multicenter comparison of the safety of oral versus intravenous acetylcysteine for treatment of acetaminophen overdose. Clin Toxicol (Phila) 48(5):424–430

Berry PA, Antoniades CG, Hussain MJ, McPhail MJ, Bernal W, Vergani D, Wendon JA (2010) Admission levels and early changes in serum interleukin-10 are predictive of poor outcome in acute liver failure and decompensated cirrhosis. Liver Int 30(5):733–740

Boetticher NC, Peine CJ, Kwo P, Abrams GA, Patel T, Aqel B, Boardman L, Gores GJ, Harmsen WS, McClain CJ, Kamath PS, Shah VH (2008) A randomized, double-blinded, placebo-controlled multicenter trial of etanercept in the treatment of alcoholic hepatitis. Gastroenterology 135(6):1953–1960

Bone-Larson CL, Hogaboam CM, Evanhoff H, Strieter RM, Kunkel SL (2001) IFN-gamma-inducible protein-10 (CXCL10) is hepatoprotective during acute liver injury through the induction of CXCR2 on hepatocytes. J Immunol 167(12):7077–7083

Bourdi M, Davies JS, Pohl LR (2011) Mispairing C57BL/6 substrains of genetically engineered mice and wild-type controls can lead to confounding results as it did in studies of JNK2 in acetaminophen and concanavalin a liver injury. Chem Res Toxicol 24(6):794–796

Bourdi M, Masubuchi Y, Reilly TP, Amouzadeh HR, Martin JL, George JW, Shah AG, Pohl LR (2002) Protection against acetaminophen-induced liver injury and lethality by interleukin 10: role of inducible nitric oxide synthase. Hepatology 35(2):289–298

Burkc AS, MacMillan-Crow LA, Hinson JA (2010) Reactive nitrogen species in acetaminophen-induced mitochondrial damage and toxicity in mouse hepatocytes. Chem Res Toxicol 23(7):1286–1292

Clements JA, Critchley JA, Prescott LF (1984) The role of sulphate conjugation in the metabolism and disposition of oral and intravenous paracetamol in man. Br J Clin Pharmacol 18(4):481–485

Corcoran GB, Wong BK (1986) Role of glutathione in prevention of acetaminophen-induced hepatotoxicity by N-acetyl-L-cysteine in vivo: studies with N-acetyl-D-cysteine in mice. J Pharmacol Exp Ther 238(1):54–61

Cover C, Liu J, Farhood A, Malle E, Waalkes MP, Bajt ML, Jaeschke H (2006) Pathophysiological role of the acute inflammatory response during acetaminophen hepatotoxicity. Toxicol Appl Pharmacol 216(1):98–107

Cover C, Mansouri A, Knight TR, Bajt ML, Lemasters JJ, Pessayre D, Jaeschke H (2005) Peroxynitrite-induced mitochondrial and endonuclease-mediated nuclear DNA damage in acetaminophen hepatotoxicity. J Pharmacol Exp Ther 315(2):879–887

Cubero FJ, Zoubek ME, Hu W, Peng J, Zhao G, Nevzorova YA, Al Masaoudi M, Bechmann LP, Boekschoten MV, Muller M, Preisinger C, Gassler N, Canbay AE, Luedde T, Davis RJ, Liedtke C, Trautwein C (2016) Combined activities of JNK1 and JNK2 in hepatocytes protect against toxic liver injury. Gastroenterology 150(4):968–981

Dagda RK, Cherra SJ 3rd, Kulich SM, Tandon A, Park D, Chu CT (2009) Loss of PINK1 function promotes mitophagy through effects on oxidative stress and mitochondrial fission. J Biol Chem 284(2):13843–13855

Dahlin DC, Miwa GT, Lu AY, Nelson SD (1984) N-acetyl-p-benzoquinone imine: a cytochrome P-450-mediated oxidation product of acetaminophen. Proc Natl Acad Sci U S A 81(5):1327–1331

Dambach DM, Watson LM, Gray KR, Durham SK, Laskin DL (2002) Role of CCR2 in macrophage migration into the liver during acetaminophen-induced hepatotoxicity in the mouse. Hepatology 35(5):1093–1103

Dara L, Johnson H, Suda J, Win S, Gaarde W, Han D, Kaplowitz N (2015) Receptor interacting protein kinase 1 mediates murine acetaminophen toxicity independent of the necrosome and not through necroptosis. Hepatology 62(6):1847–1857

Davern TJ 2nd, James LP, Hinson JA, Polson J, Larson AM, Fontana RJ, Lalani E, Munoz S, Shakil AO, Lee WM, Acute Liver Failure Study Group (2006) Measurement of serum acetaminophen-protein adducts in patients with acute liver failure. Gastroenterology 130(3):687–694

Deutsch M, Graffeo CS, Rokosh R, Pansari M, Ochi A, Levie EM, Van Heerden E, Tippens DM, Greco S, Barilla R, Tomkötter L, Zambirinis CP, Avanzi N, Gulati R, Pachter HL, Torres-Hernandez A, Eisenthal A, Daley D, Miller G (2015) Divergent effects of RIP1 or RIP3 blockade in murine models of acute liver injury. Cell Death Dis 6:e1759

Du K, Farhood A, Jaeschke H (2017) Mitochondria-targeted antioxidant Mito-tempo protects against acetaminophen hepatotoxicity. Arch Toxicol 91 (2):761–773

Du K, Ramachandran A, Jaeschke H (2016a) Oxidative stress during acetaminophen hepatotoxicity: sources, pathophysiological role and therapeutic potential. Redox Biol 10:148–156

Du K, Ramachandran A, Weemhoff JL, Chavan H, Xie Y, Krishnamurthy P, Jaeschke H (2016b) Metformin protects against acetaminophen hepatotoxicity by attenuation of mitochondrial oxidant stress and dysfunction. Toxicol Sci 154(2):214–226

Du K, Williams CD, McGill MR, Jaeschke H (2014) Lower susceptibility of female mice to acetaminophen hepatotoxicity: role of mitochondrial glutathione, oxidant stress and c-Jun N-terminal kinase. Toxicol Appl Pharmacol 281(1):58–66

Du K, Xie Y, McGill MR, Jaeschke H (2015) Pathophysiological significance of c-Jun N-terminal kinase in acetaminophen hepatotoxicity. Expert Opin Drug Metab Toxicol 11(11):1769–1779

Duan L, Davis JS, Woolbright BL, Du K, Cahkraborty M, Weemhoff J, Jaeschke H, Bourdi M (2016) Differential susceptibility to acetaminophen-induced liver injury in sub-strains of C57BL/6 mice: 6N versus 6J. Food Chem Toxicol 98(Pt B):107–118.

El-Hassan H, Anwar K, Macanas-Pirard P, Crabtree M, Chow SC, Johnson VL, Lee PC, Hinton RH, Price SC, Kass GE (2003) Involvement of mitochondria in acetaminophen-induced apoptosis and hepatic injury: roles of cytochrome c, Bax, bid, and caspases. Toxicol Appl Pharmacol 191(2):118–129

Enomoto A, Itoh K, Nagayoshi E, Haruta J, Kimura T, O'Connor T, Harada T, Yamamoto M (2001) High sensitivity of Nrf2 knockout mice to acetaminophen hepatotoxicity associated with decreased expression of ARE-regulated drug metabolizing enzymes and antioxidant genes. Toxicol Sci 59(1):169–177

Feng D, Wang Y, Wang H, Weng H, Kong X, Martin-Murphy BV, Li Y, Park O, Dooley S, Ju C, Gao B (2014) Acute and chronic effects of IL-22 on acetaminophen-induced liver injury. J Immunol 193(5):2512–2518

Fujimoto K, Kumagai K, Ito K, Arakawa S, Ando Y, Oda S, Yamoto T, Manabe S (2009) Sensitivity of liver injury in heterozygous Sod2 knockout mice treated with troglitazone or acetaminophen. Toxicol Pathol 37(2):193–200

Grasl-Kraupp B, Ruttkay-Nedecky B, Koudelka H, Bukowska K, Bursch W, Schulte-Hermann R (1995) In situ detection of fragmented DNA (TUNEL assay) fails to discriminate among apoptosis, necrosis, and autolytic cell death: a cautionary note. Hepatology 21(5):1465–1468

Gross A, Jockel J, Wei MC, Korsmeyer SJ (1998) Enforced dimerization of BAX results in its translocation, mitochondrial dysfunction and apoptosis. EMBO J 17(14):3878–3885

Gujral JS, Knight TR, Farhood A, Bajt ML, Jaeschke H (2002) Mode of cell death after acetaminophen overdose in mice: apoptosis or oncotic necrosis? Toxicol Sci 67(2):322–328

Günther C, He GW, Kremer AE, Murphy JM, Petrie EJ, Amann K, Vandenabeele P, Linkermann A, Poremba C, Schleicher U, Dewitz C, Krautwald S, Neurath MF, Becker C, Wirtz S (2016) The pseudokinase MLKL mediates programmed hepatocellular necrosis independently of RIPK3 during hepatitis. J Clin Invest 126(11):4346–4360

Hadi M, Dragovic S, van Swelm R, Herpers B, van de Water B, Russel FG, Commandeur JN, Groothuis GM (2013) AMAP, the alleged non-toxic isomer of acetaminophen, is toxic in rat and human liver. Arch Toxicol 87(1):155–1652013

Han D, Dara L, Win S, Than TA, Yuan L, Abbasi SQ, Liu ZX, Kaplowitz N (2013) Regulation of drug-induced liver injury by signal transduction pathways: critical role of mitochondria. Trends Pharmacol Sci 34(4):243–253

Hanawa N, Shinohara M, Saberi B, Gaarde WA, Han D, Kaplowitz N (2008) Role of JNK translocation to mitochondria leading to inhibition of mitochondria bioenergetics in acetaminophen-induced liver injury. J Biol Chem 283(20):13565–13577

He S, Wang L, Miao L, Wang T, Du F, Zhao L, Wang X (2009) Receptor interacting protein kinase-3 determines cellular necrotic response to TNF-alpha. Cell 137(6):1100–1111

Heard KJ, Green JL, James LP, Judge BS, Zolot L, Rhyee S, Dart RC (2011) Acetaminophen-cysteine adducts during therapeutic dosing and following overdose. BMC Gastroenterol 11:20

Hogaboam CM, Bone-Larson CL, Steinhauser ML, Lukacs NW, Colletti LM, Simpson KJ, Strieter RM, Kunkel SL (1999) Novel CXCR2-dependent liver regenerative qualities of ELR-containing CXC chemokines. FASEB J 13(12):1565–1574

Holt MP, Cheng L, Ju C (2008) Identification and characterization of infiltrating macrophages in acetaminophen-induced liver injury. J Leukoc Biol 84(6):1410–1421

Hu J, Kholmukhamedov A, Lindsey CC, Beeson CC, Jaeschke H, Lemasters JJ (2016a) Translocation of iron from lysosomes to mitochondria during acetaminophen-induced hepatocellular injury: protection by starch-desferal and minocycline. Free Radic Biol Med 97:418–426

Hu J, Ramshesh VK, McGill MR, Jaeschke H, Lemasters JJ (2016b) Low dose acetaminophen induces reversible mitochondrial dysfunction associated with transient c-Jun N-terminal kinase activation in mouse liver. Toxicol Sci 150(1):204–215

Huebener P, Pradere JP, Hernandez C, Gwak GY, Caviglia JM, Mu X, Loike JD, Jenkins RE, Antoine DJ, Schwabe RF (2015) The HMGB1/RAGE axis triggers neutrophil-mediated injury amplification following necrosis. J Clin Invest 125(2):539–550

Jaeschke H (1990) Glutathione disulfide formation and oxidant stress during acetaminophen-induced hepatotoxicity in mice in vivo: the protective effect of allopurinol. J Pharmacol Exp Ther 255(3):935–941

Jaeschke H (2006) Mechanisms of liver injury. II. Mechanisms of neutrophil-induced liver cell injury during hepatic ischemia-reperfusion and other acute inflammatory conditions. Am J Physiol Gastrointest Liver Physiol 290(6):G1083–G1088

Jaeschke H (2011) Reactive oxygen and mechanisms of inflammatory liver injury: present concepts. J Gastroenterol Hepatol 26(Suppl 1):173–179

Jaeschke H (2016) Mechanisms of acetaminophen hepatotoxicity: do we need JNK for cell death? Gastroenterology 151(2):371–372

Jaeschke H, Bajt ML (2006) Intracellular signaling mechanisms of acetaminophen-induced liver cell death. Toxicol Sci 89(1):31–41

Jaeschke H, Cover C, Bajt ML (2006) Role of caspases in acetaminophen-induced liver injury. Life Sci 78(15):1670–1676

Jaeschke H, Liu J (2007) Neutrophil depletion protects against murine acetaminophen hepatotoxicity: another perspective. Hepatology 45(6):1588–1589

Jaeschke H, McGill MR, Ramachandran A (2012a) Oxidant stress, mitochondria, and cell death mechanisms in drug-induced liver injury: lessons learned from acetaminophen hepatotoxicity. Drug Metab Rev 44(1):88–106

Jaeschke H, McGill MR, Williams CD, Ramachandran A (2011) Current issues with acetaminophen hepatotoxicity--a clinically relevant model to test the efficacy of natural products. Life Sci 88(17–18):737–745

Jaeschke H, Williams CD, Ramachandran A, Bajt ML (2012b) Acetaminophen hepatotoxicity and repair: the role of sterile inflammation and innate immunity. Liver Int 32(1):8–20

James LP, Chiew A, Abdel-Rahman SM, Letzig L, Graudins A, Day P, Roberts D (2013) Acetaminophen protein adduct formation following low-dose acetaminophen exposure: comparison of immediate-release vs extended-release formulations. Eur J Clin Pharmacol 69(4):851–857

James LP, Letzig L, Simpson PM, Capparelli E, Roberts DW, Hinson JA, Davern TJ, Lee WM (2009) Pharmacokinetics of acetaminophen-protein adducts in adults with acetaminophen overdose and acute liver failure. Drug Metab Dispos 37(8):1779–1784

James LP, McCullough SS, Knight TR, Jaeschke H, Hinson JA (2003a) Acetaminophen toxicity in mice lacking NADPH oxidase activity: role of peroxynitrite formation and mitochondrial oxidant stress. Free Radic Res 37(12):1289–1297

James LP, McCullough SS, Lamps LW, Hinson JA (2003b) Effect of N-acetylcysteine on acetaminophen toxicity in mice: relationship to reactive nitrogen and cytokine formation. Toxicol Sci 75(2):458–467

Ju C, Reilly TP, Bourdi M, Radonovich MF, Brady JN, George JW, Pohl LR (2002) Protective role of Kupffer cells in acetaminophen-induced hepatic injury in mice. Chem Res Toxicol 15(12):1504–1513

Kim I, Rodriguez-Enriquez S, Lemasters JJ (2007) Selective degradation of mitochondria by mitophagy. Arch Biochem Biophys 462(2):245–253

Kim YH, Hwang JH, Kim KS, Noh JR, Choi DH, Kim DK, Tadi S, Yim YH, Choi HS, Lee CH (2015) Metformin ameliorates acetaminophen hepatotoxicity via Gadd45β-dependent regulation of JNK signaling in mice. J Hepatol 63(1):75–82

Knight TR, Ho YS, Farhood A, Jaeschke H (2002) Peroxynitrite is a critical mediator of acetaminophen hepatotoxicity in murine livers: protection by glutathione. J Pharmacol Exp Ther 303(2):468–475

Knight TR, Kurtz A, Bajt ML, Hinson JA, Jaeschke H (2001) Vascular and hepatocellular peroxynitrite formation during acetaminophen toxicity: role of mitochondrial oxidant stress. Toxicol Sci 62(2):212–220

Knockaert L, Fromenty B, Robin MA (2011) Mechanisms of mitochondrial targeting of cytochrome P450 2E1: physiopathological role in liver injury and obesity. FEBS J 278(22):4252–4260

Kon K, Kim JS, Jaeschke H, Lemasters JJ (2004) Mitochondrial permeability transition in acetaminophen-induced necrosis and apoptosis of cultured mouse hepatocytes. Hepatology 40(2):1170–1179

Kon K, Kim JS, Uchiyama A, Jaeschke H, Lemasters JJ (2010) Lysosomal iron mobilization and induction of the mitochondrial permeability transition in acetaminophen-induced toxicity to mouse hepatocytes. Toxicol Sci 117(1):101–108

Korsmeyer SJ, Shutter JR, Veis DJ, Merry DE, Oltvai ZN (1993) Bcl-2/Bax: a rheostat that regulates an anti-oxidant pathway and cell death. Semin Cancer Biol 4(6):327–332

Kuboki S, Shin T, Huber N, Eismann T, Galloway E, Schuster R, Blanchard J, Edwards MJ, Lentsch AB (2008) Hepatocyte signaling through CXC chemokine receptor-2 is detrimental to liver recovery after ischemia/reperfusion in mice. Hepatology 48(4):1213–1223

Latchoumycandane C, Goh CW, Ong MM, Boelsterli UA (2007) Mitochondrial protection by the JNK inhibitor leflunomide rescues mice from acetaminophen-induced liver injury. Hepatology 45(2):412–421

Lauterburg BH, Corcoran GB, Mitchell JR (1983) Mechanism of action of N-acetylcysteine in the protection against the hepatotoxicity of acetaminophen in rats in vivo. J Clin Invest 71(4):980–991

Lawson JA, Farhood A, Hopper RD, Bajt ML, Jaeschke H (2000) The hepatic inflammatory response after acetaminophen overdose: role of neutrophils. Toxicol Sci 54(2):509–516

Lawson JA, Fisher MA, Simmons CA, Farhood A, Jaeschke H (1999) Inhibition of Fas receptor (CD95)-induced hepatic caspase activation and apoptosis by acetaminophen in mice. Toxicol Appl Pharmacol 156(3):179–186

Lee WM (2013) Drug-induced acute liver failure. Clin Liver Dis 17(4):575–586

Li JX, Feng JM, Wang Y, Li XH, Chen XX, Su Y, Shen YY, Chen Y, Xiong B, Yang CH, Ding J, Miao ZH (2014) The B-Raf(V600E) inhibitor dabrafenib selectively inhibits RIP3 and alleviates acetaminophen-induced liver injury. Cell Death Dis 5:e1278

Liu K, Czaja MJ (2013) Regulation of lipid stores and metabolism by lipophagy. Cell Death Differ 20(1):3–11

Liu ZX, Han D, Gunawan B, Kaplowitz N (2006) Neutrophil depletion protects against murine acetaminophen hepatotoxicity. Hepatology 43(6):1220–1230

Lundbäck P, Lea JD, Sowinska A, Ottosson L, Fürst CM, Steen J, Aulin C, Clarke JI, Kipar A, Klevenvall L, Yang H, Palmblad K, Park BK, Tracey KJ, Blom AM, Andersson U, Antoine DJ, Erlandsson Harris H (2016) A novel high mobility group box 1 neutralizing chimeric antibody attenuates drug-induced liver injury and post injury inflammation in mice. Hepatology 64(5):1699–1710

McGill MR, Jaeschke H (2013) Metabolism and disposition of acetaminophen: recent advances in relation to hepatotoxicity and diagnosis. Pharm Res 30(9):2174–2187

McGill MR, Du K, Xie Y, Bajt ML, Ding WX, Jaeschke H (2015) The role of the c-Jun N-terminal kinases 1/2 and receptor-interacting protein kinase 3 in furosemide-induced liver injury. Xenobiotica 45(5):442–449

McGill MR, Lebofsky M, Norris HR, Slawson MH, Bajt ML, Xie Y, Williams CD, Wilkins DG, Rollins DE, Jaeschke H (2013) Plasma and liver acetaminophen-protein adduct levels in mice after acetaminophen treatment: dose-response, mechanisms, and clinical implications. Toxicol Appl Pharmacol 269(3):240–249

McGill MR, Sharpe MR, Williams CD, Taha M, Curry SC, Jaeschke H (2012a) The mechanism underlying acetaminophen-induced hepatotoxicity in humans and mice involves mitochondrial damage and nuclear DNA fragmentation. J Clin Invest 122(4):1574–1583

McGill MR, Williams CD, Xie Y, Ramachandran A, Jaeschke H (2012b) Acetaminophen-induced liver injury in rats and mice: comparison of protein adducts, mitochondrial dysfunction, and oxidative stress in the mechanism of toxicity. Toxicol Appl Pharmacol 264(3):387–394

McGill MR, Yan HM, Ramachandran A, Murray GJ, Rollins DE, Jaeschke H (2011) HepaRG cells: a human model to study mechanisms of acetaminophen hepatotoxicity. Hepatology 53(3):974–982

Michael SL, Pumford NR, Mayeux PR, Niesman MR, Hinson JA (1999) Pretreatment of mice with macrophage inactivators decreases acetaminophen hepatotoxicity and the formation of reactive oxygen and nitrogen species. Hepatology 30(1):186–195

Mitchell JR, Jollow DJ, Potter WZ, Davis DC, Gillette JR, Brodie BB (1973) Acetaminophen-induced hepatic necrosis. I. Role of drug metabolism. J Pharmacol Exp Ther 187(1):185–194

Mobasher MA, González-Rodriguez A, Santamaría B, Ramos S, Martín MÁ, Goya L, Rada P, Letzig L, James LP, Cuadrado A, Martín-Pérez J, Simpson KJ, Muntané J, Valverde AM (2013) Protein tyrosine phosphatase 1B modulates GSK3β/Nrf2 and IGFIR signaling pathways in acetaminophen-induced hepatotoxicity. Cell Death Dis 4:e626

Mossanen JC, Krenkel O, Ergen C, Govaere O, Liepelt A, Puengel T, Heymann F, Kalthoff S, Lefebvre E, Eulberg D, Luedde T, Marx G, Strassburg CP, Roskams T, Trautwein C, Tacke F (2016) Chemokine (C-C motif) receptor 2-positive monocytes aggravate the early phase of acetaminophen-induced acute liver injury. Hepatology 64(5):1667–1682

Nakagawa H, Maeda S, Hikiba Y, Ohmae T, Shibata W, Yanai A, Sakamoto K, Ogura K, Noguchi T, Karin M, Ichijo H, Omata M (2008) Deletion of apoptosis signal-regulating kinase 1 attenuates acetaminophen-induced liver injury by inhibiting c-Jun N-terminal kinase activation. Gastroenterology 135(4):1311–1321

Ni HM, Bockus A, Boggess N, Jaeschke H, Ding WX (2012a) Activation of autophagy protects against acetaminophen-induced hepatotoxicity. Hepatology 55(1):222–232

Ni HM, Boggess N, McGill MR, Lebofsky M, Borude P, Apte U, Jaeschke H, Ding WX (2012b) Liver-specific loss of Atg5 causes persistent activation of Nrf2 and protects against acetaminophen-induced liver injury. Toxicol Sci 127(2):438–450

Ni HM, McGill MR, Chao X, Du K, Williams JA, Xie Y, Jaeschke H, Ding WX (2016) Removal of acetaminophen protein adducts by autophagy protects against acetaminophen-induced liver injury in mice. J Hepatol 65(2):354–362

Ni HM, Williams JA, Jaeschke H, Ding WX (2013) Zonated induction of autophagy and mitochondrial spheroids limits acetaminophen-induced necrosis in the liver. Redox Biol 1:427–432

Nieto N, Friedman SL, Cederbaum AI (2002) Stimulation and proliferation of primary rat hepatic stellate cells by cytochrome P450 2E1-derived reactive oxygen species. Hepatology 35(1):62–73

O'Malley GF, Mizrahi F, Giraldo P, O'Malley RN, Rollins D, Wilkins D (2015) Protein-derived acetaminophen-cysteine can be detected after repeated supratherapeutic ingestion of acetaminophen in the absence of hepatotoxicity. J Med Toxicol 11(3):317–330

Pumford NR, Roberts DW, Benson RW, Hinson JA (1990) Immunochemical quantitation of 3-(cystein-S-yl)acetaminophen protein adducts in subcellular liver fractions following a hepatotoxic dose of acetaminophen. Biochem Pharmacol 40(3):573–579

Qiu Y, Benet LZ, Burlingame AL (2001) Identification of hepatic protein targets of the reactive metabolites of the non-hepatotoxic regioisomer of acetaminophen, 3'-hydroxyacetanilide, in the mouse in vivo using two-dimensional gel electrophoresis and mass spectrometry. Adv Exp Med Biol 500:663–673

Ramachandran A, Lebofsky M, Weinman SA, Jaeschke H (2011) The impact of partial manganese superoxide dismutase (SOD2)-deficiency on mitochondrial oxidant stress, DNA fragmentation and liver injury during acetaminophen hepatotoxicity. Toxicol Appl Pharmacol 251(3):226–233

Ramachandran A, McGill MR, Xie Y, Ni HM, Ding WX, Jaeschke H (2013) Receptor interacting protein kinase 3 is a critical early mediator of acetaminophen-induced hepatocyte necrosis in mice. Hepatology 58(6):2099–2108

Reuben A, Tillman H, Fontana RJ, Davern T, McGuire B, Stravitz RT, Durkalski V, Larson AM, Liou I, Fix O, Schilsky M, McCashland T, Hay JE, Murray N, Shaikh OS, Ganger D, Zaman A,

Han SB, Chung RT, Smith A, Brown R, Crippin J, Harrison ME, Koch D, Munoz S, Reddy KR, Rossaro L, Satyanarayana R, Hassanein T, Hanje AJ, Olson J, Subramanian R, Karvellas C, Hameed B, Sherker AH, Robuck P, Lee WM (2016) Outcomes in adults with acute liver failure between 1998 and 2013: an observational cohort study. Ann Intern Med 164(11):724–732
Roberts DW, Lee WM, Hinson JA, Bai S, Swearingen CJ, Stravitz RT, Reuben A, Letzig L, Simpson PM, Rule J, Fontana RJ, Ganger D, Reddy KR, Liou I, Fix O, James LP (2017) An immunoassay to rapidly measure acetaminophen protein adducts accurately identifies patients with acute liver injury or failure. Clin Gastroenterol Hepatol 15(4):555-562.e3
Rosen GM, Rauckman EJ, Ellington SP, Dahlin DC, Christie JL, Nelson SD (1984) Reduction and glutathione conjugation reactions of N-acetyl-p-benzoquinone imine and two dimethylated analogues. Mol Pharmacol 25(1):151–157
Roychowdhury S, McMullen MR, Pisano SG, Liu X, Nagy LE (2013) Absence of receptor interacting protein kinase 3 prevents ethanol-induced liver injury. Hepatology 57(5):1773–1783
Rumack BH, Matthew H (1975) Acetaminophen poisoning and toxicity. Pediatrics 55(6):871–876
Ryan PM, Bourdi M, Korrapati MC, Proctor WR, Vasquez RA, Yee SB, Quinn TD, Chakraborty M, Pohl LR (2012) Endogenous interleukin-4 regulates glutathione synthesis following acetaminophen-induced liver injury in mice. Chem Res Toxicol 25(1):83–93
Saito C, Lemasters JJ, Jaeschke H (2010a) c-Jun N-terminal kinase modulates oxidant stress and peroxynitrite formation independent of inducible nitric oxide synthase in acetaminophen hepatotoxicity. Toxicol Appl Pharmacol 246(1–2):8–17
Saito C, Zwingmann C, Jaeschke H (2010b) Novel mechanisms of protection against acetaminophen hepatotoxicity in mice by glutathione and N-acetylcysteine. Hepatology 51(1):246–254
Schmidt LE, Dalhoff K (2005) Alpha-fetoprotein is a predictor of outcome in acetaminophen-induced liver injury. Hepatology 41(1):26–31
Sharma M, Gadang V, Jaeschke A (2012) Critical role for mixed-lineage kinase 3 in acetaminophen-induced hepatotoxicity. Mol Pharmacol 82(5):1001–1007
Singh R, Kaushik S, Wang Y, Xiang Y, Novak I, Komatsu M, Tanaka K, Cuervo AM, Czaja MJ (2009) Autophagy regulates lipid metabolism. Nature 458(7242):1131–1135
Tait SW, Green DR (2013) Mitochondrial regulation of cell death. Cold Spring Harb Perspect Biol 5(9)
Takeshige K, Baba M, Tsuboi S, Noda T, Ohsumi Y (1992) Autophagy in yeast demonstrated with proteinase-deficient mutants and conditions for its induction. J Cell Biol 119(2):301–311
Tirmenstein MA, Nelson SD (1989) Subcellular binding and effects on calcium homeostasis produced by acetaminophen and a nonhepatotoxic regioisomer, 3'-hydroxyacetanilide, in mouse liver. J Biol Chem 264(17):9814–9819
Vucur M, Reisinger F, Gautheron J, Janssen J, Roderburg C, Cardenas DV, Kreggenwinkel K, Koppe C, Hammerich L, Hakem R, Unger K, Weber A, Gassler N, Luedde M, Frey N, Neumann UP, Tacke F, Trautwein C, Heikenwalder M, Luedde T (2013) RIP3 inhibits inflammatory hepatocarcinogenesis but promotes cholestasis by controlling caspase-8- and JNK-dependent compensatory cell proliferation. Cell Rep 4(4):776–790
Wancket LM, Meng X, Rogers LK, Liu Y (2012) Mitogen-activated protein kinase phosphatase (Mkp)-1 protects mice against acetaminophen-induced hepatic injury. Toxicol Pathol 40(8):1095–1105
Wang H, Sun L, Su L, Rizo J, Liu L, Wang LF, Wang FS, Wang X (2014) Mixed lineage kinase domain-like protein MLKL causes necrotic membrane disruption upon phosphorylation by RIP3. Mol Cell 54(1):133–146
Wild AC, Moinova HR, Mulcahy RT (1999) Regulation of gamma-glutamylcysteine synthetase subunit gene expression by the transcription factor Nrf2. J Biol Chem 274(47):33627–33636
Williams CD, Bajt ML, Sharpe MR, McGill MR, Farhood A, Jaeschke H (2014) Neutrophil activation during acetaminophen hepatotoxicity and repair in mice and humans. Toxicol Appl Pharmacol 275(2):122–133

Williams CD, Koerner MR, Lampe JN, Farhood A, Jaeschke H (2011) Mouse strain-dependent caspase activation during acetaminophen hepatotoxicity does not result in apoptosis or modulation of inflammation. Toxicol Appl Pharmacol 257(3):449–458
Williams JA, Ding WX (2015) Targeting Pink1-Parkin-mediated mitophagy for treating liver injury. Pharmacol Res 102:264–269
Win S, Than TA, Han D, Petrovic LM, Kaplowitz N (2011) C-Jun N-terminal kinase (JNK)-dependent acute liver injury from acetaminophen or tumor necrosis factor (TNF) requires mitochondrial sab protein expression in mice. J Biol Chem 286(40):35071–35078
Win S, Than TA, Min RW, Aghajan M, Kaplowitz N (2016) c-Jun N-terminal kinase mediates mouse liver injury through a novel sab (SH3BP5)-dependent pathway leading to inactivation of intramitochondrial Src. Hepatology 63(6):1987–2003
Woolbright BL, Ramachandran A, McGill MR, Yan HM, Bajt ML, Sharpe MR, Lemasters JJ, Jaeschke H (2012) Lysosomal instability and cathepsin B release during acetaminophen hepatotoxicity. Basic Clin Pharmacol Toxicol 111(6):417–425
Woolbright BL, Jaeschke H (2015) Sterile inflammation in acute liver injury: myth or mystery? Expert Rev Gastroenterol Hepatol 9(8):1027–1029
Woolbright BL, Jaeschke H (2017) Role of the inflammasome in acetaminophen-induced liver injury and acute liver failure. J Hepatol 66(4):836–848
Xie Y, McGill MR, Cook SF, Sharpe MR, Winefield RD, Wilkins DG, Rollins DE, Jaeschke H (2015a) Time course of acetaminophen-protein adducts and acetaminophen metabolites in circulation of overdose patients and in HepaRG cells. Xenobiotica 45(10):921–929
Xie Y, McGill MR, Dorko K, Kumer SC, Schmitt TM, Forster J, Jaeschke H (2014) Mechanisms of acetaminophen-induced cell death in primary human hepatocytes. Toxicol Appl Pharmacol 279(3):266–274
Xie Y, McGill MR, Du K, Dorko K, Kumer SC, Schmitt TM, Ding WX, Jaeschke H (2015b) Mitochondrial protein adducts formation and mitochondrial dysfunction during N-acetyl-m-aminophenol (AMAP)-induced hepatotoxicity in primary human hepatocytes. Toxicol Appl Pharmacol 289(2):213–222
Xie Y, Ramachandran A, Breckenridge DG, Liles JT, Lebofsky M, Farhood A, Jaeschke H (2015c) Inhibitor of apoptosis signal-regulating kinase 1 protects against acetaminophen-induced liver injury. Toxicol Appl Pharmacol 286(1):1–9
Yan HM, Ramachandran A, Bajt ML, Lemasters JJ, Jaeschke H (2010) The oxygen tension modulates acetaminophen-induced mitochondrial oxidant stress and cell injury in cultured hepatocytes. Toxicol Sci 117(2):515–523
Yin XM, Ding WX, Gao W (2008) Autophagy in the liver. Hepatology 47(5):1773–1785
Yin XM, Wang K, Gross A, Zhao Y, Zinkel S, Klocke B, Roth KA, Korsmeyer SJ (1999) Bid-deficient mice are resistant to Fas-induced hepatocellular apoptosis. Nature 400(6747):886–891
You Q, Holt M, Yin H, Li G, Hu CJ, Ju C (2013) Role of hepatic resident and infiltrating macrophages in liver repair after acute injury. Biochem Pharmacol 86(6):836–843
Zaher H, Buters JT, Ward JM, Bruno MK, Lucas AM, Stern ST, Cohen SD, Gonzalez FJ (1998) Protection against acetaminophen toxicity in CYP1A2 and CYP2E1 double-null mice. Toxicol Appl Pharmacol 152(1):193–199
Zhang YF, He W, Zhang C, Liu XJ, Lu Y, Wang H, Zhang ZH, Chen X, Xu DX (2014) Role of receptor interacting protein (RIP)1 on apoptosis-inducing factor-mediated necroptosis during acetaminophen-evoked acute liver failure in mice. Toxicol Lett 225(3):445–453
Zhang DW, Shao J, Lin J, Zhang N, Lu BJ, Lin SC, Dong MQ, Han J (2009) RIP3, an energy metabolism regulator that switches TNF-induced cell death from apoptosis to necrosis. Science 325(5939):332–336
Zhang J, Yang Y, He W, Sun L (2016) Necrosome core machinery: MLKL. Cell Mol Life Sci 73(11–12):2153–2163

Chapter 4
Cell Death and Liver Injuries in Hepatitis C Virus Infection

Jiyoung Lee and Jing-hsiung James Ou

Abbreviations

ACC	Acetyl-CoA carboxylase
AMPK	5′AMP-activated protein kinase
Apaf-1	Apoptotic protease activating factor 1
ApoB	Apolipoprotein B
ApoE	Apolipoprotein E
ARE	Antioxidant response element
ATF6	Activating transcription factor 6
cFLIP	Cellular FADD-like interleukin-1β-converting enzyme (FLICE)-like inhibitory protein
CMA	Chaperone-mediated autophagy
CTL	Cytotoxic T lymphocyte
DAA	Direct acting antiviral
DD	Death domain
DR	Death receptor
Drp1	Dynamin-related protein 1
dsRNA	Double-stranded RNA
ECM	Extracellular matrix
ER	Endoplasmic reticulum
ERAD	ER-associated degradation
ERSE	ER stress response element
FADD	Fas-Associated protein with Death Domain
FASN	Fatty acid synthase
FFA	Free fatty acid

J. Lee • J.-h.J. Ou (✉)
Department of Molecular Microbiology and Immunology, Keck School of Medicine, University of Southern California, Los Angeles, CA, USA
e-mail: jamesou@hsc.usc.edu

W.-X. Ding, X.-M. Yin (eds.), *Cellular Injury in Liver Diseases*, Cell Death in Biology and Diseases, DOI 10.1007/978-3-319-53774-0_4

X.-M. Yin and Z. Dong (Series eds.), *Cell Death in Biology and Diseases*

HCC	Hepatocellular carcinoma
HCV	Hepatitis C virus
HMGR	HMG-CoA reductase
HMGS	HMG-CoA synthase
HSC	Hepatic stellate cells
IFN	Interferon
IFNAR	IFN-α receptor
IRE1	Inositol-requiring enzyme 1
IRES	Internal ribosome entry site
IRF-3	IFN regulatory factor 3
ISG	IFN-stimulated gene
ISRE	Interferon-stimulated response element
JNK	Jun N-terminal kinase
LC3	Microtubule-associated protein light chain 3
LD	Lipid droplet
LVP	Lipo-viro-particle
MDA5	Melanoma differentiation antigen 5
mtDNA	Mitochondrial DNA
mTOR	Mammalian target of rapamycin
MTP	Microsomal triglyceride transfer protein
NASH	Non-alcoholic steatohepatitis
ORF	Open reading frame
PAMP	Pathogen-associated molecular pattern
PARP	Poly (ADP-ribose) polymerase
PI4KA	Phosphatidylinositol-4-kinase alpha
PI4P	Phosphatidylinositol-4-phosphate
PLIN	Perilipin
PRR	Pattern recognition receptor
RID	Receptor internalization and degradation
RIG-I	Retinoic acid inducible gene-I
RNS	Reactive nitrogen species
ROS	Reactive oxygen species
SERCA	Sarcoplasmic/endoplasmic reticulum calcium ATPase
SR-BI	Scavenger receptor class B type I
SREBP-1	Sterol regulatory element binding protein 1
SVR	Sustained virological response
TAK1	TAB-transforming growth factor-β-activated kinase 1
TGF-β	Transforming growth factor-β
TLR	Toll-like receptor
TNF-α	Tumor necrosis factor-α
TRADD	TNFRSF1A-Associated via Death Domain
TRAF2	TNF receptor-associated factor 2
TRAIL	TNF-related apoptosis-inducing ligand
TRIF	Toll/interleukin-1 receptor/resistance domain-containing adaptor-inducing interferon-β

Ulk1	Unc-51-like autophagy activating kinase 1
UPR	Unfolded protein response
UTR	Untranslated region
VLDL	Very low density lipoprotein
ΔΨm	Mitochondrial membrane potential

4.1 Viral Hepatitis

Hepatitis refers to inflammatory conditions of the liver caused by various factors including alcohol, drugs, non-alcoholic fatty liver disease (NAFLD), autoimmune disease, and microbial infections. Many viruses such as cytomegalovirus, Epstein-Barr virus, yellow fever virus, rubella virus, and dengue virus can infect the liver and cause hepatitis when they cause systemic infections. However, the most common causes of viral hepatitis are hepatitis A virus (HAV), hepatitis B virus (HBV), hepatitis C virus (HCV), hepatitis delta virus (HDV), and hepatitis E virus (HEV). HAV and HEV are transmitted by the fecal-oral route and cause self-limiting acute infection, while HBV, HCV, and HDV are transmitted by the parental route via blood or other body fluids. HBV and HCV can cause chronic infection and severe liver diseases such as liver cirrhosis and hepatocellular carcinoma (HCC). HDV is a defective virus and requires the help provided by HBV for its replication. It can therefore only be co-transmitted with HBV or superinfect chronic HBV carriers. In this chapter, we will mainly focus on HCV and its effect on hepatic cell death and liver injuries.

4.2 Hepatitis C Virus

4.2.1 Hepatitis C Virus

HCV is one of the most important human pathogens. Globally, there are approximately 146 million people who are chronically infected by HCV (Majumdar et al. 2016), with up to four million new cases every year. People infected by HCV often become chronic carriers of the virus. The chronic infection by HCV may be asymptomatic. However, it can progress into severe liver diseases including fatty liver, cirrhosis, and HCC.

HCV is a small, enveloped virus with a single-stranded RNA genome. It has a very narrow host range and infects only humans and the closely related primates such as chimpanzees. This virus has a diameter of approximately 55 nm with an icosahedral capsid (Ishida et al. 2001; Kaito et al. 1994). The hepatotropic nature of HCV is due in part to the hepatocyte-restricted expression of its co-receptors, claudin-I and scavenger receptor class B type I (SR-BI). Based on the phylogenetic analysis, HCV has been grouped into six distinct genotypes and 88 subtypes, of which 21 are un-assigned (Smith et al. 2014). Between different genotypes, 60–70% genomic sequence homology is observed (Simmonds 2004), with the highest sequence variability present in the coding sequences of the hypervariable regions of viral envelope proteins and the most conserved sequence located in the 5′ untranslated region (UTR). Among different HCV subtypes, genotype 1b accounts for over 70% of overall HCV infections, making it the most common subtype worldwide. While genotypes 1a and 1b are found in the United States, genotype 1a is mostly associated with intravenous drug abuse (Alter 1999; Hnatyszyn 2005; Zein et al. 1996). HCV JFH1 strain, a genotype 2a strain, can efficiently infect and replicate in cell cultures, enabling detailed studies on the life cycle of HCV. The JFH1 virus particle has a density of 1.15–1.17 g/ml (Wakita et al. 2005).

The traditional treatment for chronic HCV infection uses pegylated interferon usually in combination with ribavirin, a guanosine analog known to have multiple antiviral effects (Leyssen et al. 2005). This treatment, however, was less than satisfactory due to the intolerable side effects and a sustained response rate of only ~50% for the combination therapy (Kelleher and Afdhal 2005). The use of direct acting antiviral (DAA) drugs in 2011 greatly improved the therapeutic outcome for HCV patients, achieving a sustained virological response (SVR) rate of >90%. DAA drugs target HCV non-structural proteins including the NS3/4A protease, NS5A, and the NS5B RNA-dependent RNA polymerase (Majumdar et al. 2016), all of which are essential for HCV replication. In contrast to DAAs, the development of HCV vaccines has encountered extreme difficulties due mainly to the high viral mutation rate, which has a frequency of approximately 10^{-3} base substitutions per site per year (Major et al. 1999).

4.2.2 HCV Genome

HCV has a positive single-stranded RNA genome of 9.6-kb. This genome encodes a single open reading frame flanked by UTRs. The HCV 5′UTR is 341 nucleotides long. It contains four highly structured and conserved domains known as domains I to IV. Domains II-IV along with the first 20–40 nucleotides of the coding sequence serve as the internal ribosome entry site (IRES) for the initiation of viral protein translation (Honda et al. 1996; Wang et al. 1995). The 3′UTR has a variable length in the vicinity of 225 nucleotides and can be divided into three regions: a 30–40-nucleotide variable sequence at the 5′-end, a long poly(U)-poly(U/UC) tract, and a highly conserved 98-nucleotide sequence at the 3′-end that contains three stem-loop structures known as SL1–3 (Kolykhalov et al. 1996; Tanaka et al. 1996). The 3′UTR interacts with the NS5B RNA-dependent RNA polymerase and plays an essential role in HCV RNA replication (Friebe and Bartenschlager 2002; Lee et al. 2004; Yi and Lemon 2003a, b). The ORF encodes a large polyprotein with a length of slightly more than 3000 amino acids. The order of the proteins encoded by the ORF from the N-terminus to the C-terminus is core, E1, E2, p7, NS2, NS3, NS4A,

NS4B, NS5A, and NS5B. The core protein forms the viral capsid, E1 and E2 are envelope proteins, p7 is an ion-channel protein, NS2 is a protease, and NS3-NS5B are nonstructural proteins required for viral RNA replication.

Since the HCV genomic RNA has a positive polarity, it will serve as the template to direct the translation of viral proteins once it is released into the cytosol after infection. The translation of HCV proteins takes place on the rough endoplasmic reticulum (ER). As mentioned above, this protein translation is mediated by the IRES. Unlike the cap-dependent mRNA translation, the translation of the cap-less HCV RNA is dependent on the direct binding of the 40S ribosome to the IRES, which then recruits other translation initiation factors such as eIF3 and eIF2 to form a 48S pre-initiation complex (Otto and Puglisi 2004). Further recruitment of eIF5B and the 60S ribosomal subunit leads to the formation of a functional 80S ribosome and the initiation of the HCV protein synthesis (Chevaliez and Pawlotsky 2006; Ji et al. 2004).

After the translation of the core protein, the internal signal sequence located at the C-terminus of the core protein targets the HCV polyprotein to the ER membrane where the polyprotein is proteolytically cleaved by host peptidases and viral proteases to give rise to mature viral proteins (Chevaliez and Pawlotsky 2006).

4.2.3 *HCV Life Cycle*

HCV binds to its co-receptors, which include CD81, SR-BI (also known as CLA-1), a tight junction protein occludin, and members of the claudin family (CLDN1, 6, and 9) on the cell surface to initiate infection of hepatocytes (Burlone and Budkowska 2009). The epidermal growth factor receptor (EGFR)/Ephrin receptor A2 (EphA2) and Niemann-Pick C1-like 1 (NPC1L1) cholesterol uptake receptor had also been identified as important host factors for HCV entry (Lupberger et al. 2011; Sainz et al. 2012). Glycosaminoglycans (GAGs) such as heparin sulfate, the lectins DC-specific intracellular adhesion molecule-3-grabbing non-integrin (DC-SIGN or CD209), and liver-specific intercellular adhesion molecule-3 (ICAM-3)-grabbing integrin (L-SIGN or CD209L) may facilitate the binding process by serving as an initial docking site (Gardner et al. 2003; Lozach et al. 2004; Morikawa et al. 2007). The low-density lipoprotein receptor (LDLR) had also been shown to be involved in HCV entry (Chung and Wasan 2004). The HCV envelope glycoproteins E1 and E2 are the viral components that play major roles in the receptor-binding event and viral entry. The initial attachment is followed by the high affinity interaction between E2 and receptors, CD81 and SR-BI and the subsequent lateral movement of these complexes to the tight junctions (Farquhar et al. 2012; Pileri et al. 1998; Scarselli et al. 2002).

Binding of HCV to the host receptors and co-receptors triggers clathrin- and dynamin-mediated endocytosis resulting in the internalization of HCV into the cells (Blanchard et al. 2006; Meertens et al. 2006). Following the fusion between endocytic vesicles and endosomes, the acidification of the endosomes triggers the fusion process of the viral envelope with the endosomal membrane, resulting in the release of HCV genome into the cytosol. Studies on glycoproteins of flaviviruses indicated that low pH could induce conformational changes of envelope proteins to expose the fusion peptide that mediates the membrane fusion (Lavillette et al. 2006; Tscherne et al. 2006).

The released HCV genomic RNA then directs the synthesis of HCV proteins. The expression of NS4B protein alone is sufficient to induce membrane remodeling for the formation of clusters of membrane vesicles known as the membranous web, which provides a microenvironment for the assembly of the HCV RNA replication complex (Egger et al. 2002; Gosert et al. 2003). The initial replication produces the negative-stranded RNA that will serve as the template for synthesis of the positive-stranded genomic RNA.

The N-terminal residues of HCV core protein possess the property of self-assembly (Klein et al. 2004, 2005). It has been shown that HCV core proteins could form capsid-like particles in *E. coli* (Kunkel et al. 2001). That particular study also demonstrated that the self-assembly of the core protein required the minimum of 124 amino acids at the N-terminus of the core protein and structured RNA molecules. Indeed, the binding of the core protein to HCV RNA required the positive-sense RNA from the 5′ end to nucleotide 2327 including the 5′UTR and a part of the ORF (Shimoike et al. 1999). Further studies indicated that the binding of core to HCV RNA was achieved through stem-loop domains I and III and the nucleotides 23–41 (Tanaka et al. 2000). Shimoike et al. further demonstrated that the binding of core to nucleotides 1–344 of the viral genomic RNA suppressed the IRES-driven translation (Shimoike et al. 1999). Taken together, the binding of the core protein to the viral genomic RNA seems to initiate the encapsidation process of the genomic RNA and simultaneously suppress the translation of HCV RNA.

HCV particles are further modified before they are released from the infected cells. One of the significant modifications is their association with lipid components. HCV particles found in both patient sera and cell culture media exhibited both low and high densities (reviewed in (Popescu and Dubuisson 2010)). HCV particles with a low density due to association with apolipoproteins and triacylglycerols were highly infectious whereas those with higher densities had lower infectivity (Andre et al. 2005; Gastaminza et al. 2006; Lindenbach et al. 2006), indicating the importance of lipid components in its infectivity. HCV assembly takes place on lipid droplet (LD), an organelle for the storage of intracellular lipids (Li et al. 2009). LDs have a core structure made up of triacylglycerols and cholesteryl esters surrounded by a single layer of phospholipids. The proximal localization of LDs to ER membranes permits an efficient transfer of the core protein from the ER to LDs. The maturation of the core protein requires the cleavage between core and E1 by a signal peptidase followed by a second cleavage for the removal of the signal sequence by a signal peptide peptidase. The second cleavage is thought to be essential for the localization of the core protein to the proximity of LD (McLauchlan et al. 2002). The D2 hydrophobic domain in the C-terminus of the core protein mediates the association of the protein to LD (Barba et al. 1997; Qiang and Jhaveri 2012). In hepatocytes, LDs are essential for the secretion of very low density lipoproteins

(VLDLs), for which the microsomal triglyceride transfer protein (MTP) and apolipoprotein B (ApoB) are indispensable (Popescu and Dubuisson 2010). Although the exact mechanism of how MTP and ApoB participate in the production of VLDL remains unclear, it is known that MTP interacts with ApoB and transfers triacylglycerols to VLDL (Sundaram and Yao 2010). Then the pre-VLDL accepts bulk lipids from LDs to become a mature form of VLDL. Studies correlating the maturation of HCV particles and the maturation of VLDL have improved our understanding of the maturation and the secretion processes of HCV particles (Gastaminza et al. 2006, 2008; Huang et al. 2007b). The inhibition of MTP activity abolished VLDL assembly as well as its secretion (Wang et al. 1999). Similarly, the inhibition of MTP prevented the secretion of infectious HCV particles from infected cells (Gastaminza et al. 2008). It had also been demonstrated that the depletion of ApoB with shRNA specific for ApoB also impaired the HCV infectivity. This observation was further supported by the finding that membrane vesicles containing HCV replication complexes isolated from infected human hepatoma cells displayed high levels of proteins involved in VLDL production, including ApoB, ApoE, and MTP (Huang et al. 2007b). Therefore, it is likely that HCV exploits the synthesis and secretion pathway of VLDL for its assembly and egress.

4.3 Immune Evasion and Chronicity of HCV

The resolution of acute HCV infection requires efficient innate and adapted immune responses. However, HCV is able to evade these immune responses to establish chronic infection in approximately 75–85% of patients it infects. Studies in recent years have significantly improved our knowledge on how HCV evades host immunity to attain persistence.

During the process of infection, microbial pathogens entering into hepatocytes will encounter various host defense mechanisms. This will culminate in the production of interferons (IFNs) and vigorous antiviral responses. There are three classes of IFNs based on the receptor complexes that they engage. Among these classes, type I IFNs are the best characterized. Type I IFNs include 13 IFN-α subtypes, IFN-β, IFN-κ, IFN-ε, IFN-o, IFN-τ, and IFN-δ. These molecules bind to the IFN-α receptor (IFNAR) 1 and 2. Upon binding of type I IFNs to the IFNAR1/IFNAR2 dimer, the signaling cascade is initiated, leading to the production of IFN-stimulated genes (ISGs) such as ISG15, the GTPase myxovirus resistance 1 (Mx1), ribonuclease L (RNaseL), 2′-5′ oligo adenylate synthetase (OAS), and protein kinase RNA-activated (PKR) (Sadler and Williams 2008).

The production of IFNs is stimulated by pattern-recognition receptors (PRRs) that sense pathogen-associated molecular patterns (PAMPs). The retinoic acid inducible gene-I (RIG-I)-like receptors (RLRs) including RIG-I, melanoma differentiation antigen 5 (MDA5), and laboratory of genetics and physiology 2 (LGP2) are PRRs that detect intracellular PAMPs. Both RIG-I and MDA5 recognize double-stranded RNA (dsRNA) to induce type I IFNs, although the specific PAMPs that they detect are different (Stetson and Medzhitov 2006). Toll-like receptors (TLRs) are also a family of PRRs and an important part of the host innate immune system. TLR1, TLR2, TLR4, TLR5, and TLR6 are expressed on the cell surface to detect PAMPs from bacteria, fungi, and viruses. TLR3, TLR7, TLR8, and TLR9 reside in endosomes and recognize intracellular viral nucleic acids. HCV could disrupt the PRR signaling pathways. The HCV NS3/4A protease blocked the RIG-I signaling by cleaving MAVS/IPS-1/VISA/Cardif (Foy et al. 2005; Li et al. 2005), an important adaptor molecule for RIG-I signaling (Lin et al. 2006). NS3/4A could also cleave Toll/interleukin-1 receptor/resistance domain-containing adaptor-inducing interferon-β (TRIF), an important adaptor that mediates TLR3 signaling for the activation of IFN regulatory factor 3 (IRF-3), thus suppressing IFN production (Li et al. 2005). In 293T cells, NS3/4A bound to TBK1 and interfered with its association with IRF-3 (Otsuka et al. 2005). The binding of NS4B to the stimulator of interferon genes (STING) in the ER membrane disrupted the association between STING and MAVS that was essential for IFN-β production (Nitta et al. 2013). In addition, NS5A was implicated in the inhibition of PKR and OAS through direct binding. NS5A could also bind to MyD88 and inhibit TLR signaling (reviewed in (Bowie and Unterholzner 2008)). The activation of PKR inhibited the cap-dependent protein translation by inactivating the translation initiation factor eIF2α and hence suppressed the translation of IFNs (Arnaud et al. 2010, 2011). It did not affect HCV protein translation, which is mediated by IRES and is cap-independent.

Although the signal transduction leading to IFN production was hampered by the expression of HCV proteins, other studies suggested the engagement of TLRs in the host defense against HCV. IFN-β produced by TLR7, which detects single-stranded RNA, could suppress HCV replication (Lee et al. 2006). Our lab also found that HCV induced the expression of tumor necrosis factor-α (TNF-α) in hepatocytes after infection (Lee et al. 2015). This induction was mediated by TLR7 and TLR8, which reside in endosomes and are activated by single-stranded RNA. The induction of TNF-α by HCV was biphasic, with the first-phase induction occurring soon after the endocytosis of HCV particles at 2 h post-infection. As this first-phase induction was prompt, TLR7 and TLR8 likely sensed a PAMP in the HCV genomic RNA either due to the disintegration of the viral particles soon after their endocytic uptake or the delivery of newly uncoated HCV RNA from the cytosol back into the endosomes. This first phase subsided at about 8 h after infection. The second-phase induction was detected at 24 h post-infection. TNF-α is required for type I interferon signaling, as the silencing of either TNF-α or its receptor TNFR1 abrogated the expression of IFN-α receptor 2 (IFNAR2) and desensitized HCV-infected cells to type I IFNs. As the expression of IFNs by HCV-infected hepatocytes was inefficient, it is likely that TNF-α expressed by HCV-infected hepatocytes supports the expression of IFNAR2 through the autocrine mechanism to sensitize infected cells to IFNs produced by other cell types such as macrophages, the main producer of type I IFNs.

Concomitant with the second-phase induction of TNF-α by HCV at 24 h post-infection was the loss of TNFR1, apparently degraded in lysosomes (Lee et al. 2015). It is known that internalized cell surface receptors after their activation may be recycled back to the cell surface. This is true for TNFR1. However, the disappearance of TNFR1 during HCV infection possibly indicated a failure in recycling of the internalized TNFR1. Human adenovirus E3-10.4K/14.5K (RIDα/β) complex is also called receptor internalization and degradation (RID) due to its ability to down-regulate cell surface receptors such as epidermal growth factor receptor (EGFR) and TNFR1 (Chin and Horwitz 2005). Chin et al. demonstrated that the binding of RID to TNFR1 led to the dynamin-dependent and endosomal/lysosomal-dependent down-regulation of TNFR1 from the cell surface (Chin and Horwitz 2005). The poliovirus protein 3A could also induce the disappearance of cell surface TNFR1 (Neznanov et al. 2001). A positive correlation between the up-regulation of TNFR1 in dendritic cells and the SVR in HCV patients had been observed (Cubillas et al. 2010). Our result indicated that silencing of TNF-α exacerbated the disappearance of TNFR1 (data not shown). How HCV induces the degradation of TNFR1 is an interesting question, and the answer to this question will provide important information for further understanding how HCV evades the host defense mechanisms.

Interestingly, although TNF-α could support IFN signaling to suppress HCV replication, it had also been shown to promote HCV infection through a paracrine mechanism to disrupt the tight junctions and promote the entry of the virus into polarized hepatocytes (Fletcher et al. 2014).

4.4 Cell Death and Liver Injury in Chronic HCV Infection

Liver diseases associated with chronic HCV infection include steatosis, cirrhosis, and HCC (Yamane et al. 2013). HCV proteins and replication can directly cause lipid accumulation, ER stress, oxidative stress, hepatic cell death, and cellular transformation. Liver injuries during chronic HCV infection can also be due to inflammatory responses as well as the deaths of infected and bystander hepatocytes targeted by cytotoxic T lymphocytes (CTLs) and other immune cells (Cecere et al. 2004; Heydtmann 2009). Prolonged hepatic cell death and regeneration are linked to the pathological progression of liver diseases during chronic HCV infection. For instance, UTP and ATP released from apoptotic bodies can activate $P2Y_2$ purinergic receptors on hepatic stellate cells (HSCs) (Elliott et al. 2009). The transdifferentiation of HSCs into invasive myofibroblast-like cells stimulates production of an excessive amount of extracellular matrix (ECM) components and type I collagen, characteristics of matrix remodeling in liver fibrosis. HCV-infected hepatocytes also produce transforming growth factor-β (TGF-β), a key pro-fibrogenic cytokine, which boosts the activation and proliferation of HSCs (Presser et al. 2013).

Mechanistic explanations of cell death associated with HCV-induced liver injury derived from in vitro and in vivo studies include the activation of programmed cell death, the production of reactive oxygen species (ROS), reactive nitrogen species (RNS), and altered lipid metabolism. Modes of cell death include apoptosis, necrosis, necroptosis, and autophagy.

4.4.1 Apoptosis: Implications in Inflammation

Apoptosis is a programmed cell death induced by various stimulants such as the fluctuation of intracellular ions, DNA damage, signal transduction, and infection. Two pathways of apoptosis are known: extrinsic and intrinsic. The extrinsic pathway of apoptosis is triggered by the stimulation of death receptors, while the intrinsic pathway is initiated by a variety of cellular stresses causing mitochondrial damages. Both pathways activate caspase cascades that merge at caspase-3 as an executioner. Caspases are cysteine proteases that cleave after specific aspartic acid residues and are activated during apoptosis. Unlike necrosis, another type of cell death, apoptosis, characterized by chromatin condensation, nuclear fragmentation, cell shrinkage, and membrane blebbing, does not spill out intracellular contents. However, massive apoptosis of cells can influence the interstitial milieu and the neighboring cells.

A positive correlation between hepatic apoptosis and the severity of inflammation has been established in chronic HCV patients. The hepatocytes of patients with chronic HCV infection had increased expression levels of Fas (CD95) and FasL (CD95L), which trigger cell death and inflammation in the liver (Ferenbach et al. 1997; Hiramatsu et al. 1994; Seidel et al. 2005). In addition, the activation of caspases and the cleavage of poly (ADP-ribose) polymerase (PARP), a hallmark of apoptosis, were detected in HCV-infected liver and positively correlated with the degree of inflammation and liver injury (Bantel et al. 2001; Nelson et al. 1997). The secretion of the pro-inflammatory cytokine TNF-α by HCV-specific CTLs was also evident in the liver infected by HCV (Ando et al. 1997; Koziel et al. 1995). Despite the distinct presence of immune responses during HCV infection, most HCV patients failed to clear the viral infection. This may be due to the loss of HCV-specific CD8+ T-lymphocytes by apoptosis during acute infection and the loss of active memory states of the lymphocytes during chronic infection (Radziewicz et al. 2008). Studies using ectopically expressed HCV proteins presented inconsistent results on the relationship between HCV and apoptosis. For instance, HCV core had been reported to be pro-apoptotic by some (Benali-Furet et al. 2005; Chou et al. 2005) and anti-apoptotic by others (Jahan et al. 2011; Nguyen et al. 2006; Saito et al. 2006). Cell culture-based HCV infection studies using the JFH-1 virus provided more consistent results, which indicated that HCV infection could directly induce apoptosis or enhance the apoptotic response to stimuli (Wakita et al. 2005).

Failure to maintain cellular homeostasis and the increase in oxidative stress, ER stress, or calcium influx can cause the depolarization of mitochondria and the release of pro-apoptotic factors such as cytochrome C and apoptotic protease activating factor 1 (Apaf-1) into the cytosol to initiate intrinsic apoptosis (Breckenridge et al. 2003; Kannan and Jain 2000). The infection of Huh7 and Huh7.5 hepatoma cells with HCV resulted in a delay in cell proliferation and the induction of apoptosis as evidenced by DNA fragmentation, cell rounding, and the activation of caspase-3 (Deng et al. 2008). It also induced the accumulation of Bax, a pro-apoptotic member of the Bcl-2 family, on the mitochondrial membrane to cause an alteration of the mitochondrial membrane potential (ΔΨm) and the release of cytochrome C into the cytosol (Deng et al. 2008). The expression of HCV proteins NS5A and core was sufficient to induce ER stress, the increase of cytosolic Ca^{2+}, and the production of ROS/RNS (Dionisio et al. 2009). HCV, however, has also developed mechanisms to suppress apoptotic cell death. It induced dynamin-related protein 1 (Drp1) to enhance mitochondrial fission and mitophagy, a form of autophagy that selectively removes damaged mitochondria, to facilitate the clearance of damaged mitochondria to shun apoptosis and the activation of the interferon-stimulated response element (ISRE) (Kim et al. 2014). This removal of damaged mitochondria also allowed the maintenance of the energy level generated by glycolysis, which supports the propagation of the virus (Kim et al. 2014).

The extrinsic apoptotic pathway involves the activation of death receptors including TNF-α receptors, Fas, and TNF-related apoptosis-inducing ligand (TRAIL) receptors. TNF-α has multiple activities including the induction of apoptosis. The TNF-α receptor TNFR1 is also known as death receptor 1 (DR1) because its C-terminal cytosolic domain contains the death domain (DD). Upon the binding of TNF-α, the DD of TNFR1 recruits TNFRSF1A-Associated via Death Domain (TRADD), a DD-containing adaptor proteins and Fas-Associated protein with Death Domain (FADD) in a sequential order. The receptor complex then activates caspases and induces apoptosis (Micheau and Tschopp 2003; Schutze et al. 2008). However, the end result of signaling through TNFR1 is actually more complicated due to the formation of two signaling complexes, TNFR1 complex I and II, after its activation (Micheau and Tschopp 2003; Vanlangenakker et al. 2011). RIP1 plays a decisive role in switching between the two temporally and spatially distinct complexes. Complex I (TNFR1-CI), formed on the plasma membrane, recruits RIP1 and cellular inhibitor of apoptosis proteins (cIAP) 1 and 2 (Haas et al. 2009). The polyubiquitination of RIP1 in the presence of cIAP1/2 allows the activation of TAB-transforming growth factor-β-activated kinase 1 (TAK1) and IκB kinase (IKK), leading to degradation of inhibitor of NF-κB (IκB) and the activation of NF-κB. NEMO, one of the IKK components, binds to RIP1 to sequester it from caspase-8. In this manner, TNFR1-CI plays an anti-apoptotic role. TNFR1 complex II (TNFR1-CII) forms upon the internalization of TNFR1 and the sequential recruitment of TRADD, RIP3, FADD, and caspase-8. The deubiquitination of RIP1 by cylindromatosis (CYLD) directs RIP1 to TNFR-CII where it is cleaved by caspase-8. This event activates caspase-dependent apoptosis. Caspase-8 also cleaves cIAP1/2 to inhibit the signaling from complex I. On the contrary, an endogenous caspase-8 inhibitor called cellular FADD-like interleukin-1β-converting enzyme (FLICE)-like inhibitory protein (cFLIP) resulted from the activation of complex I inhibits FADD/caspase-8 induced apoptosis. Thus, these two complexes mutually suppress each other to control the life and death of the cell.

Reports regarding the effects of HCV viral proteins on TNF-α-mediated apoptosis are complex and contradictory. The HCV protein NS5A inhibited TNF-α-mediated apoptosis by binding to TRADD, thus blocking further signaling from TNFR1 (Majumder et al. 2002; Miyasaka et al. 2003). In another study, the binding of NS5A with TNF receptor-associated factor 2 (TRAF2) inhibited NF-κB activation and enhanced apoptosis in response to TNF-α (Park et al. 2002). HCV core had also been shown to inhibit the caspase-8 activation induced by TNF-α by sustaining the expression of cFLIP (Saito et al. 2006). The association of the core protein with TNFR1 enhanced NF-κB activation and suppressed apoptosis (Zhu et al. 2001). Zhu et al., on the contrary, proved that HCV core facilitated FADD recruitment to TNFR1 and potentiated apoptosis by TNF-α (Zhu et al. 2001). These inconsistent results suggest that the effect of HCV proteins after TNF-α treatment depends on TNFR1-CI for cell survival and TNFR1-CII for cell death.

In contrast, cell culture-based infection studies largely agreed on the fact that HCV either promoted apoptosis or sensitized cells to apoptosis. It was shown that HCV infection up-regulated TRAIL and death receptors (DR) 4 and 5 in LH86 hepatoma cells (Zhu et al. 2007), which could sensitize cells to apoptosis. While naive Huh7 human hepatoma cells were not responsive to TRAIL, Huh7 cells containing the HCV RNA replicon underwent apoptosis when stimulated by TRAIL (Deng et al. 2012). HCV infection of hepatoma cells primed the cells for TRAIL- and TNF-α-mediated apoptosis (Jang et al. 2014; Park et al. 2012). In particular, HCV infection sensitized hepatoma cells to TNF-α-induced cell death by suppressing the NF-κB activation (Park et al. 2012).

4.4.2 Reactive Oxygen Species: Implications in Hepatic Cell Death and Tumorigenesis

Reactive oxygen species (ROS) are naturally occurring free radical derivatives of oxygen molecules produced during aerobic respiration. ROS is mainly produced by mitochondria and the ER as by-products of cellular metabolic processes. They include superoxide (O_2^-), peroxide (O_2^{2-}), hydroxyl radical ($OH^{\bullet}$), hydrogen peroxide (H_2O_2), and others. Due to the high reactivity of their unpaired electron, excessive levels of ROS can jeopardize the integrity of the cell by affecting macromolecules like proteins, lipids, and DNA. Cells can protect themselves against ROS by expressing antioxidant enzymes including superoxide dismutase (SOD), catalase, and glutathione peroxidase. They can also neutralize ROS by using antioxidants such as diet-derived vitamins A, E, B6, B12 and C, tocopherol, carotenoids, selenium, zinc, and glutathione (GSH) (Cichoz-Lach and Michalak 2014; Reshi et al. 2014). The neutralization of ROS by SOD, for example, occurs through the conversion of two ROS molecules into molecular oxygen and hydrogen peroxide. The latter is then further converted to water and oxygen by catalase. Therefore, maintaining the balance between oxidants and anti-oxidants in a physiologically tolerable range is critical for cellular homeostasis. The term oxidative stress refers to a hazardous state in a cell attained either due to the excessive production of ROS or the inefficient removal of ROS.

HCV induces oxidative stress. Patients with chronic HCV infection exhibited an elevated level of ROS (Barbaro et al. 1999; Ivanov et al. 2013; Paracha et al. 2013) and malondialdehyde (MDA), a product of lipid peroxidation (Farinati et al. 1995). The correlation between chronic HCV infection and a reduced copy number in mitochondrial DNA as a result of oxidative stress had also been reported (Yen et al. 2012). HCV infection of Huh7.5.1 hepatoma cells induced ROS and TGF-β (Lin et al. 2010). The latter likely plays important roles in the progression of hepatic fibrosis during chronic HCV infection in patients. The induction of ROS by HCV was through multiple mechanisms including the induction of ER stress and mitochondrial dysfunction. The individual expression of HCV proteins including core, E1, E2, NS4B, and NS5A core was sufficient to induce ER stress (Ploen and Hildt 2015). NS5A and core could also induce the release of Ca^{2+} from the ER by inhibiting the sarcoplasmic/endoplasmic reticulum calcium ATPase (SERCA) (Dionisio et al. 2009; Gong et al. 2001). HCV core could also enhance the activity of the mitochondrial permeability transition pore and increase the Ca^{2+} influx into mitochondria (Li et al. 2007). The mitochondrial dysfunction as a result of Ca^{2+} uptake stimulated the production of ROS in cells harboring HCV proteins (Medvedev et al. 2016). This would result in a decrease in ATP synthesis and changes in oxidative phosphorylation along with damages to mitochondrial DNA (mtDNA) to exacerbate the liver damage (Cichoz-Lach and Michalak 2014). NS3 could also trigger the ROS production via the activation of NADPH oxidase in monocytes (Bureau et al. 2001) and lymphocytes (Thoren et al. 2004). Garcia-Mediavilla et al. also demonstrated that both NS5A and the core protein expressed in hepatocyte-derived cells were able to induce ROS and RNS upon stimulation with cytokines and up-regulate iNOS gene expression (Garcia-Mediavilla et al. 2005). ROS could suppress HCV RNA replication, likely by disrupting the HCV RNA replication complex (Choi et al. 2004; Huang et al. 2007a; Medvedev et al. 2016). Perhaps for this reason, HCV proteins could also activate the cellular antioxidant pathways. HCV core, E1, E2, NS4B, and NS5A could induce genes containing the antioxidant response element (ARE) via ROS-dependent and independent mechanisms (Ivanov et al. 2011). Another study demonstrated that the AP-1 activation by NS5A via p38 MAPK and JNK led to the expression of the antioxidant MnSOD (Qadri et al. 2004). Interestingly, ROS has also been proposed to cause mutations in the viral genome and help the virus to escape host immune responses (Forns et al. 1999).

ROS affects mitochondrial permeability by altering mitochondrial membrane potential. As mentioned above, this can result in the release of cytochrome C and the activation of the caspase cascade, causing cell death. HCV-induced apoptosis was linked to Bax activation due to oxidative stress and the decrease in the levels of NF-κB and anti-apoptotic BCL-xL (Joyce et al. 2009). Pro-apoptotic Bax forms channels on the outer membrane of mitochondria for the release of cytochrome C. The observation that HCV induced apoptosis via the induction of oxidative stress was further supported by the finding that the antioxidant N-acetylcysteine (NAC) and the Jun N-terminal kinase (JNK) inhibitor, SP600125, suppressed transcriptional up-regulation of Bim, the pro-apoptotic bcl-2-like-protein 11, during HCV infection (Deng et al. 2015). The susceptibility of mtDNA to oxidative stress in hepatitis C patients (Yen et al. 2012) is presumably due to its proximity to the electron transport chain (Reshi et al. 2014). Damaged mtDNA can cause failure in cellular respiration due to impaired synthesis of mitochondrial proteins of the respiratory chain complex, which ultimately leads to apoptosis (Arciello et al. 2013).

Other pathogenic effects of oxidative stress in HCV-related fibrosis and liver cancer had been suggested. ROS can decrease the level of GSH and induce the expression of pro-fibrogenic TGF-β to promote hepatic fibrosis and inflammatory liver diseases (Poli 2000). The ectopic expression of the HCV E2 protein in HSC-T6, a hepatic stellate cell (HSC) line, induced the ROS-related activation of this cell line, which then produced collagen α and α-smooth muscle actin (α-SMA) (Ming-Ju et al. 2011). While the production of collagens and the differentiation of HSCs into myofibroblasts can contribute to the development of hepatic fibrosis, the biological significance of this finding is unclear, as HCV is not known to infect HSCs. Oxidative DNA damage can activate p53 to induce apoptosis to prevent the accumulation of genomic aberrations and the initiation of carcinogenesis. The up-regulation of 3β-hydroxysterol Δ24-reductase (DHCR24) in connection with oxidative stress in cells harboring full-length HCV genome had been reported (Tsukiyama-Kohara 2012). DHCR24, an enzyme involved in cholesterol biosynthesis, suppressed p53-induced apoptosis in these cells and thus its up-regulation by HCV implicated its involvement in HCV-induced hepatocarcinogenesis.

As mentioned above, oxidative stress can induce lipid peroxidation, which is the modification of the carbon-carbon double bonds found in polyunsaturated fatty acids (PUFAs) by free radicals to produce lipid peroxyl radicals and lipid hydroperoxides. Lipid peroxidation can be counteracted by the expression of antioxidant genes, and, if this antioxidant defense mechanism is impaired, cell death ensues. Lipid peroxidation had recently been shown to suppress the infectivity of HCV H77c (genotype 1a) and N (genotype 1b) strains by affecting NS4A and NS5B in the HCV RNA replication complex at amino acid residues within or proximal to the transmembrane domains (Yamane et al. 2014).

4.4.3 Autophagy: Implications in Cell Survival and Cancer

Autophagy is a process in which cytoplasmic contents and organelles are sequestered in membrane vesicles called autophagosomes for delivery to lysosomes for degradation and recycling. During starvation, autophagy supplies nutrients for cell survival by breaking down cellular macromolecules such as proteins, carbohydrates, and lipids. Autophagy is also important for maintaining cellular homeostasis by removing misfolded proteins to lessen the ER stress, damaged organelles, and protein aggregates.

The initiation of autophagy usually requires a complex comprised of Beclin-1, the class III phosphatidylinositol-3-kinase (PI3KC3) Vps34, Atg14, and p150. This complex produces phosphatidylinositol-3-phosphate (PI3P), which recruits the double FYVE-containing protein 1 (DFCP1) and the WD-repeat domain phosphoinositide-interacting (WIPI) protein for the formation of omegasomes and phagophores, the precursors of autophagosomes. The conjugation of Atg5 to Atg12 by

E1 ubiquitin ligase-like Atg7 and by E2 ubiquitin ligase-like Atg10 creates the Atg5-Atg12 conjugate, which then becomes associated with Atg16. The resultant E3-like complex is required for the elongation of phagophores. The microtubule-associated protein light chain 3 (LC3) is cleaved at its C-terminus by Atg4 to produce LC3-I. Atg3, Atg4, and Atg7 further modify LC3-I by conjugating LC3-I to phosphatidylethanolamine (PE). This produces the lipidated form LC3-II, which will be redirected from cytosol to outer and inner surfaces of phagophores. During translocation to phagophores, LC3-II binds to and brings unfolded or ubiquitinated proteins with it for degradation. The recruitment of LC3-II to phagophores leads to the formation of the enclosed double-membrane autophagosomes. V-ATPase, the vacuolar-type ATP-dependent proton pump, lowers the pH inside autophagosomes. Upon the fusion between autophagosome and lysosome, the acidic content in autolysosome allows the activation of lysosomal enzymes and the degradation of the cargos of autophagosomes.

Basal level autophagy under normal conditions is mostly mediated by Unc-51-like autophagy activating kinase (Ulk1). Ulk1 can transmit autophagic signals triggered by the activation of 5′adenosine monophosphate-activated protein kinase (AMPK) or the inactivation of mammalian target of rapamycin (mTOR). The ER stress induced by HCV could inactivate the Akt-mTOR signaling pathway, which led to induction of autophagy (Huang et al. 2013a).

The ER stress will induce the unfolded protein response (UPR) by activating three key enzymes: the inositol-required protein 1 (IRE1), the activating transcription factor 6 (ATF6), and the protein kinase-like ER kinase (PERK). IRE1 is an RNase that induces the splicing of the XBP1 RNA for the expression of XBP1. ATF6 is cleaved under the ER stress, which leads to the nuclear localization of its cytosolic domain. Both XBP1 and the activated ATF6 are transcription factors that can bind to the ER stress response element (ERSE) and the UPR element (UPRE) to activate the expression of genes involved in protein folding and ER-associated degradation (ERAD) of proteins. PERK phosphorylates and inactivates the translation initiation factor eIF2α to suppress the cap-dependent translation to alleviate the ER stress. It can also induce the expression of CCAAT/−enhancer-binding protein homologous protein (CHOP), which downregulates the anti-apoptotic mitochondrial protein Bcl-2 to induce apoptosis if the ER stress prolongs.

HCV can induce autophagy via the induction of ER stress and the UPR (Ke and Chen 2011; Sir et al. 2008). The silencing of IRE1, ATF6, or PERK suppressed the LC3 lipidation and HCV replication, indicating the involvement of the UPR in the induction of autophagy. Curiously, Mohi et al. reported that the induction of autophagy by HCV genotype 1b and 2a subgenomic replicons was independent of IRE1 (Mohl et al. 2012). The reason for this discrepancy of the results is unclear. Our laboratory also found that HCV NS5A and NS5B and the nascent HCV RNA colocalized with autophagosomes, indicating that the HCV RNA replication took place on autophagosomal membranes (Sir et al. 2012). The importance of the localization of the replication complex to autophagosomal membrane was further verified by the observations that the inhibition of autophagy by silencing the expression of LC3 or Atg7 suppressed HCV RNA replication and that the purified autophagosomes could mediate HCV RNA replication in vitro (Sir et al. 2012). Interestingly, the expression of HCV NS4B alone was found to be sufficient to induce autophagosomes via the formation of a complex with Rab5 and Vps34 (Su et al. 2011).

In addition to its role in HCV replication, autophagy induced by HCV had also been suggested to improve the survival of HCV-infected cells, which otherwise might undergo apoptosis caused by the ER stress (Dash et al. 2016). Indeed, impairing autophagy in HCV replicon cells containing either the subgenomic or full-genomic RNA led to the accumulation of vacuoles and cell death (Taguwa et al. 2011), and the knockdown of Beclin-1 in immortalized human hepatocytes led to the activation of caspase-mediated apoptosis (Shrivastava et al. 2011). As mentioned above, HCV had also been shown to induce mitochondrial fission and mitophagy, which attenuated apoptosis to promote HCV persistence after infection (Kim et al. 2014). The continuous protection by autophagy of HCV-infected cells carrying damaged DNA may promote tumorigenesis (Pal et al. 2010). The disruption of autophagy by silencing Beclin 1 and Atg7 also enhanced the production of IFN-α, IFN-β, 2′,5′-OAS1, and IFN-α-inducible protein 27 in immortalized human hepatocytes (IHH) infected by HCV, indicating that autophagy could also suppress host innate immunity (Shrivastava et al. 2011).

4.4.4 Lipid Metabolism: Implications in Steatosis

HCV infection can induce steatosis (Modaresi Esfeh and Ansari-Gilani 2016), which can progress to fibrosis and HCC (Ohki et al. 2008; Yoon and Hu 2006). Furthermore, hepatic steatosis in HCV patients is associated with poor responses to the combination therapy of IFN-α and ribavirin (Szanto et al. 2006; Yaginuma et al. 2006). Studies using cell cultures indicated an important role of lipids in HCV replication. HCV acquired lipoproteins during its maturation process. The term lipo-viro-particles (LVP) refers to HCV particles that are associated with lipoprotein components including triglycerides, ApoB, and ApoE (Ujino et al. 2016). This association with lipoproteins significantly enhances the infectivity of HCV particles (Gastaminza et al. 2008; Huang et al. 2007b; Ujino et al. 2016). ApoE, for example, specifically interacts with heparan sulfate on the cell surface prior to the receptor-mediated entry of HCV particles, and the blockage of this interaction totally abrogates HCV infection (Jiang et al. 2012). Lipid metabolizing pathways also assist in the morphogenesis and release of LVPs (Fukuhara et al. 2015). The morphogenesis of HCV particles is closely associated with LDs, an intracellular storage of neutral fat and cholesterol. The viral capsid assembly is initiated by the association of the HCV core protein with membrane structures surrounding LDs (Miyanari et al. 2007). The specific trafficking of HCV core to LDs requires diacylglycerol acyltransferase-1 (DGAT1) (Herker et al. 2010). Similarly, the interaction of TIP47, an LD-associated protein, with HCV genomic RNA-loaded NS5A is indispensable for the release of infectious particles (Ploen et al. 2013a, b) and HCV replication (Vogt et al. 2013). The mechanism underling HCV egress is not fully understood. However, the physical resemblance between infectious LVP and lipoproteins and the requirement of the coat protein complex II, a component in the assembly and secretion pathway of VLDL, suggest that HCV may use the lipoprotein pathway for viral release (Fukuhara et al. 2015). In addition, the finding that ApoE interacts with the viral protein NS5A offers an explanation to how ApoE participates in the assembly of viral particles (Cun et al. 2010; Huang et al. 2007b).

The ability of HCV to regulate cellular lipid metabolism allows the virus to enhance its replication and propagation. The accumulation of fat in hepatocytes can occur through several different mechanisms including enhanced lipogenesis, enhanced fatty acid uptake, changes in lipid metabolism, and a decrease in lipoprotein secretion (Negro 2006). HCV induced de novo synthesis of fatty acid by up-regulating fatty acid synthase (FASN), and the interaction between FASN and NS5B enhanced HCV replication (Huang et al. 2013b; Yang et al. 2008). The expression of the HCV core protein alone could induce the expression of lipid metabolism-related genes including sterol regulatory element binding protein 1 (SREBP-1), acetyl-CoA carboxylase (ACC), HMG-CoA reductase (HMGR), and HMG-CoA synthase (HMGS) (Yu et al. 2013).

The modulation of major regulators of nutrient homeostasis AMPK and mTOR during viral infection is not uncommon and is highly related to viral pathogenesis (Brunton et al. 2013). HCV also regulates lipid metabolism by exploitation of cellular nutritional and energy pathways. AMPK is an energy sensor and regulator of energy homeostasis inside the cell. AMPK is activated by a low intracellular energy state dictated by an increased AMP/ATP ratio. In such a case, AMPK facilitates the breakdown of nutrient storages like fat and glycogen, while AMPK inhibits the synthesis of the two to supply ATP for maintenance of basic cellular functions. The inhibition of AMPK resulted in a lipid-rich environment. In contrast, the down-regulation of the AMPK activity by phosphorylation at serine residues 485 and 491 resulted in high intracellular lipid contents, which enhanced the replication of HCV RNA (Yu et al. 2013). The forced activation of AMPK by an AMP analog, 5-aminoimidazole-4-carboxamide 1-β-D-ribofuranoside (AICAR) suppressed HCV replication, while compound C, a specific AMPK inhibitor, inhibited the effect of AICAR on HCV (Nakashima et al. 2011). The activation of AMPK by phosphorylating the LD-associated protein perilipin (PLIN) 2 led to the degradation of LDs via the chaperone-mediated autophagy (CMA) (Kaushik and Cuervo 2016).

HCV infection causes the reorganization of membranes in hepatocytes to generate the membranous web, which contains the HCV RNA replication complex and a high concentration of phosphatidylinositol-4-phosphate (PI4P), which may be important for the formation of the membranous web (Ploen and Hildt 2015; Wang and Tai 2016). The increase of PI4P in HCV-infected cells is due to the activity of HCV NS5A, which can bind to and activate phosphatidylinositol-4-kinase IIIα (PI4KA) (Wang and Tai 2016). The oxysterol-binding protein (OSBP) and the four-phosphate adaptor protein 2 (FAPP2) are two effectors of PI4P. They are lipid transfer proteins (LTPs) that are responsible for non-vesicular sterol and glycosphingolipid trafficking, respectively (Wang and Tai 2016). PI4P likely plays an important role in the recruitment of cholesterol and sphingolipids, which are important for the formation of lipid rafts for HCV RNA replication (Saxena et al. 2012).

The fat accumulation in hepatocytes induced by HCV can lead to chronic liver inflammation (Farinati et al. 1995), the resistance to IFNs (Gunduz et al. 2012), and the resistance to insulin (Miquilena-Colina et al. 2011). Hepatic steatosis of chronic HCV patients was also correlated with apoptosis as evidenced by the caspase activation and the increased ratio of Bax/Bcl-2, where Bax promotes apoptosis and Bcl-2 suppresses apoptosis (Walsh et al. 2004). Increased caspase activities were also detected in the liver biopsies as well as in the serum of chronic HCV patients with hepatic steatosis (Seidel et al. 2005). Hepatic lipoapoptosis is a term describing apoptosis induced by the toxicity from excessive fatty acids and is prominent in non-alcoholic steatohepatitis (NASH) (Wree et al. 2011). Free fatty acids (FFAs) sensitized hepatocytes to TRAIL by upregulating the expression of DR5, a TRAIL receptor, via JNK (Malhi et al. 2007). The activation of JNK by FFAs also led to the activation of pro-apoptotic Bim, a BH3 domain-only protein and Bax (Malhi et al. 2006). The induction of the steatosis state in HCV replicon cells with a mixture of the unsaturated oleic fatty acid and the saturated palmitic fatty acid enhanced HCV replication, which was accompanied by a reduced IFN response as well as a reduced cell viability (Gunduz et al. 2012).

4.5 Conclusion and Future Perspective

Host cells can sense viral infections and activate antiviral responses including autophagy, ROS, cytokines, and ISGs to suppress the replication of viruses and remove them, and if these defense mechanisms fail, cells will undergo apoptosis to prevent the continuous replication and propagation of viruses. In contrast, many viruses including HCV have also developed sophisticated mechanisms to subvert these intracellular antiviral responses to enhance their replication and persistence. HCV can disrupt PRR signaling pathways to suppress the expression of IFNs. It can also take control of the cellular lipid metabolism to enhance its replication. The host cell can respond to HCV infection by undergoing apoptosis. However, HCV has also evolved mechanism to suppress apoptosis for its persistent replication. The interactions between HCV and its host cells in the control of cell death and survival are summarized in Fig. 4.1. As shown in the figure, HCV gene products and its replication can induce the ER stress, from which two distinct signaling events are generated. On one hand, the ER stress and the inhibition of SERCA by HCV causes leakage of Ca^{2+} from the ER into the cytosol, which enters mitochondria and results in the reduction of $\Delta\Psi m$. Consequently, cytochrome C is released from the mitochondria, which leads to the initiation of intrinsic apoptosis of HCV-infected cells. On the other hand, HCV uses the ER stress and its associated UPR to induce autophagy/mitophagy to enhance its replication and to suppress the host innate immune response. This induction of autophagy/mitophagy can also enhance cell proliferation and tumorigenesis. HCV can also increase the level of ROS to cause oxidative stress, lipid peroxidation, host DNA damage, and cell death or the initiation of hepatocarcinogenesis. HCV infection can also activate TLR7/8 to induce the expression of TNF-α, which can support IFN signaling through an autocrine mechanism or promote HCV entry into naive

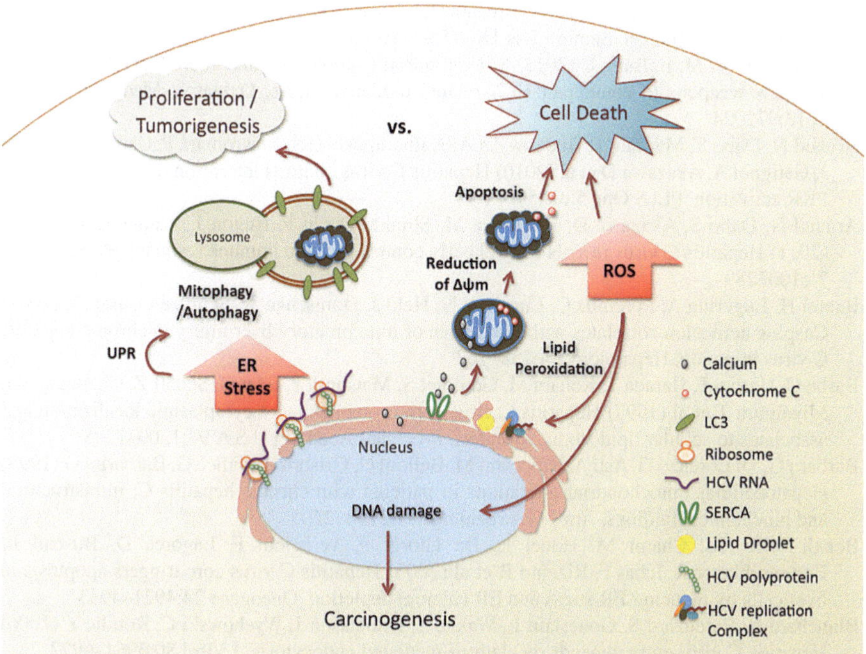

Fig. 4.1 Cell death vs. proliferation during HCV infection. HCV gene products and its replication induce the ER stress. The ER stress causes leakage of Ca^{2+} from the ER into the cytosol. The mitochondrial uptake of Ca^{2+} reduces the mitochondrial membrane potential ($\Delta\Psi$m) and releases cytochrome C to the cytosol leading to the initiation of apoptosis. Damaged mitochondria generate ROS that causes oxidative stress, lipid peroxidation, and oxidative damage of genomic DNA. Oxidative attack of mitochondrial DNA can exacerbate apoptosis, while oxidative damage of genomic DNA can lead to hepatocarcinogenesis. The ER stress can also trigger autophagy and mitophagy via UPR to remove damaged mitochondria to promote cell survival and proliferation/tumorigenesis

hepatocytes through a paracrine mechanism. TNF-α may also suppress apoptosis by activation NF-κB or induce extrinsic apoptosis via the distinct activities of TNFR1-CI and TNFRI-CII, respectively. Clearly, the interactions between HCV and its host cells are very complicated and a slight shifting of the balance can cause the cell to either undergo apoptosis for viral clearance or to survive, which will favor viral persistence and lead to hepatocarcinogenesis.

References

Alter MJ (1999) Hepatitis C virus infection in the United States. J Hepatol 31(Suppl 1):88–91

Ando K, Hiroishi K, Kaneko T, Moriyama T, Muto Y, Kayagaki N, Yagita H, Okumura K, Imawari M (1997) Perforin, Fas/Fas ligand, and TNF-alpha pathways as specific and bystander killing mechanisms of hepatitis C virus-specific human CTL. J Immunol 158:5283–5291

Andre P, Perlemuter G, Budkowska A, Brechot C, Lotteau V (2005) Hepatitis C virus particles and lipoprotein metabolism. Semin Liver Dis 25:93–104

Arciello M, Gori M, Balsano C (2013) Mitochondrial dysfunctions and altered metals homeostasis: new weapons to counteract HCV-related oxidative stress. Oxidative Med Cell Longev 2013:971024

Arnaud N, Dabo S, Maillard P, Budkowska A, Kalliampakou KI, Mavromara P, Garcin D, Hugon J, Gatignol A, Akazawa D et al (2010) Hepatitis C virus controls interferon production through PKR activation. PLoS One 5:e10575

Arnaud N, Dabo S, Akazawa D, Fukasawa M, Shinkai-Ouchi F, Hugon J, Wakita T, Meurs EF (2011) Hepatitis C virus reveals a novel early control in acute immune response. PLoS Pathog 7:e1002289

Bantel H, Lugering A, Poremba C, Lugering N, Held J, Domschke W, Schulze-Osthoff K (2001) Caspase activation correlates with the degree of inflammatory liver injury in chronic hepatitis C virus infection. Hepatology 34:758–767

Barba G, Harper F, Harada T, Kohara M, Goulinet S, Matsuura Y, Eder G, Schaff Z, Chapman MJ, Miyamura T et al (1997) Hepatitis C virus core protein shows a cytoplasmic localization and associates to cellular lipid storage droplets. Proc Natl Acad Sci U S A 94:1200–1205

Barbaro G, Di Lorenzo G, Asti A, Ribersani M, Belloni G, Grisorio B, Filice G, Barbarini G (1999) Hepatocellular mitochondrial alterations in patients with chronic hepatitis C: ultrastructural and biochemical findings. Am J Gastroenterol 94:2198–2205

Benali-Furet NL, Chami M, Houel L, De Giorgi F, Vernejoul F, Lagorce D, Buscail L, Bartenschlager R, Ichas F, Rizzuto R et al (2005) Hepatitis C virus core triggers apoptosis in liver cells by inducing ER stress and ER calcium depletion. Oncogene 24:4921–4933

Blanchard E, Belouzard S, Goueslain L, Wakita T, Dubuisson J, Wychowski C, Rouille Y (2006) Hepatitis C virus entry depends on clathrin-mediated endocytosis. J Virol 80:6964–6972

Bowie AG, Unterholzner L (2008) Viral evasion and subversion of pattern-recognition receptor signalling. Nat Rev Immunol 8:911–922

Breckenridge DG, Germain M, Mathai JP, Nguyen M, Shore GC (2003) Regulation of apoptosis by endoplasmic reticulum pathways. Oncogene 22:8608–8618

Brunton J, Steele S, Ziehr B, Moorman N, Kawula T (2013) Feeding uninvited guests: mTOR and AMPK set the table for intracellular pathogens. PLoS Pathog 9:e1003552

Bureau C, Bernad J, Chaouche N, Orfila C, Beraud M, Gonindard C, Alric L, Vinel JP, Pipy B (2001) Nonstructural 3 protein of hepatitis C virus triggers an oxidative burst in human monocytes via activation of NADPH oxidase. J Biol Chem 276:23077–23083

Burlone ME, Budkowska A (2009) Hepatitis C virus cell entry: role of lipoproteins and cellular receptors. J Gen Virol 90:1055–1070

Cecere A, Marotta F, Vangieri B, Tancredi L, Gattoni A (2004) Progressive liver injury in chronic hepatitis C infection is related to altered cellular immune response and to different citokine profile. Panminerva Med 46:171–187

Chevaliez S, and Pawlotsky JM (2006). HCV genome and life cycle. In: Tan SL (ed) Hepatitis C viruses: genomes and molecular biology, Norfolk, (UK): Horizon Bioscience; 2006. Chapter 1. Available from: https://www.ncbi.nlm.nih.gov/books/NBK1630/

Chin YR, Horwitz MS (2005) Mechanism for removal of tumor necrosis factor receptor 1 from the cell surface by the adenovirus RIDalpha/beta complex. J Virol 79:13606–13617

Choi J, Lee KJ, Zheng Y, Yamaga AK, Lai MM, Ou JH (2004) Reactive oxygen species suppress hepatitis C virus RNA replication in human hepatoma cells. Hepatology 39:81–89

Chou AH, Tsai HF, Wu YY, Hu CY, Hwang LH, Hsu PI, Hsu PN (2005) Hepatitis C virus core protein modulates TRAIL-mediated apoptosis by enhancing bid cleavage and activation of mitochondria apoptosis signaling pathway. J Immunol 174:2160–2166

Chung NS, Wasan KM (2004) Potential role of the low-density lipoprotein receptor family as mediators of cellular drug uptake. Adv Drug Deliv Rev 56:1315–1334

Cichoz-Lach H, Michalak A (2014) Oxidative stress as a crucial factor in liver diseases. World J Gastroenterol 20:8082–8091

Cubillas R, Kintner K, Phillips F, Karandikar NJ, Thiele DL, Brown GR (2010) Tumor necrosis factor receptor 1 expression is upregulated in dendritic cells in patients with chronic HCV who respond to therapy. Hepat Res Treat 2010:429243

Cun W, Jiang J, Luo G (2010) The C-terminal alpha-helix domain of apolipoprotein E is required for interaction with nonstructural protein 5A and assembly of hepatitis C virus. J Virol 84:11532–11541

Dash S, Chava S, Aydin Y, Chandra PK, Ferraris P, Chen W, Balart LA, Wu T, Garry RF (2016) Hepatitis C virus infection induces autophagy as a prosurvival mechanism to alleviate hepatic ER-stress response. Viruses 8:150

Deng L, Adachi T, Kitayama K, Bungyoku Y, Kitazawa S, Ishido S, Shoji I, Hotta H (2008) Hepatitis C virus infection induces apoptosis through a bax-triggered, mitochondrion-mediated, caspase 3-dependent pathway. J Virol 82:10375–10385

Deng Z, Yan H, Hu J, Zhang S, Peng P, Liu Q, Guo D (2012) Hepatitis C virus sensitizes host cells to TRAIL-induced apoptosis by up-regulating DR4 and DR5 via a MEK1-dependent pathway. PLoS One 7:e37700

Deng L, Chen M, Tanaka M, Ku Y, Itoh T, Shoji I, Hotta H (2015) HCV upregulates Bim through the ROS/JNK signalling pathway, leading to bax-mediated apoptosis. J Gen Virol 96:2670–2683

Dionisio N, Garcia-Mediavilla MV, Sanchez-Campos S, Majano PL, Benedicto I, Rosado JA, Salido GM, Gonzalez-Gallego J (2009) Hepatitis C virus NS5A and core proteins induce oxidative stress-mediated calcium signalling alterations in hepatocytes. J Hepatol 50:872–882

Egger D, Wolk B, Gosert R, Bianchi L, Blum HE, Moradpour D, Bienz K (2002) Expression of hepatitis C virus proteins induces distinct membrane alterations including a candidate viral replication complex. J Virol 76:5974–5984

Elliott MR, Chekeni FB, Trampont PC, Lazarowski ER, Kadl A, Walk SF, Park D, Woodson RI, Ostankovich M, Sharma P et al (2009) Nucleotides released by apoptotic cells act as a find-me signal to promote phagocytic clearance. Nature 461:282–286

Farinati F, Cardin R, De Maria N, Della Libera G, Marafin C, Lecis E, Burra P, Floreani A, Cecchetto A, Naccarato R (1995) Iron storage, lipid peroxidation and glutathione turnover in chronic anti-HCV positive hepatitis. J Hepatol 22:449–456

Farquhar MJ, Hu K, Harris HJ, Davis C, Brimacombe CL, Fletcher SJ, Baumert TF, Rappoport JZ, Balfe P, McKeating JA (2012) Hepatitis C virus induces CD81 and claudin-1 endocytosis. J Virol 86:4305–4316

Ferenbach DA, Haydon GH, Rae F, Malcomson RD, Harrison DJ (1997) Alteration in mRNA levels of fas splice variants in hepatitis C-infected liver. J Pathol 183:299–304

Fletcher NF, Sutaria R, Jo J, Barnes A, Blahova M, Meredith LW, Cosset FL, Curbishley SM, Adams DH, Bertoletti A et al (2014) Activated macrophages promote hepatitis C virus entry in a tumor necrosis factor-dependent manner. Hepatology 59:1320–1330

Forns X, Purcell RH, Bukh J (1999) Quasispecies in viral persistence and pathogenesis of hepatitis C virus. Trends Microbiol 7:402–410

Foy E, Li K, Sumpter R Jr, Loo YM, Johnson CL, Wang C, Fish PM, Yoneyama M, Fujita T, Lemon SM et al (2005) Control of antiviral defenses through hepatitis C virus disruption of retinoic acid-inducible gene-I signaling. Proc Natl Acad Sci U S A 102:2986–2991

Friebe P, Bartenschlager R (2002) Genetic analysis of sequences in the 3' nontranslated region of hepatitis C virus that are important for RNA replication. J Virol 76:5326–5338

Fukuhara T, Ono C, Puig-Basagoiti F, Matsuura Y (2015) Roles of lipoproteins and apolipoproteins in particle formation of hepatitis C virus. Trends Microbiol 23:618–629

Garcia-Mediavilla MV, Sanchez-Campos S, Gonzalez-Perez P, Gomez-Gonzalo M, Majano PL, Lopez-Cabrera M, Clemente G, Garcia-Monzon C, Gonzalez-Gallego J (2005) Differential contribution of hepatitis C virus NS5A and core proteins to the induction of oxidative and nitrosative stress in human hepatocyte-derived cells. J Hepatol 43:606–613

Gardner JP, Durso RJ, Arrigale RR, Donovan GP, Maddon PJ, Dragic T, Olson WC (2003) L-SIGN (CD 209L) is a liver-specific capture receptor for hepatitis C virus. Proc Natl Acad Sci U S A 100:4498–4503

Gastaminza P, Kapadia SB, Chisari FV (2006) Differential biophysical properties of infectious intracellular and secreted hepatitis C virus particles. J Virol 80:11074–11081
Gastaminza P, Cheng G, Wieland S, Zhong J, Liao W, Chisari FV (2008) Cellular determinants of hepatitis C virus assembly, maturation, degradation, and secretion. J Virol 82:2120–2129
Gong G, Waris G, Tanveer R, Siddiqui A (2001) Human hepatitis C virus NS5A protein alters intracellular calcium levels, induces oxidative stress, and activates STAT-3 and NF-kappa B. Proc Natl Acad Sci U S A 98:9599–9604
Gosert R, Egger D, Lohmann V, Bartenschlager R, Blum HE, Bienz K, Moradpour D (2003) Identification of the hepatitis C virus RNA replication complex in Huh-7 cells harboring subgenomic replicons. J Virol 77:5487–5492
Gunduz F, Aboulnasr FM, Chandra PK, Hazari S, Poat B, Baker DP, Balart LA, Dash S (2012) Free fatty acids induce ER stress and block antiviral activity of interferon alpha against hepatitis C virus in cell culture. Virol J 9:143
Haas TL, Emmerich CH, Gerlach B, Schmukle AC, Cordier SM, Rieser E, Feltham R, Vince J, Warnken U, Wenger T et al (2009) Recruitment of the linear ubiquitin chain assembly complex stabilizes the TNF-R1 signaling complex and is required for TNF-mediated gene induction. Mol Cell 36:831–844
Herker E, Harris C, Hernandez C, Carpentier A, Kaehlcke K, Rosenberg AR, Farese RV Jr, Ott M (2010) Efficient hepatitis C virus particle formation requires diacylglycerol acyltransferase-1. Nat Med 16:1295–1298
Heydtmann M (2009) Macrophages in hepatitis B and hepatitis C virus infections. J Virol 83:2796–2802
Hiramatsu N, Hayashi N, Katayama K, Mochizuki K, Kawanishi Y, Kasahara A, Fusamoto H, Kamada T (1994) Immunohistochemical detection of Fas antigen in liver tissue of patients with chronic hepatitis C. Hepatology 19:1354–1359
Hnatyszyn HJ (2005) Chronic hepatitis C and genotyping: the clinical significance of determining HCV genotypes. Antivir Ther 10:1–11
Honda M, Ping LH, Rijnbrand RC, Amphlett E, Clarke B, Rowlands D, Lemon SM (1996) Structural requirements for initiation of translation by internal ribosome entry within genome-length hepatitis C virus RNA. Virology 222:31–42
Huang H, Chen Y, Ye J (2007a) Inhibition of hepatitis C virus replication by peroxidation of arachidonate and restoration by vitamin E. Proc Natl Acad Sci U S A 104:18666–18670
Huang H, Sun F, Owen DM, Li W, Chen Y, Gale M Jr, Ye J (2007b) Hepatitis C virus production by human hepatocytes dependent on assembly and secretion of very low-density lipoproteins. Proc Natl Acad Sci U S A 104:5848–5853
Huang H, Kang R, Wang J, Luo G, Yang W, Zhao Z (2013a) Hepatitis C virus inhibits AKT-tuberous sclerosis complex (TSC), the mechanistic target of rapamycin (MTOR) pathway, through endoplasmic reticulum stress to induce autophagy. Autophagy 9:175–195
Huang JT, Tseng CP, Liao MH, Lu SC, Yeh WZ, Sakamoto N, Chen CM, Cheng JC (2013b) Hepatitis C virus replication is modulated by the interaction of nonstructural protein NS5B and fatty acid synthase. J Virol 87:4994–5004
Ishida S, Kaito M, Kohara M, Tsukiyama-Kohora K, Fujita N, Ikoma J, Adachi Y, Watanabe S (2001) Hepatitis C virus core particle detected by immunoelectron microscopy and optical rotation technique. Hepatol Res 20:335–347
Ivanov AV, Smirnova OA, Ivanova ON, Masalova OV, Kochetkov SN, Isaguliants MG (2011) Hepatitis C virus proteins activate NRF2/ARE pathway by distinct ROS-dependent and independent mechanisms in HUH7 cells. PLoS One 6:e24957
Ivanov AV, Bartosch B, Smirnova OA, Isaguliants MG, Kochetkov SN (2013) HCV and oxidative stress in the liver. Viruses 5:439–469
Jahan S, Khaliq S, Siddiqi MH, Ijaz B, Ahmad W, Ashfaq UA, Hassan S (2011) Anti-apoptotic effect of HCV core gene of genotype 3a in Huh-7 cell line. Virol J 8:522
Jang JY, Kim SJ, Cho EK, Jeong SW, Park EJ, Lee WC, Lee SH, Kim SG, Kim YS, Kim HS et al (2014) TRAIL enhances apoptosis of human hepatocellular carcinoma cells sensitized by hepatitis C virus infection: therapeutic implications. PLoS One 9:e98171

Ji H, Fraser CS, Yu Y, Leary J, Doudna JA (2004) Coordinated assembly of human translation initiation complexes by the hepatitis C virus internal ribosome entry site RNA. Proc Natl Acad Sci U S A 101:16990–16995

Jiang J, Cun W, Wu X, Shi Q, Tang H, Luo G (2012) Hepatitis C virus attachment mediated by apolipoprotein E binding to cell surface heparan sulfate. J Virol 86:7256–7267

Joyce MA, Walters KA, Lamb SE, Yeh MM, Zhu LF, Kneteman N, Doyle JS, Katze MG, Tyrrell DL (2009) HCV induces oxidative and ER stress, and sensitizes infected cells to apoptosis in SCID/Alb-uPA mice. PLoS Pathog 5:e1000291

Kaito M, Watanabe S, Tsukiyama-Kohara K, Yamaguchi K, Kobayashi Y, Konishi M, Yokoi M, Ishida S, Suzuki S, Kohara M (1994) Hepatitis C virus particle detected by immunoelectron microscopic study. J Gen Virol 75(Pt 7):1755–1760

Kannan K, Jain SK (2000) Oxidative stress and apoptosis. Pathophysiology 7:153–163

Kaushik S, Cuervo AM (2016) AMPK-dependent phosphorylation of lipid droplet protein PLIN2 triggers its degradation by CMA. Autophagy 12:432–438

Ke PY, Chen SS (2011) Activation of the unfolded protein response and autophagy after hepatitis C virus infection suppresses innate antiviral immunity in vitro. J Clin Invest 121:37–56

Kelleher TB, Afdhal N (2005) Maintenance therapy for chronic hepatitis C. Curr Gastroenterol Rep 7:50–53

Kim SJ, Syed GH, Khan M, Chiu WW, Sohail MA, Gish RG, Siddiqui A (2014) Hepatitis C virus triggers mitochondrial fission and attenuates apoptosis to promote viral persistence. Proc Natl Acad Sci U S A 111:6413–6418

Klein KC, Polyak SJ, Lingappa JR (2004) Unique features of hepatitis C virus capsid formation revealed by de novo cell-free assembly. J Virol 78:9257–9269

Klein KC, Dellos SR, Lingappa JR (2005) Identification of residues in the hepatitis C virus core protein that are critical for capsid assembly in a cell-free system. J Virol 79:6814–6826

Kolykhalov AA, Feinstone SM, Rice CM (1996) Identification of a highly conserved sequence element at the 3' terminus of hepatitis C virus genome RNA. J Virol 70:3363–3371

Koziel MJ, Dudley D, Afdhal N, Grakoui A, Rice CM, Choo QL, Houghton M, Walker BD (1995) HLA class I-restricted cytotoxic T lymphocytes specific for hepatitis C virus. Identification of multiple epitopes and characterization of patterns of cytokine release. J Clin Invest 96:2311–2321

Kunkel M, Lorinczi M, Rijnbrand R, Lemon SM, Watowich SJ (2001) Self-assembly of nucleocapsid-like particles from recombinant hepatitis C virus core protein. J Virol 75: 2119–2129

Lavillette D, Bartosch B, Nourrisson D, Verney G, Cosset FL, Penin F, Pecheur EI (2006) Hepatitis C virus glycoproteins mediate low pH-dependent membrane fusion with liposomes. J Biol Chem 281:3909–3917

Lee H, Shin H, Wimmer E, Paul AV (2004) Cis-acting RNA signals in the NS5B C-terminal coding sequence of the hepatitis C virus genome. J Virol 78:10865–10877

Lee J, Wu CC, Lee KJ, Chuang TH, Katakura K, Liu YT, Chan M, Tawatao R, Chung M, Shen C et al (2006) Activation of anti-hepatitis C virus responses via toll-like receptor 7. Proc Natl Acad Sci U S A 103:1828–1833

Lee J, Tian Y, Chan ST, Kim JY, Cho C, Ou JH (2015) TNF-alpha induced by hepatitis C virus via TLR7 and TLR8 in hepatocytes supports interferon signaling via an autocrine mechanism. PLoS Pathog 11:e1004937

Leyssen P, Balzarini J, De Clercq E, Neyts J (2005) The predominant mechanism by which ribavirin exerts its antiviral activity in vitro against flaviviruses and paramyxoviruses is mediated by inhibition of IMP dehydrogenase. J Virol 79:1943–1947

Li K, Foy E, Ferreon JC, Nakamura M, Ferreon AC, Ikeda M, Ray SC, Gale M Jr, Lemon SM (2005) Immune evasion by hepatitis C virus NS3/4A protease-mediated cleavage of the toll-like receptor 3 adaptor protein TRIF. Proc Natl Acad Sci U S A 102:2992–2997

Li Y, Boehning DF, Qian T, Popov VL, Weinman SA (2007) Hepatitis C virus core protein increases mitochondrial ROS production by stimulation of Ca2+ uniporter activity. FASEB J 21: 2474–2485

Li ZZ, Berk M, McIntyre TM, Feldstein AE (2009) Hepatic lipid partitioning and liver damage in nonalcoholic fatty liver disease: role of stearoyl-CoA desaturase. J Biol Chem 284:5637–5644

Lin R, Lacoste J, Nakhaei P, Sun Q, Yang L, Paz S, Wilkinson P, Julkunen I, Vitour D, Meurs E et al (2006) Dissociation of a MAVS/IPS-1/VISA/Cardif-IKKepsilon molecular complex from the mitochondrial outer membrane by hepatitis C virus NS3-4A proteolytic cleavage. J Virol 80:6072–6083

Lin W, Tsai WL, Shao RX, Wu G, Peng LF, Barlow LL, Chung WJ, Zhang L, Zhao H, Jang JY et al (2010) Hepatitis C virus regulates transforming growth factor beta1 production through the generation of reactive oxygen species in a nuclear factor kappaB-dependent manner. Gastroenterology 138:2509–2518. 2518 e2501.

Lindenbach BD, Meuleman P, Ploss A, Vanwolleghem T, Syder AJ, McKeating JA, Lanford RE, Feinstone SM, Major ME, Leroux-Roels G et al (2006) Cell culture-grown hepatitis C virus is infectious in vivo and can be recultured in vitro. Proc Natl Acad Sci U S A 103:3805–3809

Lozach PY, Amara A, Bartosch B, Virelizier JL, Arenzana-Seisdedos F, Cosset FL, Altmeyer R (2004) C-type lectins L-SIGN and DC-SIGN capture and transmit infectious hepatitis C virus pseudotype particles. J Biol Chem 279:32035–32045

Lupberger J, Zeisel MB, Xiao F, Thumann C, Fofana I, Zona L, Davis C, Mee CJ, Turek M, Gorke S et al (2011) EGFR and EphA2 are host factors for hepatitis C virus entry and possible targets for antiviral therapy. Nat Med 17:589–595

Major ME, Mihalik K, Fernandez J, Seidman J, Kleiner D, Kolykhalov AA, Rice CM, Feinstone SM (1999) Long-term follow-up of chimpanzees inoculated with the first infectious clone for hepatitis C virus. J Virol 73:3317–3325

Majumdar A, Kitson MT, Roberts SK (2016) Systematic review: current concepts and challenges for the direct-acting antiviral era in hepatitis C cirrhosis. Aliment Pharmacol Ther 43: 1276–1292

Majumder M, Ghosh AK, Steele R, Zhou XY, Phillips NJ, Ray R, Ray RB (2002) Hepatitis C virus NS5A protein impairs TNF-mediated hepatic apoptosis, but not by an anti-FAS antibody, in transgenic mice. Virology 294:94–105

Malhi H, Bronk SF, Werneburg NW, Gores GJ (2006) Free fatty acids induce JNK-dependent hepatocyte lipoapoptosis. J Biol Chem 281:12093–12101

Malhi H, Barreyro FJ, Isomoto H, Bronk SF, Gores GJ (2007) Free fatty acids sensitise hepatocytes to TRAIL mediated cytotoxicity. Gut 56:1124–1131

McLauchlan J, Lemberg MK, Hope G, Martoglio B (2002) Intramembrane proteolysis promotes trafficking of hepatitis C virus core protein to lipid droplets. EMBO J 21:3980–3988

Medvedev R, Ploen D, Hildt E (2016) HCV and oxidative stress: implications for HCV life cycle and HCV-associated pathogenesis. Oxidative Med Cell Longev 2016:9012580

Meertens L, Bertaux C, Dragic T (2006) Hepatitis C virus entry requires a critical postinternalization step and delivery to early endosomes via clathrin-coated vesicles. J Virol 80:11571–11578

Micheau O, Tschopp J (2003) Induction of TNF receptor I-mediated apoptosis via two sequential signaling complexes. Cell 114:181–190

Ming-Ju H, Yih-Shou H, Tzy-Yen C, Hui-Ling C (2011) Hepatitis C virus E2 protein induce reactive oxygen species (ROS)-related fibrogenesis in the HSC-T6 hepatic stellate cell line. J Cell Biochem 112:233–243

Miquilena-Colina ME, Lima-Cabello E, Sanchez-Campos S, Garcia-Mediavilla MV, Fernandez-Bermejo M, Lozano-Rodriguez T, Vargas-Castrillon J, Buque X, Ochoa B, Aspichueta P et al (2011) Hepatic fatty acid translocase CD36 upregulation is associated with insulin resistance, hyperinsulinaemia and increased steatosis in non-alcoholic steatohepatitis and chronic hepatitis C. Gut 60:1394–1402

Miyanari Y, Atsuzawa K, Usuda N, Watashi K, Hishiki T, Zayas M, Bartenschlager R, Wakita T, Hijikata M, Shimotohno K (2007) The lipid droplet is an important organelle for hepatitis C virus production. Nat Cell Biol 9:1089–1097

Miyasaka Y, Enomoto N, Kurosaki M, Sakamoto N, Kanazawa N, Kohashi T, Ueda E, Maekawa S, Watanabe H, Izumi N et al (2003) Hepatitis C virus nonstructural protein 5A inhibits tumor necrosis factor-alpha-mediated apoptosis in Huh7 cells. J Infect Dis 188:1537–1544

Modaresi Esfeh J, Ansari-Gilani K (2016) Steatosis and hepatitis C. Gastroenterol Rep (Oxf) 4:24–29

Mohl BP, Tedbury PR, Griffin S, Harris M (2012) Hepatitis C virus-induced autophagy is independent of the unfolded protein response. J Virol 86:10724–10732

Morikawa K, Zhao Z, Date T, Miyamoto M, Murayama A, Akazawa D, Tanabe J, Sone S, Wakita T (2007) The roles of CD81 and glycosaminoglycans in the adsorption and uptake of infectious HCV particles. J Med Virol 79:714–723

Nakashima K, Takeuchi K, Chihara K, Hotta H, Sada K (2011) Inhibition of hepatitis C virus replication through adenosine monophosphate-activated protein kinase-dependent and -independent pathways. Microbiol Immunol 55:774–782

Negro F (2006) Mechanisms and significance of liver steatosis in hepatitis C virus infection. World J Gastroenterol 12:6756–6765

Nelson DR, Lim HL, Marousis CG, Fang JW, Davis GL, Shen L, Urdea MS, Kolberg JA, Lau JY (1997) Activation of tumor necrosis factor-alpha system in chronic hepatitis C virus infection. Dig Dis Sci 42:2487–2494

Neznanov N, Kondratova A, Chumakov KM, Angres B, Zhumabayeva B, Agol VI, Gudkov AV (2001) Poliovirus protein 3A inhibits tumor necrosis factor (TNF)-induced apoptosis by eliminating the TNF receptor from the cell surface. J Virol 75:10409–10420

Nguyen H, Sankaran S, Dandekar S (2006) Hepatitis C virus core protein induces expression of genes regulating immune evasion and anti-apoptosis in hepatocytes. Virology 354:58–68

Nitta S, Sakamoto N, Nakagawa M, Kakinuma S, Mishima K, Kusano-Kitazume A, Kiyohashi K, Murakawa M, Nishimura-Sakurai Y, Azuma S et al (2013) Hepatitis C virus NS4B protein targets STING and abrogates RIG-I-mediated type I interferon-dependent innate immunity. Hepatology 57:46–58

Ohki T, Tateishi R, Sato T, Masuzaki R, Imamura J, Goto T, Yamashiki N, Yoshida H, Kanai F, Kato N et al (2008) Obesity is an independent risk factor for hepatocellular carcinoma development in chronic hepatitis C patients. Clin Gastroenterol Hepatol 6:459–464

Otsuka M, Kato N, Moriyama M, Taniguchi H, Wang Y, Dharel N, Kawabe T, Omata M (2005) Interaction between the HCV NS3 protein and the host TBK1 protein leads to inhibition of cellular antiviral responses. Hepatology 41:1004–1012

Otto GA, Puglisi JD (2004) The pathway of HCV IRES-mediated translation initiation. Cell 119:369–380

Pal S, Polyak SJ, Bano N, Qiu WC, Carithers RL, Shuhart M, Gretch DR, Das A (2010) Hepatitis C virus induces oxidative stress, DNA damage and modulates the DNA repair enzyme NEIL1. J Gastroenterol Hepatol 25:627–634

Paracha UZ, Fatima K, Alqahtani M, Chaudhary A, Abuzenadah A, Damanhouri G, Qadri I (2013) Oxidative stress and hepatitis C virus. Virol J 10:251

Park KJ, Choi SH, Lee SY, Hwang SB, Lai MM (2002) Nonstructural 5A protein of hepatitis C virus modulates tumor necrosis factor alpha-stimulated nuclear factor kappa B activation. J Biol Chem 277:13122–13128

Park J, Kang W, Ryu SW, Kim WI, Chang DY, Lee DH, Park do Y, Choi YH, Choi K, Shin EC et al (2012) Hepatitis C virus infection enhances TNFalpha-induced cell death via suppression of NF-kappaB. Hepatology 56:831–840

Pileri P, Uematsu Y, Campagnoli S, Galli G, Falugi F, Petracca R, Weiner AJ, Houghton M, Rosa D, Grandi G et al (1998) Binding of hepatitis C virus to CD81. Science 282:938–941

Ploen D, Hildt E (2015) Hepatitis C virus comes for dinner: how the hepatitis C virus interferes with autophagy. World J Gastroenterol 21:8492–8507

Ploen D, Hafirassou ML, Himmelsbach K, Sauter D, Biniossek ML, Weiss TS, Baumert TF, Schuster C, Hildt E (2013a) TIP47 plays a crucial role in the life cycle of hepatitis C virus. J Hepatol 58:1081–1088

Ploen D, Hafirassou ML, Himmelsbach K, Schille SA, Biniossek ML, Baumert TF, Schuster C, Hildt E (2013b) TIP47 is associated with the hepatitis C virus and its interaction with Rab9 is required for release of viral particles. Eur J Cell Biol 92:374–382

Poli G (2000) Pathogenesis of liver fibrosis: role of oxidative stress. Mol Asp Med 21:49–98
Popescu CI, Dubuisson J (2010) Role of lipid metabolism in hepatitis C virus assembly and entry. Biol Cell 102:63–74
Presser LD, McRae S, Waris G (2013) Activation of TGF-beta1 promoter by hepatitis C virus-induced AP-1 and Sp1: role of TGF-beta1 in hepatic stellate cell activation and invasion. PLoS One 8:e56367
Qadri I, Iwahashi M, Capasso JM, Hopken MW, Flores S, Schaack J, Simon FR (2004) Induced oxidative stress and activated expression of manganese superoxide dismutase during hepatitis C virus replication: role of JNK, p38 MAPK and AP-1. Biochem J 378:919–928
Qiang G, Jhaveri R (2012) Lipid droplet binding of hepatitis C virus core protein genotype 3. ISRN Gastroenterol 2012:176728
Radziewicz H, Ibegbu CC, Hon H, Osborn MK, Obideen K, Wehbi M, Freeman GJ, Lennox JL, Workowski KA, Hanson HL et al (2008) Impaired hepatitis C virus (HCV)-specific effector CD8+ T cells undergo massive apoptosis in the peripheral blood during acute HCV infection and in the liver during the chronic phase of infection. J Virol 82:9808–9822
Reshi ML, Su YC, Hong JR (2014) RNA viruses: ROS-mediated cell death. Int J Cell Biol 2014:467452
Sadler AJ, Williams BR (2008) Interferon-inducible antiviral effectors. Nat Rev Immunol 8:559–568
Sainz B Jr, Barretto N, Martin DN, Hiraga N, Imamura M, Hussain S, Marsh KA, Yu X, Chayama K, Alrefai WA et al (2012) Identification of the Niemann-pick C1-like 1 cholesterol absorption receptor as a new hepatitis C virus entry factor. Nat Med 18:281–285
Saito K, Meyer K, Warner R, Basu A, Ray RB, Ray R (2006) Hepatitis C virus core protein inhibits tumor necrosis factor alpha-mediated apoptosis by a protective effect involving cellular FLICE inhibitory protein. J Virol 80:4372–4379
Saxena V, Lai CK, Chao TC, Jeng KS, Lai MM (2012) Annexin A2 is involved in the formation of hepatitis C virus replication complex on the lipid raft. J Virol 86:4139–4150
Scarselli E, Ansuini H, Cerino R, Roccasecca RM, Acali S, Filocamo G, Traboni C, Nicosia A, Cortese R, Vitelli A (2002) The human scavenger receptor class B type I is a novel candidate receptor for the hepatitis C virus. EMBO J 21:5017–5025
Schutze S, Tchikov V, Schneider-Brachert W (2008) Regulation of TNFR1 and CD95 signalling by receptor compartmentalization. Nat Rev Mol Cell Biol 9:655–662
Seidel N, Volkmann X, Langer F, Flemming P, Manns MP, Schulze-Osthoff K, Bantel H (2005) The extent of liver steatosis in chronic hepatitis C virus infection is mirrored by caspase activity in serum. Hepatology 42:113–120
Shimoike T, Mimori S, Tani H, Matsuura Y, Miyamura T (1999) Interaction of hepatitis C virus core protein with viral sense RNA and suppression of its translation. J Virol 73:9718–9725
Shrivastava S, Raychoudhuri A, Steele R, Ray R, Ray RB (2011) Knockdown of autophagy enhances the innate immune response in hepatitis C virus-infected hepatocytes. Hepatology 53:406–414
Simmonds P (2004) Genetic diversity and evolution of hepatitis C virus—15 years on. J Gen Virol 85:3173–3188
Sir D, Chen WL, Choi J, Wakita T, Yen TS, Ou JH (2008) Induction of incomplete autophagic response by hepatitis C virus via the unfolded protein response. Hepatology 48:1054–1061
Sir D, Kuo CF, Tian Y, Liu HM, Huang EJ, Jung JU, Machida K, Ou JH (2012) Replication of hepatitis C virus RNA on autophagosomal membranes. J Biol Chem 287:18036–18043
Smith DB, Bukh J, Kuiken C, Muerhoff AS, Rice CM, Stapleton JT, Simmonds P (2014) Expanded classification of hepatitis C virus into 7 genotypes and 67 subtypes: updated criteria and genotype assignment web resource. Hepatology 59:318–327
Stetson DB, Medzhitov R (2006) Type I interferons in host defense. Immunity 25:373–381
Su WC, Chao TC, Huang YL, Weng SC, Jeng KS, Lai MM (2011) Rab5 and class III phosphoinositide 3-kinase Vps34 are involved in hepatitis C virus NS4B-induced autophagy. J Virol 85:10561–10571
Sundaram M, Yao Z (2010) Recent progress in understanding protein and lipid factors affecting hepatic VLDL assembly and secretion. Nutr Metab (Lond) 7:35

Szanto P, Grigorescu M, Dumitru I, Serban A (2006) Steatosis in hepatitis C virus infection. Response to anti-viral therapy. J Gastrointestin Liver Dis 15:117–124
Taguwa S, Kambara H, Fujita N, Noda T, Yoshimori T, Koike K, Moriishi K, Matsuura Y (2011) Dysfunction of autophagy participates in vacuole formation and cell death in cells replicating hepatitis C virus. J Virol 85:13185–13194
Tanaka T, Kato N, Cho MJ, Sugiyama K, Shimotohno K (1996) Structure of the 3' terminus of the hepatitis C virus genome. J Virol 70:3307–3312
Tanaka Y, Shimoike T, Ishii K, Suzuki R, Suzuki T, Ushijima H, Matsuura Y, Miyamura T (2000) Selective binding of hepatitis C virus core protein to synthetic oligonucleotides corresponding to the 5' untranslated region of the viral genome. Virology 270:229–236
Thoren F, Romero A, Lindh M, Dahlgren C, Hellstrand K (2004) A hepatitis C virus-encoded, nonstructural protein (NS3) triggers dysfunction and apoptosis in lymphocytes: role of NADPH oxidase-derived oxygen radicals. J Leukoc Biol 76:1180–1186
Tscherne DM, Jones CT, Evans MJ, Lindenbach BD, McKeating JA, Rice CM (2006) Time- and temperature-dependent activation of hepatitis C virus for low-pH-triggered entry. J Virol 80:1734–1741
Tsukiyama-Kohara K (2012) Role of oxidative stress in hepatocarcinogenesis induced by hepatitis C virus. Int J Mol Sci 13:15271–15278
Ujino S, Nishitsuji H, Hishiki T, Sugiyama K, Takaku H, Shimotohno K (2016) Hepatitis C virus utilizes VLDLR as a novel entry pathway. Proc Natl Acad Sci U S A 113:188–193
Vanlangenakker N, Bertrand MJ, Bogaert P, Vandenabeele P, Vanden Berghe T (2011) TNF-induced necroptosis in L929 cells is tightly regulated by multiple TNFR1 complex I and II members. Cell Death Dis 2:e230
Vogt DA, Camus G, Herker E, Webster BR, Tsou CL, Greene WC, Yen TS, Ott M (2013) Lipid droplet-binding protein TIP47 regulates hepatitis C virus RNA replication through interaction with the viral NS5A protein. PLoS Pathog 9:e1003302
Wakita T, Pietschmann T, Kato T, Date T, Miyamoto M, Zhao Z, Murthy K, Habermann A, Krausslich HG, Mizokami M et al (2005) Production of infectious hepatitis C virus in tissue culture from a cloned viral genome. Nat Med 11:791–796
Walsh MJ, Vanags DM, Clouston AD, Richardson MM, Purdie DM, Jonsson JR, Powell EE (2004) Steatosis and liver cell apoptosis in chronic hepatitis C: a mechanism for increased liver injury. Hepatology 39:1230–1238
Wang H, Tai AW (2016) Mechanisms of cellular membrane reorganization to support hepatitis C virus replication. Viruses 8:142
Wang C, Le SY, Ali N, Siddiqui A (1995) An RNA pseudoknot is an essential structural element of the internal ribosome entry site located within the hepatitis C virus 5' noncoding region. RNA 1:526–537
Wang Y, Tran K, Yao Z (1999) The activity of microsomal triglyceride transfer protein is essential for accumulation of triglyceride within microsomes in McA-RH7777 cells. A unified model for the assembly of very low density lipoproteins. J Biol Chem 274:27793–27800
Wree A, Kahraman A, Gerken G, Canbay A (2011) Obesity affects the liver - the link between adipocytes and hepatocytes. Digestion 83:124–133
Yaginuma R, Ikejima K, Okumura K, Kon K, Suzuki S, Takei Y, Sato N (2006) Hepatic steatosis is a predictor of poor response to interferon alpha-2b and ribavirin combination therapy in Japanese patients with chronic hepatitis C. Hepatol Res 35:19–25
Yamane D, McGivern DR, Masaki T, Lemon SM (2013) Liver injury and disease pathogenesis in chronic hepatitis C. Curr Top Microbiol Immunol 369:263–288
Yamane D, McGivern DR, Wauthier E, Yi M, Madden VJ, Welsch C, Antes I, Wen Y, Chugh PE, McGee CE et al (2014) Regulation of the hepatitis C virus RNA replicase by endogenous lipid peroxidation. Nat Med 20:927–935
Yang W, Hood BL, Chadwick SL, Liu S, Watkins SC, Luo G, Conrads TP, Wang T (2008) Fatty acid synthase is up-regulated during hepatitis C virus infection and regulates hepatitis C virus entry and production. Hepatology 48:1396–1403

Yen HH, Shih KL, Lin TT, Su WW, Soon MS, Liu CS (2012) Decreased mitochondrial deoxyribonucleic acid and increased oxidative damage in chronic hepatitis C. World J Gastroenterol 18:5084–5089

Yi M, Lemon SM (2003a) 3' nontranslated RNA signals required for replication of hepatitis C virus RNA. J Virol 77:3557–3568

Yi M, Lemon SM (2003b) Structure-function analysis of the 3' stem-loop of hepatitis C virus genomic RNA and its role in viral RNA replication. RNA 9:331–345

Yoon EJ, Hu KQ (2006) Hepatitis C virus (HCV) infection and hepatic steatosis. Int J Med Sci 3:53–56

Yu JW, Sun LJ, Liu W, Zhao YH, Kang P, Yan BZ (2013) Hepatitis C virus core protein induces hepatic metabolism disorders through down-regulation of the SIRT1-AMPK signaling pathway. Int J Infect Dis 17:e539–e545

Zein NN, Rakela J, Krawitt EL, Reddy KR, Tominaga T, Persing DH (1996) Hepatitis C virus genotypes in the United States: epidemiology, pathogenicity, and response to interferon therapy. Collaborative study group. Ann Intern Med 125:634–639

Zhu N, Ware CF, Lai MM (2001) Hepatitis C virus core protein enhances FADD-mediated apoptosis and suppresses TRADD signaling of tumor necrosis factor receptor. Virology 283:178–187

Zhu H, Dong H, Eksioglu E, Hemming A, Cao M, Crawford JM, Nelson DR, Liu C (2007) Hepatitis C virus triggers apoptosis of a newly developed hepatoma cell line through antiviral defense system. Gastroenterology 133:1649–1659

Chapter 5
Hepatocyte Lethal and Nonlethal Lipotoxic Injury

Petra Hirsova, Samar H. Ibrahim, Harmeet Malhi, and Gregory J. Gores

Abbreviations

CAD	Caspase-activated DNase
CHOP	CAAT/enhancer binding homologous protein
CXCL10	Chemokine (C-X-C motif) ligand 10
DAMP	Damage-associated molecular pattern
ER	Endoplasmic reticulum
EV	Extracellular vesicle
IRE1α	Inositol-requiring enzyme-1α
JNK	c-Jun N-terminal kinase
LPC	Lysophosphatidylcholine
MLK3	Mixed lineage kinase 3
NAFLD	Nonalcoholic fatty liver disease
NASH	Nonalcoholic steatohepatitis
ROCK1	Rho-associated, coiled-coil-containing protein kinase 1
TGFβ	Transforming growth factor β
TNFR1	Tumor necrosis factor receptor 1
TRAIL	Tumor necrosis factor-related apoptosis-inducing ligand
TRAIL-R	Tumor necrosis factor related apoptosis inducing ligand receptor

P. Hirsova • S.H. Ibrahim • H. Malhi • G.J. Gores, M.D (✉)
Division of Gastroenterology and Hepatology, Mayo Clinic,
200 First Street SW, Rochester, MN 55905, USA
e-mail: gores.gregory@mayo.edu

W.-X. Ding, X.-M. Yin (eds.), *Cellular Injury in Liver Diseases*, Cell Death in Biology and Diseases, DOI 10.1007/978-3-319-53774-0_5

X.-M. Yin and Z. Dong (Series eds.), *Cell Death in Biology and Diseases*

5.1 Introduction

Nonalcoholic fatty liver disease (NAFLD) is a major public health problem, associated with national increases in the prevalence of obesity and diabetes mellitus. NAFLD often occurs in the context of the metabolic syndrome associated with obesity, diabetes, hypertension, and elevated triglycerides (Rinella 2015). Insulin resistance is a cardinal feature of the metabolic syndrome. Insulin resistance results in enhanced adipose tissue lipolysis releasing an excess of free fatty acids into the circulation (de Almeida et al. 2002). NAFLD can be subdivided into patients with isolated fatty liver (hepatic steatosis) and those who have nonalcoholic steatohepatitis. Isolated fatty liver likely represents a hepatocyte detoxification process where the surfeit of circulating free fatty acids are extracted by the hepatocyte and detoxified by esterification into neutral triglycerides. These neutral triglycerides are then stored in lipid droplets that are inert. Indeed, preclinical studies in mice deficient for a critical acyltransferase, DGAT2, an enzyme responsible for fatty acid esterification into triglycerides, have increased liver injury when fed NASH-inducing diet (Yamaguchi et al. 2007). This liver injury occurs in the absence of hepatic lipid accumulation. A subset of patients with hepatic steatosis develop an inflammatory condition of the liver. This inflammatory condition termed nonalcoholic steatohepatitis (NASH) can be associated with progressive liver injury, fibrosis, and even cirrhosis. These patients may develop all the sequelae of cirrhosis, including portal hypertension and its various complications, and hepatocellular carcinoma (Rinella 2015). The mechanisms whereby hepatic inflammation occurs in obesity-associated NASH remain unclear, but appear to be related to lethal and nonlethal hepatocyte injury.

The lethal and nonlethal hepatocyte injury is due to toxic lipid signaling, a phenomenon termed lipotoxicity. To date, nonesterified fatty acids or free fatty acids have been predominantly implicated in lipotoxicity. Of the circulating free fatty acids, oleate and palmitate predominate (de Almeida et al. 2002). Oleate is a monounsaturated fatty acid and is nontoxic, whereas palmitate is a saturated fatty acid and is toxic (Malhi and Gores 2008). Palmitate is also a precursor to intracellular generation of lysophosphatidylcholine (LPC), which mimics palmitate toxicity (Han et al. 2008; Kakisaka et al. 2012a), and a precursor for de novo synthesis of C16:0 ceramide, which is increasingly linked to liver injury (Turpin et al. 2014; Raichur et al. 2014). Herein, we will define lethal and nonlethal hepatocyte injury in models employing palmitate and LPC, and how they likely contribute to NASH.

5.2 Lethal Hepatocyte Injury by Lipoapoptosis

Cellular apoptosis mediated by toxic lipids is termed lipoapoptosis. The role of hepatocyte lipoapoptosis in human NASH has been extensively studied. Apoptosis is increased in human NASH as assessed by the terminal deoxynucleotidyl transferase dUTP nick end labeling (TUNEL) assay (Feldstein et al. 2003a). This study demonstrated that the increase in TUNEL-positive cells correlates with the grade of inflammation and stage of fibrosis in NASH patients. These data suggest that hepatocyte apoptosis in the context of NASH either promotes the inflammation in NASH or it is a strong biomarker for lipotoxic events. Hence, the study of hepatocyte lipoapoptosis is quite germane to human NASH.

Free fatty acids induce lethal hepatocyte injury by triggering hepatocyte cell death by lipoapoptosis. The liver richly expresses death receptors, including Fas, tumor necrosis factor receptor 1 (TNFR1), and tumor necrosis factor-related apoptosis-inducing ligand (TRAIL) receptors 1 and 2 (Guicciardi and Gores 2009). Ligation of the Fas receptor with Fas ligand induces rapid and profound hepatocyte apoptosis. Indeed, animals receiving a Fas agonist succumb to acute and fatal liver injury within hours (Ogasawara et al. 1993). Mice fed an obesity-inducing diet rich in fat develop hepatic steatosis and are hypersensitive to Fas agonists as they rapidly undergo Fas-mediated liver injury at ligand concentrations inadequate to cause liver failure in lean animals (Feldstein et al. 2003b). Likewise, mice fed an obesity-inducing diet with accompanying hepatic steatosis are also sensitive to TRAIL receptor-mediated liver injury (Hirsova et al. 2013). In contrast to Fas, normal hepatocytes are resistant to TRAIL, but steatosis converts the hepatocyte from being TRAIL insensitive to TRAIL sensitive. Enhanced TRAIL sensitivity is, in part, due to an upregulation of TRAIL receptor 2 (TRAIL-R2) and also induction of other proapoptotic proteins (see below). The role of tumor necrosis factor alpha (TNFα) and TNFR1 signaling in fatty liver disease remain controversial and complex. However, in some models, TNFR1 knockout mice are resistant to the induction of hepatic steatosis and liver injury (Feldstein et al. 2004). Thus, all three death receptors may contribute to lipoapoptosis mediated by toxic lipids.

In addition to being activated by ligands, death receptors may also be activated by ligand-independent mechanisms. This has been extensively examined for TRAIL receptor signaling. The TRAIL receptor may be aggregated in the plasma membrane following exposure to lipotoxic molecules such as palmitate; likely this lipid promotes TRAIL receptor oligomerization in microdomains on the plasma membrane leading to proapoptotic signaling, which eventually culminates in caspase-dependent cell death (Cazanave et al. 2011). Studies suggest that similar events may occur in intracellular membranes, such as endoplasmic reticulum (ER), and that ER stress, which also occurs in NASH, may promote proapoptotic TRAIL receptor signaling at intracellular membrane sites, including the ER (Fig. 5.1) (Lu et al. 2014).

In support of a role for TRAIL receptor signaling in NASH, a variety of studies have been conducted in TRAIL receptor-deficient mice. When TRAIL receptor knockout mice are fed an obesity-inducing diet rich in fat, fructose, and cholesterol, which mimics human NASH, the mice are protected from liver injury (Idrissova et al. 2015). For example, the TRAIL receptor-deficient mice fed such a diet have decreased serum alanine transaminase values, markers of hepatocyte apoptosis, and hepatic fibrosis. These data suggest that TRAIL receptor signaling plays a key role in promoting NASH. The studies were, however, conducted in mice with germline deletion of the TRAIL receptor. Conditional knockout animal studies will be necessary

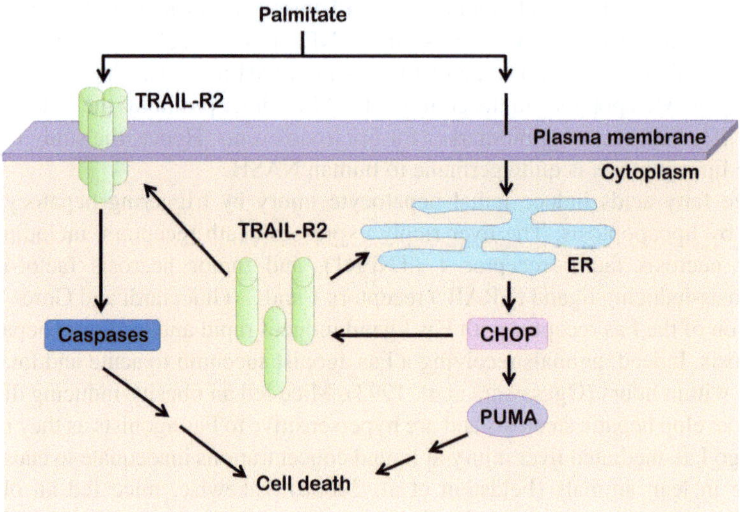

Fig. 5.1 Apoptotic signaling induced by palmitate treatment of hepatocytes. Hepatocyte treatment with palmitate results in ligand-independent clustering and activation of the death receptor TRAIL-R2 (tumor necrosis factor-like apoptosis-inducing ligand receptor 2), causing caspase-dependent cell death. Palmitate also induces endoplasmic reticulum (ER) stress, which upregulates proapoptotic BH3-only protein PUMA and TRAIL-R2 via transcription factor CHOP (CAAT/enhancer binding homologous protein)

to truly elucidate which cell types participate in this process. For example, TRAIL receptor signaling, in addition to being proapoptotic, can be proinflammatory (Hirsova et al. 2016; Gao et al. 2015), and perhaps TRAIL receptor signaling in immune cells also contributes to the NASH phenotype. Also, whether TRAIL receptor-mediated hepatic injury in mice is ligand-dependent or ligand-independent is unclear. Studies will also have to be performed in TRAIL ligand-deficient animals to fully understand TRAIL receptor signaling in obesity-associated liver injury.

In addition to modulation of death receptor proteins, hepatocyte apoptosis is also modulated by the proapoptotic and antiapoptotic members of the Bcl-2 protein family. The multidomain antiapoptotic members of this family include Bcl-2, Bcl-XL, and Mcl-1. The multidomain proapoptotic members include Bax and Bak plus BH3-only proteins. The BH3-only proteins share only the BH3 homology domain of the other members of the Bcl-2 family. The BH3-only proteins are the biosensors of proapoptotic signaling. These members neutralize the antiapoptotic proteins and activate Bax and Bak to induce cell death.

The Bcl-2 protein family has been extensively examined in vitro in hepatocytes treated with the saturated fatty acid palmitate. Palmitate induces Bim activation by upregulating its transcription via FoxO3A (Barreyro et al. 2007). Also, palmitate increases PUMA expression, a process involving the transcription factors CHOP and AP1 (Cazanave et al. 2010). CHOP and AP1 heterodimerize to enhance transcription of the potent apoptotic inducer PUMA; indeed PUMA-deficient hepatocytes are resistant to palmitate-induced lipoapoptosis (Cazanave et al. 2009). Presumably, the activation and induction of Bim and PUMA sensitize the hepatocytes to death receptor-induced apoptosis (Fig. 5.1). This concept, which has not been directly tested, could be examined by rendering PUMA-deficient cells steatotic with fatty acids and then treating them with a TRAIL agonist.

5.3 Lipoapoptosis and Liver Injury

The most nefarious consequence of liver injury is hepatic fibrosis. An exuberant wound-healing response results in a spider web-like fibrosis network causing cirrhosis. Advanced cirrhosis results in chronic liver failure. The mechanisms by which apoptosis may promote fibrosis are complex and multifactorial. Apoptotic hepatocytes can be engulfed by stellate cells, which reside in the liver sinusoids. This engulfment results in the generation of the profibrogenic transforming growth factor β (TGFβ) (Canbay et al. 2003b). The engulfment of apoptotic bodies by stellate cells also activates these cells to acquire a myofibroblast phenotype with increased expression of alpha-smooth muscle actin and secretion of collagen. Apoptotic bodies can also be engulfed by liver resident macrophages, known as Kupffer cells. Upon engulfment of apoptotic bodies, Kupffer cells also generate the profibrogenic cytokine TGFβ (Canbay et al. 2003a). More importantly, they also increase expression of Fas ligand, which then in turn may induce hepatocyte apoptosis; this results in a feed-forward cycle whereby apoptosis activates Kupffer cells, which then generate Fas ligand, causing more hepatocyte apoptosis. This feed-forward loop likely exists in NASH, contributing to hepatic damage.

NASH has also been associated with the development of hepatocellular carcinoma, a consequence of liver inflammation and fibrosis. There are two potential models to explain this process. One is that hepatocyte apoptosis results in compensatory hepatocyte proliferation. The proliferating hepatocytes are then susceptible to mutagenic events in an inflammatory milieu, resulting in malignant transformation. Inflammatory signaling is important in this paradigm of malignant transformation as demonstrated by studies linking hepatic inflammation to NASH and hepatocellular carcinoma in a mouse model where hepatocyte expression of urokinase plasminogen activator is driven by major urinary protein promoter (Nakagawa et al. 2014). In this model, endoplasmic reticulum stress, which is a feature of NASH, promoted hepatic inflammation and hepatocellular carcinoma. An alternative concept is that limited proapoptotic signaling results in the release of caspase-activated DNase (CAD) from hepatocyte mitochondria (Fig. 5.2). The endonuclease activity of CAD then results in limited DNA damage. The DNA damage results in chromosomal instability, a hallmark of malignant transformation. Indeed, two recent papers suggest that CAD-associated chromosomal instability can lead to cellular transformation and tumor formation (Ichim et al. 2015; Liu et al. 2015).

In summary, hepatocyte apoptosis by death receptor signaling can promote inflammation through Kupffer cell engulfment of apoptotic bodies, hepatic fibrosis through

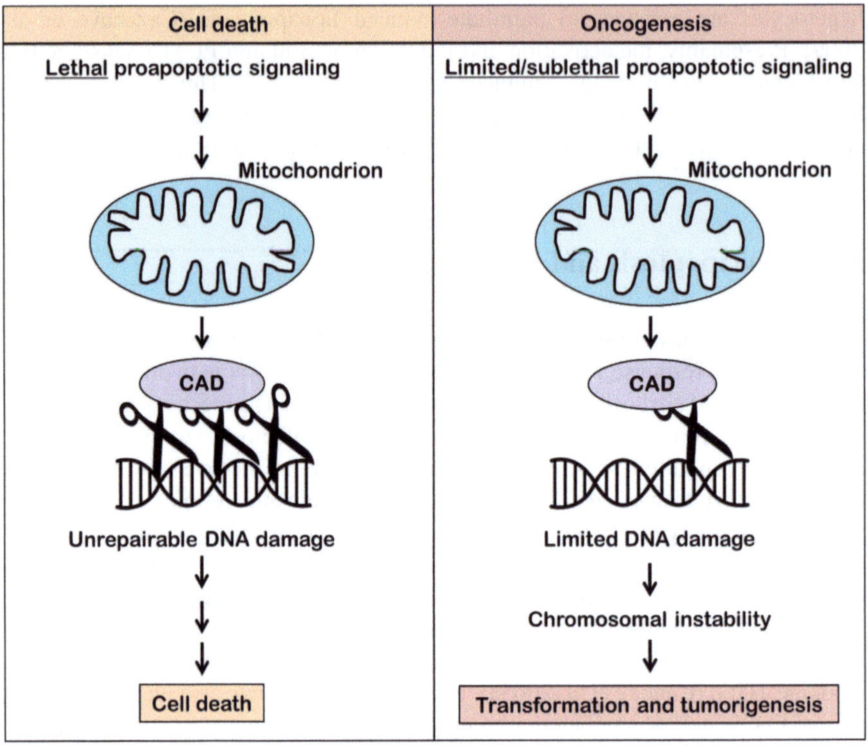

Fig. 5.2 Lethal versus sublethal proapoptotic signaling. Lethal proapoptotic signaling results in the release of caspase-activated DNase (CAD) from hepatocyte mitochondria, endonuclease activity of which results in unrepairable DNA damage and cell death. In contrast, during sublethal proapoptotic signaling the CAD endonuclease activity causes only limited DNA damage. This DNA damage results in chromosomal instability, which can lead to cellular transformation and tumor formation

engulfment of apoptotic bodies by stellate cells, and malignant transformation through limited hepatocyte DNA damage with associated chromosomal instability. Whether these processes are sufficiently robust in vivo to cause cirrhosis and liver cancer remains unclear.

5.4 Nonlethal Hepatocyte Injury

We will review two forms of nonlethal hepatocyte injury where recent advances have elucidated novel signaling mechanisms between injured hepatocytes and other cell types in the liver. One is the concept of the undead hepatocyte and the other is the release of extracellular vesicles induced by nonlethal, proapoptotic signaling in hepatocytes.

In developmental systems, proapoptotic stimuli may result in "damaged but undead" cells. These undead cells are viable and have aberrant production of proinflammatory and profibrogenic mediators (Fuchs and Steller 2011). This has been modeled in isolated systems and in hepatocytes where a key caspase, such as caspase 9, is deleted from hepatocytes that are then treated with palmitate (Kakisaka et al. 2012b). The caspase 9-deficient cells are unable to die following palmitate treatment but do display an aberrant phenotype. This aberrant phenotype is associated with release of sonic hedgehog, which is known to be profibrogenic in the liver in multiple preclinical models (Omenetti et al. 2011). Based on these observations, we speculated that the ballooned hepatocyte in NASH is actually an undead cell. The presence of ballooned hepatocytes in NASH is an advanced histopathologic marker in this disease (Kleiner et al. 2005). However, other than its appearance as a NASH biomarker, the pathogenicity of this cell remains unclear. Immunohistochemical studies suggest that there is an absence of caspase 9 in ballooned hepatocytes and they may actually mimic the undead cell observed in developmental systems (Kakisaka et al. 2012b). This concept has not been further explored except for this single report. However, a conditional liver deletion of the various executional caspases (e.g., caspase 3 or 9) could be used to explore these concepts in vivo.

One of the dilemmas is, in addition to the mechanisms above, how proapoptotic signaling promotes inflammation. It is well recognized that cell necrosis can result in the release of damage-associated molecular pattern (DAMP) proteins, which activate inflammatory cells through a multitude of processes, including activation of DAMP recognition receptors (Schaefer 2014). On the other hand, apoptotic bodies are membrane-bound and are thought not to release DAMP proteins into the interstitial space. Therefore, other than their engulfment by stellate cells and macrophages, it is unclear how apoptotic cells would be proinflammatory. It may be that apoptotic cells are simply a marker for widespread proapoptotic but nonlethal signaling. This hypothesis postulates that proapoptotic signaling is not sufficient to induce cell death in the majority of cells, but only in a minority. This minority cell population would undergo cell death by apoptosis, which by itself may not promote high level proinflammatory activity. Rather, the presence of apoptotic cells may serve as a biomarker for widespread, nonlethal proapoptotic signaling (Fig. 5.3). The nonlethal proapoptotic signaling processes could result in the release of proinflammatory extracellular vesicles.

There has been a relative explosion of information on the release of proinflammatory extracellular vesicles (EVs) by hepatocytes in the last couple of years. Therefore, we will review these data in some detail. EVs may be released from intracellular multivesicular bodies and as such are constrained in size. These small vesicles are often called exosomes because of their biogenesis and origin from the multivesicular bodies (Raposo and Stoorvogel 2013). In addition, blebbing of the plasma membrane may result in the release of small vesicles of plasma membrane origin, which by convention are termed microvesicles. Regardless of their cellular source, EVs have been demonstrated in a variety of systems to be proinflammatory. Nonlethal proapoptotic signaling clearly results in the release of EVs. These vesicles

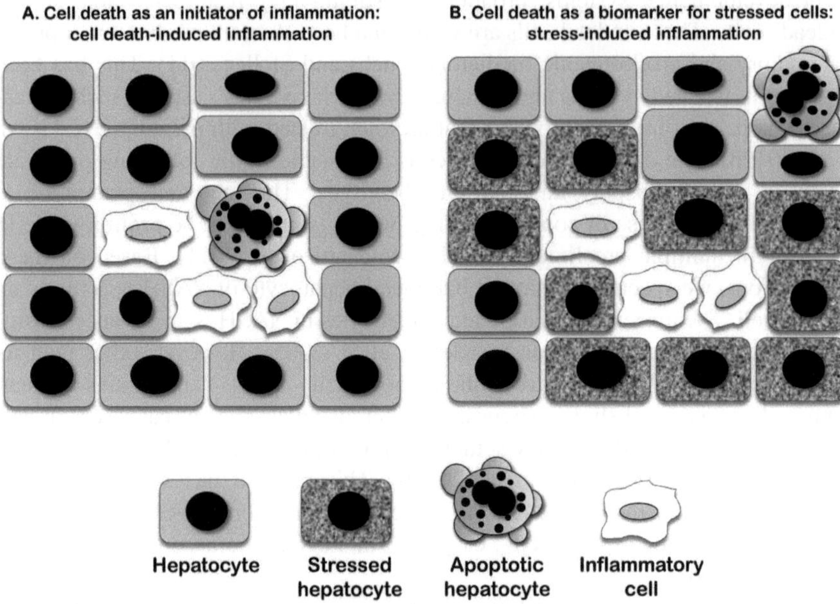

Fig. 5.3 Two concepts linking hepatocyte cell death and inflammation in NASH. (**a**). Hepatocyte cell death can directly promote inflammation via apoptotic bodies engulfed by macrophages. (**b**) Apoptotic cells can also be a marker for widespread proapoptotic but nonlethal signaling. The proapoptotic signaling is not sufficient to induce cell death in the majority of cells, but only in a minority. This minority cell population undergoing apoptosis may not induce significant proinflammatory response. Rather, the presence of apoptotic hepatocytes may serve as a biomarker for widespread, nonlethal proapoptotic signaling occurring in neighboring stressed cells. The nonlethal proapoptotic signaling could promote inflammation via release of proinflammatory extracellular vesicles from these stressed hepatocytes

are chemotactic for macrophages, and may also activate macrophages. The role EVs in recruiting and activating macrophages is highly relevant to NASH. Circulating monocytes are recruited into the liver in NASH (Miura et al. 2012; Osborn and Olefsky 2012). Once within the hepatic parenchyma, these monocyte-derived macrophages are activated to release inflammatory cytokines. Indeed, macrophages are considered to be a pivotal inflammatory cell type in NASH. There are now a couple of papers examining macrophage biology in the context of EV biology.

EV release from hepatocytes is significantly enhanced when the cells are treated with palmitate or LPC (Hirsova et al. 2016; Ibrahim et al. 2015; Kakazu et al. 2016; Povero et al. 2013). The size of these EVs is approximately the same as EVs released by unstressed cells. The release of these vesicles is, in part, stimulated by stress kinases such as mixed lineage kinase 3, which is a sensor of toxic lipids (Ibrahim et al. 2015). Mixed lineage kinase 3 activates c-Jun N-terminal kinase (JNK), which is a known proapoptotic kinase. A recent publication demonstrates that the EVs released from hepatocytes treated with LPC are enriched in the chemokine (C-X-C

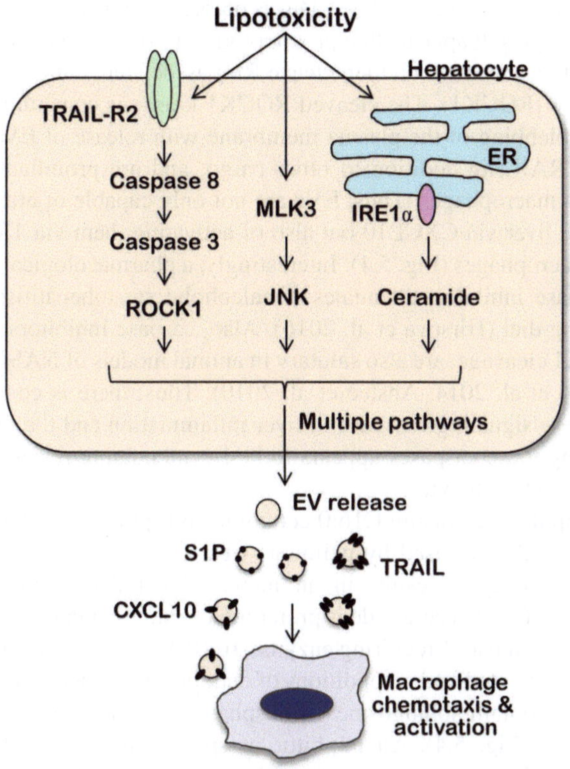

Fig. 5.4 Toxic lipids promote release of proinflammatory EVs by hepatocytes. Toxic lipids, such as palmitate and its metabolite lysophosphatidylcholine, enhance EV release by mechanisms dependent on TRAIL-R2 (tumor necrosis factor-like apoptosis-inducing ligand receptor 2) signaling cascade, stress kinase mixed lineage kinase 3 (MLK3), and the ER stress sensor IRE1α (inositol-requiring protein 1 alpha). The hepatocyte-derived EVs contain chemokines, such as chemokine (C-X-C motif) ligand 10 (CXCL10) and sphingosine 1-phosphate (S1P), which mediate monocyte/macrophage chemotaxis, and TNF-related apoptosis-inducing ligand (TRAIL), which contributes to macrophage activation

motif) ligand 10 (CXCL10) (Ibrahim et al. 2015). Indeed, deficiency of mixed lineage kinase 3 or its pharmacological inhibitors decreased the release of CXCL10-bearing vesicles (Fig. 5.4). CXCL10 is a potent chemokine for both macrophages and neutrophils (Petrovic-Djergovic et al. 2015; Hartl et al. 2008). The mechanism by which lipotoxic lipids induce CXCL10 expression and packaging into EVs is unclear, but a subject of intense investigation. Of considerable importance is the fact that in preclinical models, the genetic deficiency of CXCL10 attenuates nonalcoholic steatohepatitis (Zhang et al. 2014).

Nonlethal but proapoptotic TRAIL receptor signaling may also promote a release of EVs from the plasma membrane of hepatocytes (Hirsova et al. 2016). When primary mouse hepatocytes are treated with palmitate or LPC, there is a striking increase in the release of EVs. This is markedly reduced in TRAIL receptor-deficient hepatocytes. It appears that proapoptotic TRAIL receptor signaling results in activation of caspases which then cleave Rho-associated, coiled-coil-containing protein kinase 1 (ROCK1). The cleaved ROCK1 kinase is constitutively activated and promotes blebbing of the plasma membrane with release of EVs. These vesicles contain TRAIL, in addition to other cargo, and are proinflammatory when incubated with macrophages. Thus, EVs are not only capable of attracting macrophages into the liver via CXCL10 but also of activating them via TRAIL receptor signaling in macrophages (Fig. 5.4). Interestingly, a pharmacological agent fasudil, a ROCK1 kinase inhibitor, attenuates nonalcoholic steatohepatitisin mice fed a NASH-inducing diet (Hirsova et al. 2016). Also, caspase inhibitors, which would prevent ROCK1 cleavage, are also salutary in animal models of NASH (Witek et al. 2009; Barreyro et al. 2014; Anstee et al. 2010). Thus, there is considerable data implicating these signaling cascades in liver inflammation and the therapeutic target of inhibiting these caspases appears to be by attenuating release of EVs from lipotoxic stressed hepatocytes.

Lipotoxic lipids, specifically C16:0 ceramide and sphingosine 1-phosphate, are also enriched in EVs released by palmitate-treated cells. For example, palmitate treatment of hepatocytes results in an increased cellular ceramide generation (Kakazu et al. 2016). This ceramide is preferentially shuttled into EVs via a mechanism dependent on inositol-requiring enzyme-1α (IRE1α), one of the three unfolded protein sensors activated under conditions of endoplasmic reticulum stress. In addition, these EVs contain sphingosine 1-phosphate, a potent chemotactic molecule for macrophages (Fig. 5.4). An inhibitor of sphingosine 1-phosphate signaling attenuates macrophage chemotaxis toward lipotoxic EVs in vitro and also attenuates liver injury in mice fed a NASH-inducing diet (unpublished observation). Palmitate-induced endoplasmic reticulum stress is a well-recognized component of lipotoxicity; these studies link palmitate-induced IRE1α activation to proinflammatory macrophage recruitment to the liver. In this regard, in addition to carrying cytokines and chemokines, the lipid composition of the EVs may also be quite proinflammatory.

5.5 Conclusion

It would appear that the lethal lipotoxic injury via apoptosis may contribute to NASH. But more recent concepts suggest that nonlethal proapoptotic signaling by the release of extracellular vesicles actually may be more important in generating inflammation. The associated inflammation is largely macrophage-related; the recruited and activated macrophages then, in turn, damage hepatocytes resulting in a feed-forward cycle. Interruption of these pathways, for example by inhibiting EV generation or release, may be therapeutic in NASH. These data suggest a myriad of pharmacologic approaches to treating a disease that remains difficult to treat and a major public health problem.

Acknowledgments This work is supported in part by NIH grants DK41876 (GJG), DK97178 and DK107402 (HM), and KL2TR000136-09 (SHI). Support was also provided to P. Hirsova by the American Liver Foundation and Edward C. Kendall Research Fellowship Award (Mayo Clinic).

References

Anstee QM, Concas D, Kudo H, Levene A, Pollard J, Charlton P et al (2010) Impact of pan-caspase inhibition in animal models of established steatosis and non-alcoholic steatohepatitis. J Hepatol 53(3):542–550. doi:10.1016/j.jhep.2010.03.016. [Research Support, Non-U.S. Gov't]

Barreyro FJ, Holod S, Finocchietto PV, Camino AM, Aquino JB, Avagnina A et al (2014) The pan-caspase inhibitor Emricasan (IDN-6556) decreases liver injury and fibrosis in a murine model of non-alcoholic steatohepatitis. Liver Int. doi:10.1111/liv.12570

Barreyro FJ, Kobayashi S, Bronk SF, Werneburg NW, Malhi H, Gores GJ (2007) Transcriptional regulation of Bim by FoxO3A mediates hepatocyte lipoapoptosis. J Biol Chem 282(37):27141–27154. doi:10.1074/jbc.M704391200. M704391200 [pii]

Canbay A, Feldstein AE, Higuchi H, Werneburg N, Grambihler A, Bronk SF et al (2003a) Kupffer cell engulfment of apoptotic bodies stimulates death ligand and cytokine expression. Hepatology 38((5)):1188–1198. doi:10.1053/jhep.2003.50472. [Research Support, Non-U.S. Gov't Research Support, U.S. Gov't, P.H.S]

Canbay A, Taimr P, Torok N, Higuchi H, Friedman S, Gores GJ (2003b) Apoptotic body engulfment by a human stellate cell line is profibrogenic. Lab Invest 83((5)):655–663. [Research Support, Non-U.S. Gov't Research Support, U.S. Gov't, P.H.S.]

Cazanave SC, Elmi NA, Akazawa Y, Bronk SF, Mott JL, Gores GJ (2010) CHOP and AP-1 cooperatively mediate PUMA expression during lipoapoptosis. *Am Journal Physiol Gastrointest liver physiol* 299((1)):G236–G243. doi:10.1152/ajpgi.00091.2010. [Research Support, N.I.H., Extramural Research Support, Non-U.S. Gov't]

Cazanave SC, Mott JL, Bronk SF, Werneburg NW, Fingas CD, Meng XW et al (2011) Death receptor 5 signaling promotes hepatocyte lipoapoptosis. J Biol Chem 286(45):39336–39348. doi:10.1074/jbc.M111.280420. M111.280420 [pii]

Cazanave SC, Mott JL, Elmi NA, Bronk SF, Werneburg NW, Akazawa Y et al (2009) JNK1-dependent PUMA expression contributes to hepatocyte lipoapoptosis. J Biol Chem 284(39):26591–26602. doi:10.1074/jbc.M109.022491. M109.022491 [pii]

de Almeida IT, Cortez-Pinto H, Fidalgo G, Rodrigues D, Camilo ME (2002) Plasma total and free fatty acids composition in human non-alcoholic steatohepatitis. Clin Nutr 21(3):219–223. [Research Support, Non-U.S. Gov't]

Feldstein AE, Canbay A, Angulo P, Taniai M, Burgart LJ, Lindor KD et al (2003a) Hepatocyte apoptosis and fas expression are prominent features of human nonalcoholic steatohepatitis. Gastroenterology 125(2):437–443. doi:S0016508503009077 [pii]

Feldstein AE, Canbay A, Guicciardi ME, Higuchi H, Bronk SF, Gores GJ (2003b) Diet associated hepatic steatosis sensitizes to Fas mediated liver injury in mice. J Hepatol 39(6):978–983. doi:S0168827803004604 [pii]

Feldstein AE, Werneburg NW, Canbay A, Guicciardi ME, Bronk SF, Rydzewski R et al (2004) Free fatty acids promote hepatic lipotoxicity by stimulating TNF-alpha expression via a lysosomal pathway. Hepatology 40(1):185–194. doi:10.1002/hep.20283. [Research Support, Non U.S. Gov't Research Support, U.S. Gov't, P.H.S.].

Fuchs Y, Steller H (2011) Programmed cell death in animal development and disease. Cell 147(4):742–758. doi:10.1016/j.cell.2011.10.033. [Research Support, Non-U.S. Gov't Review]

Gao J, Wang D, Liu D, Liu M, Ge Y, Jiang M et al (2015) Tumor necrosis factor-related apoptosis-inducing ligand induces the expression of proinflammatory cytokines in macrophages and re-educates tumor-associated macrophages to an antitumor phenotype. Mol Biol Cell 26(18): 3178–3189. doi:10.1091/mbc.E15-04-0209

Guicciardi ME, Gores GJ (2009) Life and death by death receptors. FASEB J 23(6):1625–1637. doi:10.1096/fj.08-111005. [Research Support, N.I.H., Extramural Research Support, Non U.S. Gov't Review]

Han MS, Park SY, Shinzawa K, Kim S, Chung KW, Lee JH et al (2008) Lysophosphatidylcholine as a death effector in the lipoapoptosis of hepatocytes. J Lipid Res 49(1):84–97. doi:10.1194/jlr.M700184-JLR200. [Research Support, Non-U.S. Gov't]

Hartl D, Krauss-Etschmann S, Koller B, Hordijk PL, Kuijpers TW, Hoffmann F et al (2008) Infiltrated neutrophils acquire novel chemokine receptor expression and chemokine responsiveness in chronic inflammatory lung diseases. J Immunol 181(11):8053–8067

Hirsova P, Ibrahim SH, Bronk SF, Yagita H, Gores GJ (2013) Vismodegib suppresses TRAIL-mediated liver injury in a mouse model of nonalcoholic steatohepatitis. PLoS One 8(7):e70599. doi:10.1371/journal.pone.0070599. [Research Support, N.I.H., Extramural]

Hirsova P, Ibrahim SH, Krishnan A, Verma VK, Bronk SF, Werneburg NW et al (2016) Lipid-induced signaling causes release of inflammatory extracellular vesicles from hepatocytes. Gastroenterology 150:956–967. doi:10.1053/j.gastro.2015.12.037

Ibrahim SH, Hirsova P, Tomita K, Bronk SF, Werneburg NW, Harrison SA et al (2015) Mixed lineage kinase 3 mediates release of C-X-C motif ligand 10-bearing chemotactic extracellular vesicles from lipotoxic hepatocytes. Hepatology 63:731–744. doi:10.1002/hep.28252

Ichim G, Lopez J, Ahmed SU, Muthalagu N, Giampazolias E, Delgado ME et al (2015) Limited mitochondrial permeabilization causes DNA damage and genomic instability in the absence of cell death. Mol Cell 57(5):860–872. doi:10.1016/j.molcel.2015.01.018. [Research Support, Non-U.S. Gov't]

Idrissova L, Malhi H, Werneburg NW, LeBrasseur NK, Bronk SF, Fingas C et al (2015) TRAIL receptor deletion in mice suppresses the inflammation of nutrient excess. J Hepatol 62(5): 1156–1163. doi:10.1016/j.jhep.2014.11.033. [Research Support, N.I.H., Extramural Research Support, Non-U.S. Gov't]

Kakazu E, Mauer AS, Yin M, Malhi H (2016) Hepatocytes release ceramide-enriched pro-inflammatory extracellular vesicles in an IRE1alpha-dependent manner. J Lipid Res 57(2):233–245. doi:10.1194/jlr.M063412

Kakisaka K, Cazanave SC, Fingas CD, Guicciardi ME, Bronk SF, Werneburg NW et al (2012a) Mechanisms of lysophosphatidylcholine-induced hepatocyte lipoapoptosis. Am J Physiol Gastrointest Liver physiol 302(1):G77–G84. doi:10.1152/ajpgi.00301.2011. ajpgi.00301.2011 [pii]

Kakisaka K, Cazanave SC, Werneburg NW, Razumilava N, Mertens JC, Bronk SF et al (2012b) A hedgehog survival pathway in 'undead' lipotoxic hepatocytes. J Hepatol 57(4):844–851. doi:10.1016/j.jhep.2012.05.011. [Research Support, N.I.H., Extramural Research Support, Non-U.S. Gov't]

Kleiner DE, Brunt EM, Van Natta M, Behling C, Contos MJ, Cummings OW et al (2005) Design and validation of a histological scoring system for nonalcoholic fatty liver disease. Hepatology 41(6):1313–1321. doi:10.1002/hep.20701. [Multicenter Study Research Support, N.I.H., Extramural Research Support, U.S. Gov't, P.H.S. Validation Studies]

Liu X, He Y, Li F, Huang Q, Kato TA, Hall RP et al (2015) Caspase-3 promotes genetic instability and carcinogenesis. Mol Cell 58(2):284–296. doi:10.1016/j.molcel.2015.03.003. Research Support, N.I.H., Extramural Research Support, Non-U.S. Gov't Research Support, U.S. Gov't, Non-P.H.S.

Lu M, Lawrence DA, Marsters S, Acosta-Alvear D, Kimmig P, Mendez AS et al (2014) Cell death. Opposing unfolded-protein-response signals converge on death receptor 5 to control apoptosis. Science 345(6192):98–101. doi:10.1126/science.1254312. [Research Support, Non-U.S. Gov't]

Malhi H, Gores GJ (2008) Molecular mechanisms of lipotoxicity in nonalcoholic fatty liver disease. Seminars Liver Disease 28(4):360–369. doi:10.1055/s-0028-1091980. [Research Support, N.I.H., Extramural Research Support, Non-U.S. Gov't Review]

Miura K, Yang L, van Rooijen N, Ohnishi H, Seki E (2012) Hepatic recruitment of macrophages promotes nonalcoholic steatohepatitis through CCR2. American journal of physiology

Gastrointestinal and liver physiology 302(11):G1310–G1321. doi:10.1152/ajpgi.00365.2011. [Research Support, N.I.H., Extramural Research Support, Non-U.S. Gov't].

Nakagawa H, Umemura A, Taniguchi K, Font-Burgada J, Dhar D, Ogata H et al (2014) ER stress cooperates with hypernutrition to trigger TNF-dependent spontaneous HCC development. Cancer Cell 26(3):331–343. doi:10.1016/j.ccr.2014.07.001. [Research Support, N.I.H., Extramural Research Support, Non-U.S. Gov't Research Support, U.S. Gov't, Non-P.H.S.]

Ogasawara J, Watanabe-Fukunaga R, Adachi M, Matsuzawa A, Kasugai T, Kitamura Y et al (1993) Lethal effect of the anti-Fas antibody in mice. Nature 364(6440):806–809. doi:10.1038/364806a0. [Research Support, Non-U.S. Gov't]

Omenetti A, Choi S, Michelotti G, Diehl AM (2011) Hedgehog signaling in the liver. J Hepatol 54(2):366–373. doi:10.1016/j.jhep.2010.10.003. [Research Support, N.I.H., Extramural Review]

Osborn O, Olefsky JM (2012) The cellular and signaling networks linking the immune system and metabolism in disease. Nat Med 18(3):363–374. doi:10.1038/nm.2627. nm.2627 [pii]

Petrovic-Djergovic D, Popovic M, Chittiprol S, Cortado H, Ransom RF, Partida-Sanchez S (2015) CXCL10 induces the recruitment of monocyte-derived macrophages into kidney, which aggravate puromycin aminonucleoside nephrosis. Clin Exp Immunol 180(2):305–315. doi:10.1111/cei.12579. [Research Support, N.I.H., Extramural Research Support, Non-U.S. Gov't]

Povero D, Eguchi A, Niesman IR, Andronikou N, de Mollerat du Jeu X, Mulya A et al (2013) Lipid-induced toxicity stimulates hepatocytes to release angiogenic microparticles that require Vanin-1 for uptake by endothelial cells. Sci Signal 6(296):ra88. doi:10.1126/scisignal.2004512. 6/296/ra88 [pii]

Raichur S, Wang ST, Chan PW, Li Y, Ching J, Chaurasia B et al (2014) CerS2 haploinsufficiency inhibits beta-oxidation and confers susceptibility to diet-induced steatohepatitis and insulin resistance. Cell Metab 20(4):687–695. doi:10.1016/j.cmet.2014.09.015. [Research Support, Non-U.S. Gov't]

Raposo G, Stoorvogel W (2013) Extracellular vesicles: exosomes, microvesicles, and friends. J Cell Biol 200(4):373–383. doi:10.1083/jcb.201211138. [Research Support, Non-U.S. Gov't Review]

Rinella ME (2015) Nonalcoholic fatty liver disease: a systematic review. [Review]. JAMA 313(22):2263–2273. doi:10.1001/jama.2015.5370

Schaefer L (2014) Complexity of danger: the diverse nature of damage-associated molecular patterns. J Biol Chem 289(51):35237–35245. doi:10.1074/jbc.R114.619304. [Research Support, Non-U.S. Gov't Review]

Turpin SM, Nicholls HT, Willmes DM, Mourier A, Brodesser S, Wunderlich CM et al (2014) Obesity-induced CerS6-dependent C16:0 ceramide production promotes weight gain and glucose intolerance. Cell Metab 20(4):678–686. doi:10.1016/j.cmet.2014.08.002. [Research Support, Non-U.S. Gov't]

Witek RP, Stone WC, Karaca FG, Syn WK, Pereira TA, Agboola KM et al (2009) Pan-caspase inhibitor VX-166 reduces fibrosis in an animal model of nonalcoholic steatohepatitis. Hepatology 50(5):1421–1430. doi:10.1002/hep.23167. [Research Support, Non-U.S. Gov't]

Yamaguchi K, Yang L, McCall S, Huang J, Yu XX, Pandey SK et al (2007) Inhibiting triglyceride synthesis improves hepatic steatosis but exacerbates liver damage and fibrosis in obese mice with nonalcoholic steatohepatitis. Hepatology 45(6):1366–1374. doi:10.1002/hep.21655. [Research Support, N.I.H., Extramural]

Zhang X, Shen J, Man K, Chu ES, Yau TO, Sung JC et al (2014) CXCL10 plays a key role as an inflammatory mediator and a non-invasive biomarker of non-alcoholic steatohepatitis. J Hepatol 61(6):1365–1375. doi:10.1016/j.jhep.2014.07.006. [Research Support, Non-U.S. Gov't]

Chapter 6
Cell Death in Alcohol-Induced Liver Injury

Xiaojuan Chao, Shaogui Wang, and Wen-Xing Ding

Abbreviations

ADH	Alcohol dehydrogenase
AIF	Apoptosis inducing factor
ALD	Alcoholic liver disease
ALDH1	Aldehyde dehydrogenase 1
BH	BCL-2 homology
CB2	Cannabinoid receptor 2
cIAP1/2	Cellular inhibitor of apoptosis proteins 1 and 2
Cyp2E1	Cytochrome P450 family 2, subfamily E, polypeptide 1
CYLD	Cylindromatosis
DAMPs	Damage-associated molecular patterns
DISC	Death-inducing signaling complex
DR-4	TRAIL-receptor 1
DR-5	TRAIL receptor 2/death receptor 5
ER	Endoplasmic reticulum
FAEE	Fatty acid ethyl ester
FASLG	Fas ligand
FAP-1	Fas-associated phosphatase 1
$FLIP_L$	FLICE (FADD-like IL-1β-converting enzyme)-like inhibitory protein large
FADD	Fas-associated protein with a death domain
GPX4	Glutathione peroxidase 4
GSH	Glutathione

X. Chao • S. Wang • W.-X. Ding, Ph.D. (✉)
Department of Pharmacology, Toxicology and Therapeutics, University of Kansas Medical Center, MS 1018, 3901 Rainbow Blvd, Kansas City, KS 66160, USA
e-mail: wxding@kumc.edu

W.-X. Ding, X.-M. Yin (eds.), *Cellular Injury in Liver Diseases*, Cell Death in Biology and Diseases, DOI 10.1007/978-3-319-53774-0_6

X.-M. Yin and Z. Dong (Series eds.), *Cell Death in Biology and Diseases*

Gsdmd	Gasdermin D
HCC	Hepatocellular carcinoma
HMGB1	High mobility group box 1
HSC	Hepatic stellate cells
IAP	Inhibitor-of-apoptosis
IAPs	Inhibitor of apoptosis proteins
IKK	IκB kinase
KO	Knockout
LPS	Lipopolysaccharide/endotoxin
MLKL	Mixed lineage kinase domain-like protein
MPT	Mitochondrial permeability transition
NAD	Nicotinamide adenine dinucleotide
NASH	Non-alcoholic steatohepatitis
NLRs	NOD-like receptors
NOD	Nucleotide-binding oligomerization domain
PGC-1α	PPARγ co-activator-1α
Pink1	Tensin homolog-induced putative kinase 1
PPARγ	Peroxisome proliferator-activated receptor γ
PSMA2	Proteasome subunit alpha type-2
PSMC1	Proteasome 26S subunit ATPase 1
RIP	Receptor-interacting protein kinase
RIPK1	Receptor-interacting protein kinase 1
ROS	Reactive oxygen species
Smac/DIABLO	Second mitochondria-derived activator of caspase
tBid	Truncated form of Bid
TFEB	Transcription factor EB
TNFR1	TNF-α receptor 1
TRADD	TNFR-associated death domain
TRAF2	TNFR-associated factor 2
TAK1	Transforming growth factor β-activated kinase 1
TAB2	TAK1-binding protein 2
TfR1	Transferrin receptor 1
TNFR1	TNF-α receptor 1
TNF-α	Tumor necrosis factor-α
VDAC	Voltage-dependent anion channel

6.1 Introduction

Alcoholic liver disease (ALD) is a major health problem and a significant source of liver injury worldwide. The pathogenesis of ALD is characterized as steatosis, fibrosis, acute alcoholic hepatitis, cirrhosis, which eventually lead to hepatocellular carcinoma (HCC) (Gao and Bataller 2011; Nagy et al. 2016; Williams et al. 2014). In the past decades, careful research works from many labs demonstrate that cell death including apoptosis and necrosis contributes to the pathogenesis of ALD (Nagy et al. 2016; Wang et al. 2016b). Liver injury and disease can be triggered by a diverse set of metabolic, toxic, and inflammatory insults, which results in the activation of apoptosis and necrosis/necroptosis in liver cells. Liver contains multiple different cell types including parenchymal and non-parenchymal cells. Various liver diseases including ALD can be due to cell death of both cell types.

Cells can also adapt to metabolic changes and stresses induced by alcohol, which limit the extent of intracellular lipid accumulation, mitochondrial damage, and cell death. One such adaptive mechanism is autophagy, which is a conserved lysosomal degradation pathway. We previously showed that autophagy is induced by acute alcohol although it seems that autophagy is impaired during chronic alcohol exposure (Ding et al. 2010). In this book chapter, we summarize the progress on the role and mechanisms of alcohol-induced cell death and autophagy in hepatocytes. The death and autophagy of non-parenchymal cells is beyond the topic of this chapter.

6.2 Alcohol Metabolism and Oxidative Stress

Accumulated studies revealed that the metabolism of alcohol plays a critical role in alcohol-induced activation of mitochondrial apoptotic pathway and subsequent apoptosis. Ethanol is metabolized through two major oxidative and two minor non-oxidative pathways.

First, alcohol is metabolized into highly reactive acetaldehyde by alcohol dehydrogenase (ADH). Then acetaldehyde is further metabolized by cytosolic aldehyde dehydrogenase 1 (ALDH1) and mitochondrial ALDH2 into more harmless acetate (Crabb et al. 2004). The metabolism of ethanol through this process increases the conversion of nicotinamide adenine dinucleotide (NAD^+) into its reduced form, NADH. The increased ratio of NADH/NAD^+ alters cellular redox status and decreases NAD^+-dependent enzyme activities. Moreover, the increased ratio of NADH/NAD^+ also promotes excess flow of electrons in the mitochondrial respiratory chain, resulting in accumulation and leakage of electrons at the mitochondria respiratory chain complexes I and III to produce reactive oxygen species (ROS) (Bailey and Cunningham 2002). Alcohol exposure also reduces mitochondrial DNA and ribosomes, which lead to reduced mitochondrial protein and ATP synthesis (Coleman and Cunningham 1991; Mansouri et al. 1999). Moreover, chronic alcohol exposure decreases hepatic mitochondrial respiration (state III) and increases sensitivity to Ca^{2+}-mediated mitochondrial permeability transition induction resulting in mitochondrial-mediated apoptosis (King et al. 2014). In addition, chronic alcohol exposure decreases mitochondrial maximal oxygen consumption rate and in turn increases the susceptibility of hepatocytes to alcohol-induced hypoxia and liver injury (Zelickson et al. 2011). Second, ethanol is metabolized by cytochrome P450 family 2, subfamily E, polypeptide 1 (Cyp2E1), and catalase. Acute or chronic ethanol exposure induces Cyp2E1, leading to increasing ROS generation (Lu and Cederbaum 2008). Alcohol exposure also reduces antioxidant enzymes including catalase, superoxide dismutase, and glutathione peroxidase in

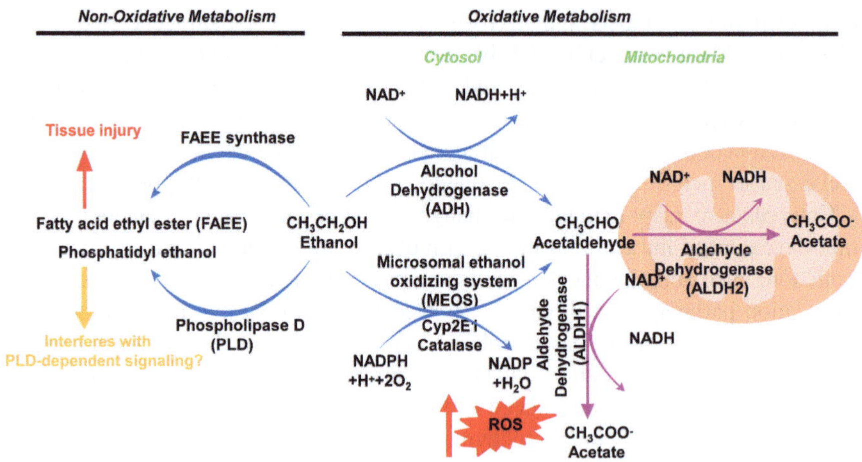

Fig. 6.1 Oxidative and non-oxidative metabolisms of alcohol. Oxidative metabolism of alcohol is mainly mediated via ADH, ALDH, Cyp2E1, and catalase. Non-oxidative metabolism of alcohol involves FAEE synthase (*for detail see the text*)

liver (Bourogaa et al. 2013). Alcohol-induced ROS production can damage mitochondrial proteins and mitochondria DNA, induce mitochondrial depolarization and onset of mitochondrial permeability transition (MPT) (Hoek et al. 2002). Onset of MPT then leads to more mitochondrial depolarization and subsequent more ROS production, which forms a vicious feed forward loop (Zorov et al. 2000). Third, in addition to the oxidative metabolism, a small portion of ethanol can also be metabolized via two non-oxidative pathways. Ethanol can directly interact with fatty acid and generates fatty acid ethyl ester (FAEE) through FAEE synthase (Zelner et al. 2013), which exacerbates alcohol-induced injury in various tissues including liver (Wu et al. 2006), pancreas (Wu et al. 2008; Werner et al. 2002), and heart (Beckemeier and Bora 1998; Wu et al. 2006, 2008;). FAEE induces mitochondria damage by binding to mitochondria membrane and uncoupling oxidative phosphorylation (Lange and Sobel 1983). In addition, ethanol also reacts with phospholipase D (PLD) to generate phosphatidyl ethanol. Due to its poor metabolism, phosphatidyl ethanol can accumulate to detectable levels following chronic consumption of large amounts of alcohol. Currently, the effects of phosphatidyl ethanol on cellular functions are not clear yet (Zakhari 2006). The oxidative and non-oxidative metabolisms of ethanol are summarized in Fig. 6.1.

6.3 Apoptosis, Necrosis, and Necroptosis in ALD

Cells can die in different modes including apoptosis, necrosis, and necroptosis. Different cell death modes are mainly based on their distinctive morphological and biochemical changes. Apoptosis or programmed cell death is characterized by nuclear fragmentation, chromatin condensation, and cellular shrinkage. Biochemically, apoptosis is generally dependent on activation of caspases although caspase-independent apoptosis may also occur. Apoptotic cells can break apart into membrane-enclosed apoptotic bodies, which contain intact organelles that are later phagocytosed by immune cells, such as macrophages/Kupffer cells in the liver without inducing an inflammatory response.

The characters of necrosis include cell swelling, membrane rupture, and release of cell contents that leads to a subsequent inflammatory response (Malhi et al. 2006, 2010). However, recent evidence suggests that necrosis can also be highly regulated, which involves the RIP1-RIP3-MLKL (mixed lineage kinase domain-like protein)-mediated necrotic cascade, a process that is termed necroptosis or programmed necrosis (Degterev et al. 2005; Zhang et al. 2009; Cho et al. 2009; He et al. 2009). Although necroptosis is similar in nature to necrosis, it is a caspase-independent programmed form of cell death. Necroptosis is also initiated by death receptors, similar to the extrinsic apoptotic pathway. Both apoptosis and necroptosis have been suggested to play a role in the pathogenesis of ALD, as discussed below in detail.

6.4 Mechanisms of Alcohol-Induced Apoptosis in Hepatocytes

In general, apoptosis can either be triggered through intrinsic (mitochondria) or extrinsic (death receptor) pathway. Both intrinsic and extrinsic apoptotic pathways appear to play a role in alcohol-induced cell death in hepatocytes.

6.5 Activation of Mitochondrial (Intrinsic) Apoptotic Pathway by Alcohol

Various apoptotic stimuli, including DNA damage, oxidative stress, or deprivation of hormone or growth factor, can activate the mitochondrial apoptotic pathway or the intrinsic pathway (Yin and Ding 2003). Following the activation of the intrinsic apoptotic pathway, a number of apoptotic factors, such as cytochrome *c*, second mitochondria-derived activator of caspase (Smac/DIABLO), HtrA2/Omi, apoptosis inducing factor (AIF), and endonuclease G, are released from mitochondria. Cytosolic cytochrome c then binds to Apaf-1 and caspase-9 to form a heptameric complex known as "Apoptosome," which subsequently activates caspase-9 in the presence of dATP (Li et al. 1997). Activated caspase-9 further cleaves and activates downstream executioner caspases, such as caspase-3, -6, and -7. Cells with the deficiency of Apaf1, caspase-9, or caspase-3 are resistant to apoptosis following intrinsic apoptotic stimuli, highlighting the importance of this pathway. Concomitantly,

Smac/DIABLO is also released into the cytosol, where it binds to the inhibitor of apoptosis proteins (IAPs), such as XIAP, to abolish their inhibitory effects on caspases. More than 400 caspase substrates have been identified that are cleaved by activated executioner caspases resulting in the characteristic morphological features of apoptosis such as DNA fragmentation/condensation, externalization of phosphatidylserine, and formation of apoptotic bodies.

The Bcl-2 family proteins are a group of evolutionarily conserved proteins that regulate apoptosis mainly on mitochondria. Bcl-2 family proteins consist of antiapoptotic and proapoptotic members, and all members possess conserved α-helices with sequence conservation clustered in BCL-2 homology (BH) domains. Antiapoptotic proteins such as Bcl-2 and Mcl-2 have all segments of BH1–BH4 domain, whereas proapoptotic proteins lack the first α-helical BH4 domain and are further subdivided into "multidomain" and "BH3-only" molecules. The multidomain proapoptotic Bcl-2 family proteins Bax and Bak form a requisite gateway on regulating the intrinsic mitochondrial apoptotic pathway. Bax is a cytosolic protein whereas Bak resides at the mitochondria membrane. Upon an intrinsic death signal, Bax translocates from cytosol to mitochondria and inserts into the outer mitochondrial membrane by forming homo-oligomers. Bak also undergoes a conformation change and oligomerization resulting in the permeabilization of the outer mitochondrial membrane followed by releasing of apoptotic proteins from the inter-membrane space (Wei et al. 2001). The propapoptotic BH3-only members such as Noxa and Puma only have amphipathic α-helical BH3 region. The BH3-only proteins are regulated at the transcriptional regulation or posttranslational modification and reside upstream in the pathway in response to death signals. The BH3-only proteins still require downstream Bax and Bak to trigger apoptosis, whereas antiapoptotic proteins such as Bcl-xL and Mcl-1 inhibit apoptosis by binding and sequestering "BH3-only" proteins to prevent the activation of Bax and Bak (Scorrano and Korsmeyer 2003).

Chronic ethanol feeding has been shown to increase the expression of Bcl-xL and Bax in the mouse livers. In cultured AML12 cells (immortalized mouse hepatocytes), it was reported that Bee venom (*Apis mellifera*) increased the expression of Bcl-2 but decreased the expression of Bax and protected against ethanol-induced cell death (Kim et al. 2010). In primary cultured rat hepatocytes, ethanol treatment induced Bax translocation to mitochondria and interacted with mitochondrial voltage-dependent anion channel (VDAC), one component of the MPT, to trigger cytochrome c release and apoptosis (Adachi et al. 2004). In addition to the liver, it was also reported that alcohol induced apoptosis in mouse adipocytes resulting in adipose tissue inflammation, which was inhibited in Bid-deficient mice (Sebastian et al. 2011).

Both oxidative and non-oxidative metabolisms of ethanol can lead to mitochondria dysfunction and induction of MPT, resulting in the release of mitochondria apoptotic factors such as cytochrome c and Smac/DIABLO. Ethanol induces MPT in primary cultured rat hepatocytes, which is inhibited by cyclosporin A, a MPT inhibitor, and by several antioxidants (Higuchi et al. 2001). More recently, Zhong et al. developed an intravital confocal/multiphoton microscopy approach, which can directly monitor the changes of mitochondrial membrane potential in mouse livers after acute alcohol administration. Using this method, they found that acute alcohol gavage induced reversible hepatic mitochondrial depolarization and onset of MPT, which was dependent on ethanol metabolism (Zhong et al. 2014). The reversibility of depolarized mitochondria in mouse liver after acute alcohol exposure is likely due to the adaptation of mitochondria to ethanol-induced damage. In line with this notion, it has been reported that acute or chronic alcohol exposure alters mitochondria structures (e.g., enlarged mitochondria) and functions in both animal models and human alcoholics livers (Garcia-Ruiz et al. 2013). Mitophagy and PGC-1α-mediated mitochondrial biogenesis are two critical adaptive mechanisms that are activated in liver cells in response to acute or chronic alcohol-induced mitochondrial damage and metabolic stress (Williams et al. 2015; Williams and Ding 2015b; Han et al. 2012). Mitochondria are highly dynamic organelles and constantly undergoing fission and fusion, a subject has not been studied in alcohol-induced liver injury. Nevertheless, a balance between alcohol-induced mitochondrial damage and their removal as well as mitochondrial repair/biogenesis is a critical factor to determine the cell fate. Maladaptation will result in accumulated damaged mitochondria and subsequent apoptosis, and liver injury.

6.6 Activation of Death Receptor (Extrinsic) Apoptotic Pathway by Alcohol

The death receptor pathway (or the extrinsic pathway) is mainly initiated when death receptor ligands bind to the death receptors. Death receptors are cell surface cytokine receptors, and mainly belong to the TNF receptor super-family protein. Death receptors share sequence homology at the cysteine-rich extracellular domains and additional sequence homology at the intracellular death domain. The death receptors expressed on hepatocytes include TNF-receptor 1 (TNFR1), Fas, TRAIL-receptor 1 (DR-4), and TRAIL-receptor 2 (DR5), and the ligands of which are tumor necrosis factor-α (TNF-α), FasL, and TRAIL, respectively (Yin and Ding 2003).

Different ligands bind to their receptors may induce different intracellular events. The Fas receptors are trimerized after FasL or the agonistic antibody (such as Jo2) binds to Fas, resulting in the clustering of the death domain in the intracellular portion of the Fas receptors, which further recruits Fas-associated protein with death domain (FADD), caspase-8, and the cellular FLICE/caspase-8-like inhibitory protein (cFLIP). The resulting complex is called death-inducing signaling complex (DISC). When caspase-8 or caspase-10 is recruited to DISC, they are activated through the so-called proximity-induced dimerization model, in which caspase-8 is dimerized and activated. For type I cells such as lymphocytes, activated caspase-8 may directly cleave downstream effector caspases such as caspase-3 to trigger apoptosis (Yin and Ding 2003). However, for Type II cells such as hepatocytes, the direct activation of caspase-3 by caspase-8 is not sufficient to induce apoptosis likely due to the abundant of inhibitor-of-apoptosis (IAP) molecules such as XIAP (Yin and

Ding 2003; Ding and Yin 2004). Activated caspase-8 then cleaves Bid, a BH3-only proapoptotic Bcl-2 family protein, to generate a truncated form of Bid (tBid). tBid then translocates to mitochondria and activates multidomain BAX and BAK to trigger the release of mitochondria apoptotic factors, such as cytochrome c and Smac/DIABLO, which further activates caspase-9 at the "apoptosome" and subsequent caspase-3 via caspase-9-mediated cleavage of caspase-3 (Yin et al. 1999b). In addition, Smac/DIABLO releases the inhibition of XIAP on caspase activation (Li et al. 1997; Du et al. 2000; Verhagen et al. 2000).

Activation of TNF-α receptor regulates cell death and inflammation, which is involved in acute and chronic liver diseases and in liver cancer. The process of TNF-α-induced caspase activation and apoptosis is a little bit more complicated and different. After TNF-α binds to TNF-α receptor 1 (TNFR1), it forms the TNFR complex I by recruiting downstream factors, including TNFR-associated death domain (TRADD), receptor interacting protein kinase 1 (RIPK1), TNFR-associated factor 2 (TRAF2), and cellular inhibitor of apoptosis proteins 1 and 2 (cIAP1/2) (Vucic et al. 2011; Zhou et al. 2012). RIPK1 is ubiquitinated by the E3 ligases cIAP1/2 and a secondary E3 ligase complex of LUBAC (linear ubiquitin chain assembly complex). Once ubiquitinated, RIPK1 then recruits transforming growth factor β-activated kinase 1 (TAK1) and the ubiquitin-binding partners TAB2 (TAK1-binding protein 2) and TAB3, which forms a platform for the downstream IκB kinase (IKK) complex to activate the NF-κB pathway. The transcription factor NF-κB (of which mammalian cells have five NF-κB factors including RelA/p65, RelB, C-Rel, p50/p105, and p52/p100) then translocates to nuclei and upregulates the expression of genes for cell survival and inflammation (Zhou et al. 2012; Vucic et al. 2011). RIPK1 can also be de-ubiquitinated by cylindromatosis (CLYD), and de-ubiquitinated RIPK1 then recruits TRADD, the FADD, $FLIP_L$, and caspase-8 to form the pro-death complex II resulting in caspase-8 activation. $FLIP_L$ is transcriptionally regulated by NF-κB, which has two death effector domains but has no cysteine protease activity, and thus acts as a negative regulator for caspase-8 activation. In hepatocytes, TNF-α-activated caspase-8 cleaves Bid and tBid translocates to mitochondria resulting in the activation of the mitochondrial-apoptotic pathway (Ding and Yin 2004).

Death receptor activation has been implicated in alcohol-induced liver injury. Acute alcohol consumption increased Fas ligand-mediated apoptosis in mouse livers, and supplemental with zinc inhibited alcohol-induced apoptosis (Lambert et al. 2003). Moreover, chronic alcohol consumption increased the expression of microRNA 21 (miR-21), which is a negative regulator of the expression of Fas ligand (TNF superfamily, member six) (FASLG) and death receptor 5 (DR5) in the mouse livers. When miR-21 was inhibited by using a specific Vivo-Morpholino, ethanol-induced liver injury was greatly exacerbated due to increased expression of DR5 and FASLG in mouse livers. Thus, it is likely that miR-21 may act as an adaptive protective mechanism against alcohol-induced apoptosis in the liver (Francis et al. 2014). In addition to Fas-mediated apoptosis, TNF-α-mediated cell death is also implicated in the pathogenesis of ALD (Yin et al. 1999a; Nagy et al. 2016). It is known that acute or chronic alcohol exposure increases gut-intestinal permeability and elevated systemic levels of gut-derived endotoxins and other microbial products (Hartmann et al. 2015). Endotoxin (LPS) is a strong activator of Kupffer cells (the resident macrophage in the liver), which leads to the production of TNF-α and TNF-α-mediated apoptosis in the liver. In addition to hepatocytes, alcohol can also induce apoptosis in adipocyte resulting in inflammation in adipose tissue. Bid-deficient mice are resistant to alcohol-induced apoptosis in both the adipose and liver tissues. Intriguingly, Bid-deficient mice are not protected from alcohol-induced liver injury and steatosis (Sebastian et al. 2011; Roychowdhury et al. 2012). These findings suggest that apoptosis may not be the major form of cell death in alcohol-exposed liver. Indeed, VX166, a pan-caspase inhibitor, fails to protect against alcohol-induced liver injury and steatosis (Roychowdhury et al. 2012). Taken together, while it is clear that hepatocyte apoptosis contributes to alcohol-induced liver injury through both mitochondrial and death receptor-mediated apoptotic pathways, it may not be critical in alcohol-induced liver injury. Other forms of cell death such as necrosis or pyroptosis may be more critical in the pathogenesis of ALD in addition to apoptosis.

6.7 Necrosis and Necroptosis in ALD

As we discussed above, upon TNF-α binding to its receptor TNFR1, it forms several complexes by recruiting different components to either trigger NF-κB pathway or induce apoptosis depending on the levels of ubiquitinated of RIPK1. Highly ubiquitinated RIPK1 activates NF-κB-mediated cell survival and inflammatory pathways, whereas de-ubiquitinated RIPK1 form the pro-death complex II. Complex II includes TRAF2, TRADD, RIPK1, FADD, c-FLIP, and caspase 8, resulting in the activation of caspase-8. Under normal apoptotic conditions, activated caspase-8 cleaves RIPK3 and RIPK1 to inactivate their functions, suggesting that apoptosis can generally suppress necroptosis (Vandenabeele et al. 2010). However, under the conditions of genetic deletion of FADD or caspase-8 or by pharmacological inhibition of caspase-8, RIPK1 interacts with RIPK3 via their RIP homotypic interaction motif (RHIM) and forms an amyloid-like structure, which is stabilized by phosphorylated RIPK1 and RIPK3. Activated RIPK3 then recruits and phosphorylates MLKL protein to form necrosome. Phosphorylated MLKL in turn oligomerizes and translocates to plasma membranes resulting in eventual membrane rupture or activating ion channels resulting in necrosis (Weinlich and Green 2014; Vanden Berghe et al. 2014). More recently, it is shown that once on the plasma membrane, MLKL can activate cell-surface proteases of the disintegrin and metalloprotease (ADAM) family to promote necroptosis, inflammation, and cell migration (Cai et al. 2016). Interestingly, when cIAPs are absent or depleted, RIPK1, RIPK3, FADD, caspase-8, and c-FLIP$_L$ form a large complex known as the ripoptosome (also called complex IIb), which activates caspase-8 and triggers apoptosis (Dillon et al. 2014; Mandal et al. 2014; Tenev et al. 2011). The assembly of ripoptosomes does not involve autocrine TNF-α, which is negatively regulated by cIAP1, cIAP2, XIAP, and c-FLIP$_L$

(Tenev et al. 2011). Notably, RIPK1 knockout mice die perinatally, exhibiting apoptosis in multiple tissues, whereas RIPK3 knockout mice are viable without obvious phenotypes (Weinlich and Green 2014). These findings suggest that RIPK1 may have a paradoxical cell survival role. Intriguingly, the kinase-dead RIPK3^{D161N} mice die at E10.5, which are protected by germline deletion of caspase-8 but not by the kinase-dead RIPK1^{D138N}, suggesting that RIPK3^{D161N} promotes lethal RIPK1 and caspase-8-dependent apoptosis (Newton et al. 2014). In contrast, kinase-dead RIPK1^{D138N} mice are viable and healthy. Moreover, small-molecule RIPK3 kinase inhibitors inhibit necroptosis but trigger apoptosis by activating the Ripoptosome (Mandal et al. 2014). Taken together, these genetic findings suggest that the formation of the ripoptosome may not require RIPK1 and RIPK3 kinase activity but requires their scaffolding functions (Newton et al. 2014). Therefore, it appears that RIPK1 and RIPK3 may have multiple roles in regulating apoptosis, necroptosis, cell survival, and inflammation (Fig. 6.2).

Recent evidence suggests that RIPK3-mediated necroptosis contributes to the pathogenesis of ALD (Roychowdhury et al. 2013; Wang et al. 2016a). RIP3 was induced by ethanol feeding in mouse livers. Consistent with this, ALD patients had increased hepatic expression of RIPK3 compared to healthy controls. We recently found that alcohol regulates the levels of RIPK3 at the posttranslation via inhibition of the ubiquitin proteasome system but not the transcription level in the liver. Specifically, we found that chronic feeding plus binge alcohol (Gao-binge alcohol) treatment decreased protein levels of proteasome subunit alpha type-2 (PSMA2) and proteasome 26S subunit ATPase 1 (PSMC1) resulting in reduced hepatic proteasomal activity. Pharmacological inhibition of proteasome by the proteasome inhibitor bortezomib or genetic depletion of PSMC1 in mouse livers all leads to the accumulation of hepatic RIP3. These data suggest that RIPK3 is degraded through ubiquitin proteasome system, and impaired proteasomal function induced by Gao-binge alcohol treatment may account for the hepatic accumulation of RIPK3. Notably, we also found increased RIPK3 expression and decreased PSMC1 expression in the livers of human ALD (Wang et al. 2016a). Importantly, RIPK3 knockout (KO) mice had decreased liver injury, steatosis, and inflammation compared with wild-type mice after alcohol administration (either chronic feeding or Gao-binge alcohol), verifying the role of RIPK3 and necroptosis in mediating ethanol-induced liver injury and progression of ALD (Roychowdhury et al. 2013; Wang et al. 2016a). Intriguingly, inhibition of RIPK1 kinase activity by 7-Cl-O-Nec-1, a potent specific RIPK1 inhibitor, suppressed Gao-binge alcohol-induced hepatic inflammation but did not protect against steatosis and liver injury. These findings suggest that alcohol-induced RIPK3-mediated necroptosis may be independent of RIPK1 kinase activity. Therefore, it seems that RIPK1 is more important in regulating inflammation, whereas RIPK3 seems to be more critical in alcohol-induced steatosis and cell death. Both RIPK1 and RIPK3 kinase dead mice are now available, future studies to use these kinase dead mice will help to further clarify the role of the kinase activity of RIPK1 or RIPK3 in alcohol-induced liver injury.

Notably, the necroptosis components of RIPK1, RIPK3 and MLKL are predominantly expressed in immune cells, and their expression levels in the liver are

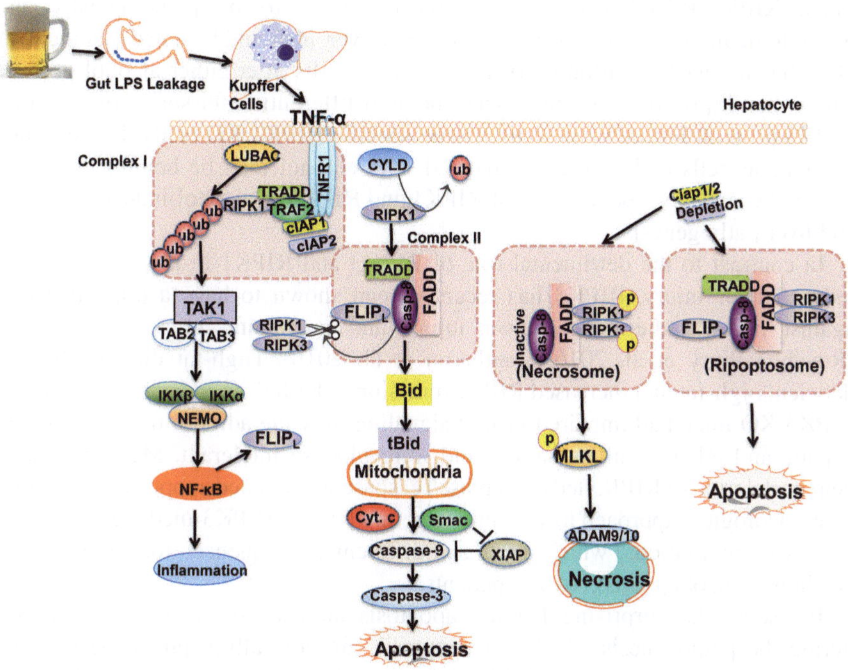

Fig. 6.2 Possible pathways leading to alcohol-induced apoptosis and necrosis. Alcohol consumption increases gut permeability resulting in elevated influx of LPS into the liver. LPS activates Kupffer cells to produce TNF-α. TNF-α binds to its receptor (TNFR1), which further recruits TRADD, RIPK1, cIAP1, cIAP2, TRAF2, and LUBAC to form Complex I, resulting in RIPK1 poly-ubiquitnation. Ubiquitinated RIPK1 then activates TAK1 and IKK complex resulting in the activation of NF-κB, which regulates inflammation and blocks apoptosis. When RIPK1 is de-ubiquitinated by CYLD, it further recruits TRADD, FADD, caspase-8, and c-FLIP$_L$ to form Complex II, resulting in caspase-8 activation. Activated caspase-8 cleaves Bid to activate the mitochondrial apoptotic pathway and trigger mitochondrial Cyto c and SMAC release. Released Cyto c promotes the activation of caspase-9 and caspase-3, whereas Smac inhibits XIAP to remove its inhibition on caspases. Activated caspase-3 then leads to hepatocyte apoptosis. Activated caspase-8 also cleaves RIPK1 and RIKP3 to inactivate RIPK1-RIPK3-mediated necroptosis. When cIAPs are depleted and caspase-8 is inhibited, RIPK1 and RIPK3 interact with each other via RHIM domains to form the amyloid-like necrosome. Auto- and transphosphorylated RIPK1 and RIPK3 then further recruit and phosphorylate downstream MLKL. Phosphorylated MLKL translocates to the plasma membrane and activates ADAM to trigger necroptosis. In the absence of cIAPs, RIPK1, RIPK3, TRADD, caspase-8, and FLIP$_L$ form a complex called the ripoptosome, which induces caspase-8 activation and apoptosis independent of RIPK1 and RIPK3 kinase activity

relatively low (He et al. 2009; Dara et al. 2015; Wang et al. 2016a). In addition to alcohol, acetaminophen treatment also increased hepatic RIPK3 proteins, although RIPK3 KO mice are only resistant to APAP-induced liver injury at the early phase and lost the protection at the late phase of APAP-induced hepatotoxicity (Ramachandran et al. 2013). In addition, knockdown of hepatic RIPK1 by using a RIP1 antisense in mice also attenuated APAP-induced liver injury (Dara et al. 2015).

Thus, RIPK1-RIPK3-mediated necroptosis seems to be pathologically and physiologically relevant in liver diseases. However, it should be noted that the liver often has increased infiltration of inflammatory cells after either alcohol or acetaminophen exposure, which makes it difficult to differentiate the source of increased RIPK3 protein. Future work to use tissue-specific RIPK1 KO or RIPK3 KO mice in immune cells or hepatocytes that fed with alcohol may be helpful to further clarify the cell type-specific role of RIPK1 and RIPK3 in alcohol-induced necrosis and liver pathogenesis.

In contrast to the detrimental role of RIPK3 and RIPK1 in alcohol or APAP-induced liver injury, RIPK3 has recently been shown to have a protective role against high-fat die-induced liver injury and preventing glucose intolerance (Roychowdhury et al. 2016; Gautheron et al. 2016). High-fat diet or choline-deficient high-fat diet increased RIP3 expression in both liver and adipocyte tissues. RIPK3 KO mice had impaired insulin signaling in white adipose tissue, increased hepatic and adipose inflammation, and were glucose intolerant. Mechanistically, genetic deletion of RIPK3 led to increased caspase-8-mediated apoptosis. Therefore, pharmacological approach to systemic targeting RIPK1-RIPK3-mediated necroptosis has to be cautious, which may have different consequences for ALD or non-alcoholic steatohepatitis (NASH) patients.

It is somewhat surprising that both apoptosis and necrosis occur simultaneously during the pathogenesis of ALD if apoptosis can normally suppress necrosis by caspase-mediated cleavage of RIPK3. Liver has the unique zonation that differs in the levels of oxygen, nutrient, and metabolic enzymes. During APAP-induced liver injury, autophagy was induced only by the zone areas that are adjacent to the necrotic areas (Ni et al. 2013). It is possible that apoptosis and necrosis may also occur in different zones of the liver after alcohol consumption. Using immunohistochemistry co-staining for apoptotic (such as activated caspase-3) and necrotic markers (such as RIPK3 and phosphorylated MLKL) may help to test this hypothesis. ALD is a chronic liver disease ranging from steatosis, fibrosis, cirrhosis, steatohepatitis, and liver cancer, and it is also likely that different cell death modes may occur in different stages of liver pathogenesis. For instance, apoptosis could mainly occur in the early phase of ALD, such as in steatosis. In contrast, necrosis may occur predominantly in the later stages of ALD, such as in alcoholic hepatitis. Future work to elucidate these possibilities will be very important to further understand the role of apoptosis and necrosis/necroptosis in ALD.

6.8 Other Forms of Regulated Programmed Cell Death in ALD

6.8.1 Pyroptosis

The term "pyropotosis" was first coined in 2001 and originated from the Greek roots *pyro* refers to fire or fever whereas *ptosis* denotes a falling. Pyroptosis is a proinflammatory form of regulated cell death that depends on the activation of pro-inflammatory caspases that are not required for apoptosis to occur (Miao et al. 2010). So far, pyroptosis is mainly documented in immune cells such as macrophages, dendritic cells, and neutrophils likely due to the high expression levels of inflammatory caspases in immune cells. Pyroptosis is associated with cell swelling and rapid plasma membrane lysis due to the formation of pores on the plasma membrane, which depends on the activation of inflammatory caspases including caspase-1 or casppase-11 (mouse) and its human orthologs caspase-4 and -5 (Vande Walle and Lamkanfi 2016). Pyroptosis shares some similar morphological features with apoptosis such as nuclear DNA fragmentation and positive annexin V staining. The underlying mechanism of DNA fragmentation for pyroptosis does not depend on CAD, the DNase activated by apoptotic caspases. Caspase-1 is activated at complexes termed inflammasome, which include cytosolic nucleotide-binding oligomerization domain (NOD)-like receptors (NLRs), or the pyrin domain containing protein AIM2. Activated caspase-1 proteolytically cleaves proIL-1β and proIL-18 to generate cleaved matured active IL-1β and IL-18. The exact mechanisms for how pyroptosis occurs are still largely unknown, but recent evidence suggests that cleavage of gasdermin D (Gsdmd) by caspase-1 and -11 may play an important role in pyroptosis (Liu et al. 2016; Ding et al. 2016; Chen et al. 2016; Sborgi et al. 2016; Kayagaki et al. 2015; Shi et al. 2015). The cleaved N-terminal fragment of Gsdmd p30 (also called Gsdmd-NT) localized to the lipid bilayer, which has a pore-forming function resulting in the permeabilization of the plasma membrane and releasing of the cellular contents (Ding et al. 2016; Liu et al. 2016). Gsdmd-NT kills cells from within the cells and does not harm the neighboring cells due to its lipid-binding preferences. Consequently, pyroptosis promotes the release of proinflammatory cytokines such as IL-1β in addition to other inflammatory factors or damage-associated molecular patterns (DAMPs), including nuclear protein high mobility group box 1 (HMGB1) (Vande Walle and Lamkanfi 2016).

Chronic alcohol exposure has been reported to activate inflammasome in mouse livers (Petrasek et al. 2012). In this study, Petrasek et al. found that mice fed with alcohol diet for 4 weeks activated caspase-1 in bone marrow-derived Kupffer cells and increased serum levels of IL-1β. Direct administration of physiological doses of IL-1β induced flammasome activation in bone marrow-derived Kupffer cells. Moreover, a recombinant IL-1 receptor antagonist blocked IL-1 signaling and markedly attenuated alcohol-induced liver inflammation, steatosis, and liver injury. While pyroptosis was not examined in this study, it is likely that alcohol may also induce caspase-1-mediated pyropotsis in the liver (Petrasek et al. 2012). Futures studies are needed to determine whether Gsdmd would also be cleaved in alcohol-fed mouse livers.

6.8.2 *Ferroptosis*

Ferroptosis is an emerging new form of programmed necrosis that is morphologically, biochemically, and genetically distinct from apoptosis, necrosis, and autophagy (Yang and Stockwell 2016). Ferroptosis was first discovered from an unbiased screening for small molecules that could selectively kill RAS mutant tumor cells. Two small molecules, erastin and RSL3, were identified that could selectively kill oncogenic RAS mutant tumor cells. This type of cell death was not apoptosis or necroptosis because inhibition of caspases or RIPK1 could not rescue the cell death. Instead, this type of cell death depended on intracellular iron but not other metals, and was termed ferroptosis (Dixon et al. 2012). Ferroptosis is associated with intracellular glutathione depletion, increased intracellular iron, and increased lipid peroxidation, resulting in the plasma membrane rupture. Subsequent studies revealed that ferroptosis is driven by loss of activity of the lipid repair enzyme glutathione peroxidase 4 (GPX4) (Yang et al. 2014). Erastin triggers ferroptosis by inhibiting cysteine transporter on the plasma membrane although erastin can also bind with mitochondrial outer membrane proteins VDAC1 and VDAC2, but the later seems not relevant to ferroptosis (Dixon et al. 2012). In contrast, RSL3 induces ferroptosis by inhibiting the enzyme activity of GPX4 (Yang and Stockwell 2016). Either lipophilic antioxidants or iron chelators can strongly suppress ferroptosis, revealing the important roles of ROS and cellular iron in ferroptosis (Yang and Stockwell 2016). Moreover, ferrastatin-1 and Liproxstatin-1, two small molecule inhibitors, specifically blocked ferroptosis but not apoptosis or necroptosis (Yang and Stockwell 2016).

Hepatic iron overload and increased oxidative stress have often been reported in patients with ALD (Williams et al. 2014; Nagy et al. 2016; Kohgo et al. 2007). In addition, it was reported that hepatic iron content in alcoholic cirrhosis patients correlated with their mortality (Ganne-Carrie et al. 2000). Moreover, iron chelation inhibited TNF-α expression (Xiong et al. 2004, 2008), attenuated alcohol-induced liver lipid peroxidation and steatosis in intragastric ethanol fed rats (Sadrzadeh et al. 1994). Conversely, cotreatment with alcohol and iron exacerbated alcohol-induced liver injury in intragastric alcohol fed rats (Tsukamoto et al. 1995). Mechanistically, alcohol-induced hepatic iron accumulation was thought to be due to increased hepatocellular uptake of iron via transferrin receptor 1 (TfR1) and increased intestinal absorption of iron, and down-regulation of hepcidin expression (Kohgo et al. 2008). While the evidence that alcohol consumption increased hepatic iron accumulation and lipid peroxidation is compelling, the role of ferroptosis in the pathogenesis of ALD has not been determined.

6.9 Autophagy and Cell Death in ALD

Accumulating evidence indicates that apoptosis, necroptosis, autophagy, and other forms of programmed cell death can mutually regulate each other (Kroemer et al. 2010; Wang et al. 2016b; Nagy et al. 2016). Autophagy is an evolutionary conserved lysosomal degradation pathway that involves the formation of the double-membrane autophagosomes, which carry the enveloped cargos and fuse with lysosomes to form autolysosomes where the autophagic cargos are degraded. Under most circumstances, autophagy acts as a pro-survival mechanism and plays a critical role in normal liver physiology and liver diseases (Czaja et al. 2013; Yin et al. 2008). However, under certain specific contexts, autophagy may also cause cell death as so-called autophagic cell death or a specific type of cell death called autosis (Liu and Levine 2015; Sakamaki and Ryan 2016). In response to stress or cell death stimuli, autophagy may help to remove misfolded toxic protein aggregates, damaged organelles such as mitochondria and lysosomes, and provide nutrient fuel for ATP production. All of which can lead to the protection against apoptosis or necrosis.

In the context of ALD, works from our lab and others have demonstrated that pharmacological activation of autophagy protects, whereas pharmacological or genetic inhibition of autophagy exacerbates alcohol-induced liver injury (Ding et al. 2010; Lin et al. 2013; Ding et al. 2011b). We found that induction of autophagy can help to remove excess lipid droplets and thus attenuates alcohol-induced steatosis (Ding et al. 2010, 2011a). Autophagy removes protein aggregates and relieves ER stress in cultured cells (Ding et al. 2007). Thus, autophagy may also help to relieve alcohol-induced ER stress although this topic has not been examined in hepatocytes that are exposed to alcohol. In addition, we also found that induction of autophagy can help to remove alcohol-induced damaged mitochondria via mitophagy through the Pink1 (tensin homolog-induced putative kinase 1)-Parkin pathway (Williams and Ding 2015a, b). Pink1 is a mitochondrial resident protein kinase and Parkin is an evolutionarily conserved E3 ligase, which is recruited to depolarized damaged mitochondria by Pink1. Once Parkin is recruited to mitochondria, it promotes the ubiquitination of mitochondrial outer membrane proteins and subsequent mitochondrial degradation by mitophagy (Ni et al. 2012). Following Gao-binge alcohol administration, Parkin translocated from cytosol to mitochondria in hepatocytes. Parkin KO mice had greater mitochondrial damage, steatosis, and liver injury compared to wild-type mice after Gao-binge alcohol (Williams et al. 2015). Similar protective role of mitophagy against alcohol-induced liver injury was also reported in chronic ethanol fed rats for 10 weeks (Eid et al. 2013). Autophagy is a very dynamic process and autophagy activity could be influenced by acute or chronic ethanol exposure. While we previously found that mice subjected to acute alcohol administration (mimic the alcohol binge condition) had decreased mTOR and increased autophagic flux in mouse livers, chronic alcohol or Gao-binge alcohol impairs autophagy (Ding et al. 2010; Thomes et al. 2015). We found that Gao-binge alcohol increased hepatic mTOR and impairs transcription factor EB (TFEB)-mediated lysosomal biogenesis resulting in insufficient autophagy (unpublished results, Ding Lab). Ongoing research in our lab is aimed to modulate the TFEB pathway to attenuate or prevent ALD.

In addition to hepatocytes, autophagy in the liver non-parenchymal cells such as hepatic stellate cells (HSC) and macrophages may also play a critical role in alcohol-induced liver injury. It has been reported that autophagy in HSC may promote liver fibrosis by increasing lipid droplet degradation via lipophagy resulting in HSC activation. Stellate cell-specific Atg7 KO mice are resistant to CCl_4-induced fibrosis in vivo *(*Hernandez-Gea et al. 2012*)*. It has been reported that chronic alcohol feeding in rats increased ER stress and activated autophagy in rat HSC, which might promote chronic ethanol-induced fibrosis (Hernandez-Gea et al. 2013). In contrast to HSC, macrophage-specific Atg5 KO mice are more susceptible to CCl_4-induced fibrosis and endotoxin-induced liver injury (Lodder et al. 2015; Ilyas et al. 2016). Moreover, macrophage autophagy can also inhibit inflammasome-mediated IL-1β generation and secretion and limit acute toxin-induced liver injury in mice (Ilyas et al. 2016). Cannabinoid receptor 2 (CB2) is a G protein coupled receptor and predominantly expresses at immune cells such as macrophages and protects against inflammatory liver diseases. Recently, it was found that a CB2 agonist JWH-133 protected against Gao-binge alcohol-induced steatosis and inflammation in mice. Interestingly, JWH-133 increased autophagic flux in macrophages and its protection was lost in macrophage-specific Atg5 KO mice (Denaes et al. 2016). These findings suggest that macrophage autophagy may also protect against alcohol-induced liver injury.

Increasing evidence now supports a mutual regulation among autophagy, apoptosis, necroptosis, and ferroptosis. As discussed above, on one hand, autophagy protects against apoptosis or necrosis by removing toxic protein aggregates and damaged organelles. On the other hand, apoptosis can also suppress autophagy by directly inducing caspase-mediated cleavage and inactivation of essential autophagy proteins such as Beclin 1 (Li et al. 2011). However, under certain specific contexts, autophagy can paradoxically promote apoptosis. For instance, autophagy degrades FAP-1 (Fas-associated phosphatase 1), a negative regulator of Fas, which promotes Fas ligand-induced apoptosis in Type I cells that do not require mitochondrial activation (Gump et al. 2014). In contrast, autophagy inhibits apoptosis in Type II cells that require mitochondrial activation for amplifying the apoptotic signals or on treatment with TRAIL in either Type I or II cells. These findings suggest that autophagy may either promote or inhibit apoptosis depends on the cell type or specific apoptotic stimuli. In addition to apoptosis, autophagy can also mutually regulate necroptosis. In MAP3K7-deficient tumor cells, TRAIL predominantly induced necroptosis but not apoptosis. In this model, it was found that RIPK1 and MLKL were recruited to the autophagosomal membranes likely via the autophagy substrate protein p62 since p62 directly interacts with RIPK1 via its ZZ domain (Goodall et al. 2016a, b). Inhibition of the formation and assembly of early autophagosomal membrane such as the knockdown of Atg5, Atg7, or Beclin 1, but not the late autolysosomal functions such as the use of chloroquine or bafilomycin A1, prevented TRAIL-induced cell death in MAP3K7-deficient cells (Goodall et al. 2016b). These results suggest that autophagosomal membranes may provide scaffold platforms for the RIPK1-RIPK3-MLKL necrosome formation and promote necroptosis. Conversely, necroptosis machinery may also regulate autophagy. TFEB, the master regulator of lysosomal biogenesis and autophagy-related gene expression, is mainly regulated at the posttranslational level and mTOR and ERK can phosphorylate TFEB and inactivate its function. It has been reported that RIPK1 can activate ERK and inhibit TFEB resulting in the repression of basal autophagy (Yonekawa et al. 2015).

As discussed above, physiological levels of autophagy act as a cell survival mechanism for the maintenance of cellular homeostasis during various stress conditions;

however, accumulation evidence also suggests that autophagy may also cause cell death when excessive autophagy is induced under certain specific conditions. Autosis is a new form of cell death that is induced by autophagy-inducing peptide Tat-Beclin-1 (a potent autophagy-inducing cell permeable peptide mimic Beclin-1) or glucose starvation in vitro cultured cells or in rat neurons subjected to cerebral hypoxia-ischemia. Autosis is characterized by enhanced cell substrate adhesion, enlarged perinuclear space and dilation and fragmentation of ER, which depends on autophagy-related genes but not apoptosis and necroptosis genes (Liu et al. 2013). In addition to depending on autophagy-related genes, pharmacological or genetic inactivation of Na^+, K^+-ATPase inhibits autosis in vitro and in vivo. More recently, another line of evidence that autophagy contributes to non-apoptotic cell death has been added on the regulation of ferroptosis. As discussed above, ferroptosis is a programmed necrosis that depends on the cellular iron, glutathione (GSH) depletion, and lipid peroxidation. It is known that intracellular iron binds with ferritin that serves as the major intracellular iron storage. Interestingly, autophagy selectively degrades ferritin (termed ferritinophagy) via the cargo receptor protein NCOA4 resulting in increased levels of labile iron and reactive oxygen species and ferroptosis, which can be blocked by knockdown of autophagy-related genes or NCOA4 (Gao et al. 2016). Whether autosis or ferritinophagy-mediated ferroptosis would play a role in alcohol-induced liver injury remains unknown and needs further future investigations.

Taken together, it is clear that there is a complex mutual regulatory network among autophagy, apoptosis, necroptosis, and other forms of cell death (Fig. 6.3). After alcohol exposure, a cell's fate is decided by the balance of cell survival (autophagy) versus damage (apoptosis, necroptosis, autosis, and ferroptosis). Disruption of the balance from cell survival autophagy toward cell injury (apoptosis, necroptosis, autosis, and ferroptosis) will eventually lead to liver injury and ALD after alcohol exposure.

6.10 Concluding Remarks and Future Perspective

ALD is a major chronic liver disease and health problem in the United States and worldwide. No successful treatments for ALD are available despite decades of research efforts. Alcohol consumption can activate cell death and cell adaptive survival pathways, such as apoptosis, necrosis, and autophagy, in the liver. The cell survival autophagy and different modes of cell death can mutually regulate each other, and the balance between cell death and autophagy may eventually decide the pathogenesis of ALD. In addition to apoptosis and necrosis/necroptosis and autophagy in ALD, emerging new forms of programmed cell death such as ferroptosis, pyroptosis, and autosis and their contributions in ALD still need to be further investigated. Because all the different forms of cell death and autophagy mutually regulate each other, combined approaches for targeting all of them might be an ideal strategy to treat ALD. Future works are definitely needed to explore these approaches and investigate the beneficial effects against the pathogenesis of ALD.

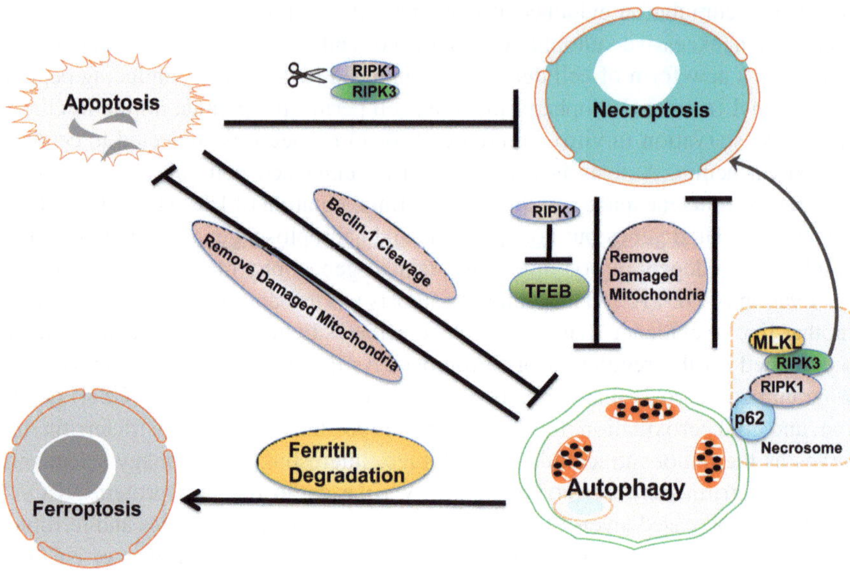

Fig. 6.3 Possible interplay among apoptosis, necrosis, ferroptosis, and autophagy. Apoptosis blocks necroptosis through caspase-8-mediated cleavage and inactivation of RIPK1 and RIPK3. Apoptosis also inhibits autophagy by caspase-mediated cleavage and inactivation of Beclin 1. Autophagosomal membranes may provide a platform for the assembly and activation of necrosome to promote necroptosis. Conversely, the necroptosis protein RIPK1 negatively regulates autophagy by phosphorylating and inactivating TFEB. Autophagy inhibits both apoptosis and necrosis by removing damaged mitochondria. Autophagy may promote ferroptosis via autophagic degradation of ferritin to increase intracellular free iron

Acknowledgements The research in the Ding Lab was supported in part by the NIH/NIAAA funds R01 AA020518, NIH R01 DK102142, National Center for Research Resources P20RR021940, and the National Institute of General Medical Sciences P20 GM103549. All the authors have no conflict of interest to claim.

References

Adachi M, Higuchi H, Miura S, Azuma T, Inokuchi S, Saito H, Kato S, Ishii H (2004) Bax interacts with the voltage-dependent anion channel and mediates ethanol-induced apoptosis in rat hepatocytes. Am J Physiol Gastrointest Liver Physiol 287:G695–G705

Bailey SM, Cunningham CC (2002) Contribution of mitochondria to oxidative stress associated with alcoholic liver disease. Free Radic Biol Med 32:11–16

Beckemeier ME, Bora PS (1998) Fatty acid ethyl esters: potentially toxic products of myocardial ethanol metabolism. J Mol Cell Cardiol 30:2487–2494

Bourogaa E, Nciri R, Mezghani-Jarraya R, Racaud-Sultan C, Damak M, El Feki A (2013) Antioxidant activity and hepatoprotective potential of Hammada scoparia against ethanol-induced liver injury in rats. J Physiol Biochem 69:227–237

Cai Z, Zhang A, Choksi S, Li W, Li T, Zhang XM, Liu ZG (2016) Activation of cell-surface proteases promotes necroptosis, inflammation and cell migration. Cell Res 26:886–900

Chen X, He WT, Hu L, Li J, Fang Y, Wang X, Xu X, Wang Z, Huang K, Han J (2016) Pyroptosis is driven by non-selective gasdermin-D pore and its morphology is different from MLKL channel-mediated necroptosis. Cell Res 26:1007–1020

Cho YS, Challa S, Moquin D, Genga R, Ray TD, Guildford M, Chan FK (2009) Phosphorylation-driven assembly of the RIP1-RIP3 complex regulates programmed necrosis and virus-induced inflammation. Cell 137:1112–1123

Coleman WB, Cunningham CC (1991) Effect of chronic ethanol consumption on hepatic mitochondrial transcription and translation. Biochim Biophys Acta 1058:178–186

Crabb DW, Matsumoto M, Chang D, You M (2004) Overview of the role of alcohol dehydrogenase and aldehyde dehydrogenase and their variants in the genesis of alcohol-related pathology. Proc Nutr Soc 63:49–63

Czaja MJ, Ding WX, Donohue TM Jr, Friedman SL, Kim JS, Komatsu M, Lemasters JJ, Lemoine A, Lin JD, Ou JH, Perlmutter DH, Randall G, Ray RB, Tsung A, Yin XM (2013) Functions of autophagy in normal and diseased liver. Autophagy 9:1131–1158

Dara L, Johnson H, Suda J, Win S, Gaarde W, Han D, Kaplowitz N (2015) Receptor interacting protein kinase 1 mediates murine acetaminophen toxicity independent of the necrosome and not through necroptosis. Hepatology 62(6):1847–1857

Degterev A, Huang Z, Boyce M, Li Y, Jagtap P, Mizushima N, Cuny GD, Mitchison TJ, Moskowitz MA, Yuan J (2005) Chemical inhibitor of nonapoptotic cell death with therapeutic potential for ischemic brain injury. Nat Chem Biol 1:112–119

Denaes T, Lodder J, Chobert MN, Ruiz I, Pawlotsky JM, Lotersztajn S, Teixeira-Clerc F (2016) The cannabinoid receptor 2 protects against alcoholic liver disease via a macrophage autophagy-dependent pathway. Sci Rep 6:28806

Dillon CP, Weinlich R, Rodriguez DA, Cripps JG, Quarato G, Gurung P, Verbist KC, Brewer TL, Llambi F, Gong YN, Janke LJ, Kelliher MA, Kanneganti TD, Green DR (2014) RIPK1 blocks early postnatal lethality mediated by caspase-8 and RIPK3. Cell 157:1189–1202

Ding J, Wang K, Liu W, She Y, Sun Q, Shi J, Sun H, Wang DC, Shao F (2016) Pore-forming activity and structural autoinhibition of the gasdermin family. Nature 535:111–116

Ding WX, Li M, Chen X, Ni HM, Lin CW, Gao W, Lu B, Stolz DB, Clemens DL, Yin XM (2010) Autophagy reduces acute ethanol-induced hepatotoxicity and steatosis in mice. Gastroenterology 139:1740–1752

Ding WX, Li M, Yin XM (2011a) Selective taste of ethanol-induced autophagy for mitochondria and lipid droplets. Autophagy 7:248–249

Ding WX, Manley S, Ni HM (2011b) The emerging role of autophagy in alcoholic liver disease. Exp Biol Med (Maywood) 236:546–556

Ding WX, Ni HM, Gao W, Yoshimori T, Stolz DB, Ron D, Yin XM (2007) Linking of autophagy to ubiquitin-proteasome system is important for the regulation of endoplasmic reticulum stress and cell viability. Am J Pathol 171:513–524

Ding WX, Yin XM (2004) Dissection of the multiple mechanisms of TNF-alpha-induced apoptosis in liver injury. J Cell Mol Med 8:445–454

Dixon SJ, Lemberg KM, Lamprecht MR, Skouta R, Zaitsev EM, Gleason CE, Patel DN, Bauer AJ, Cantley AM, Yang WS, Morrison B 3rd, Stockwell BR (2012) Ferroptosis: an iron-dependent form of nonapoptotic cell death. Cell 149:1060–1072

Du C, Fang M, Li Y, Li L, Wang X (2000) Smac, a mitochondrial protein that promotes cytochrome c-dependent caspase activation by eliminating IAP inhibition. Cell 102:33–42

Eid N, Ito Y, Maemura K, Otsuki Y (2013) Elevated autophagic sequestration of mitochondria and lipid droplets in steatotic hepatocytes of chronic ethanol-treated rats: an immunohistochemical and electron microscopic study. J Mol Histol 44:311–326

Francis H, McDaniel K, Han Y, Liu X, Kennedy L, Yang F, McCarra J, Zhou T, Glaser S, Venter J, Huang L, Levine P, Lai JM, Liu CG, Alpini G, Meng F (2014) Regulation of the extrinsic apoptotic pathway by microRNA-21 in alcoholic liver injury. J Biol Chem 289:27526–27539

Ganne-Carrie N, Christidis C, Chastang C, Ziol M, Chapel F, Imbert-Bismut F, Trinchet JC, Guettier C, Beaugrand M (2000) Liver iron is predictive of death in alcoholic cirrhosis: a multivariate

study of 229 consecutive patients with alcoholic and/or hepatitis C virus cirrhosis: a prospective follow up study. Gut 46:277–282
Gao B, Bataller R (2011) Alcoholic liver disease: pathogenesis and new therapeutic targets. Gastroenterology 141:1572–1585
Gao M, Monian P, Pan Q, Zhang W, Xiang J, Jiang X (2016) Ferroptosis is an autophagic cell death process. Cell Res 26:1021–1032
Garcia-Ruiz C, Kaplowitz N, Fernandez-Checa JC (2013) Role of mitochondria in alcoholic liver disease. Curr Pathobiol Rep 1:159–168
Gautheron J, Vucur M, Schneider AT, Severi I, Roderburg C, Roy S, Bartneck M, Schrammen P, Diaz MB, Ehling J, Gremse F, Heymann F, Koppe C, Lammers T, Kiessling F, Van Best N, Pabst O, Courtois G, Linkermann A, Krautwald S, Neumann UP, Tacke F, Trautwein C, Green DR, Longerich T, Frey N, Luedde M, Bluher M, Herzig S, Heikenwalder M, Luedde T (2016) The necroptosis-inducing kinase RIPK3 dampens adipose tissue inflammation and glucose intolerance. Nat Commun 7:11869
Goodall ML, Cramer SD, Thorburn A (2016a) Autophagy complexes cell death by necroptosis. Oncotarget 7(32):50818–50819
Goodall ML, Fitzwalter BE, Zahedi S, Wu M, Rodriguez D, Mulcahy-Levy JM, Green DR, Morgan M, Cramer SD, Thorburn A (2016b) The autophagy machinery controls cell death switching between apoptosis and necroptosis. Dev Cell 37:337–349
Gump JM, Staskiewicz L, Morgan MJ, Bamberg A, Riches DW, Thorburn A (2014) Autophagy variation within a cell population determines cell fate through selective degradation of Fap-1. Nat Cell Biol 16:47–54
Han D, Ybanez MD, Johnson HS, McDonald JN, Mesropyan L, Sancheti H, Martin G, Martin A, Lim AM, Dara L, Cadenas E, Tsukamoto H, Kaplowitz N (2012) Dynamic adaptation of liver mitochondria to chronic alcohol feeding in mice: biogenesis, remodeling, and functional alterations. J Biol Chem 287:42165–42179
Hartmann P, Seebauer CT, Schnabl B (2015) Alcoholic liver disease: the gut microbiome and liver cross talk. Alcohol Clin Exp Res 39:763–775
He S, Wang L, Miao L, Wang T, Du F, Zhao L, Wang X (2009) Receptor interacting protein kinase-3 determines cellular necrotic response to TNF-alpha. Cell 137:1100–1111
Hernandez-Gea V, Ghiassi-Nejad Z, Rozenfeld R, Gordon R, Fiel MI, Yue Z, Czaja MJ, Friedman SL (2012) Autophagy releases lipid that promotes fibrogenesis by activated hepatic stellate cells in mice and in human tissues. Gastroenterology 142:938–946
Hernandez-Gea V, Hilscher M, Rozenfeld R, Lim MP, Nieto N, Werner S, Devi LA, Friedman SL (2013) Endoplasmic reticulum stress induces fibrogenic activity in hepatic stellate cells through autophagy. J Hepatol 59:98–104
Higuchi H, Adachi M, Miura S, Gores GJ, Ishii H (2001) The mitochondrial permeability transition contributes to acute ethanol-induced apoptosis in rat hepatocytes. Hepatology 34:320–328
Hoek JB, Cahill A, Pastorino JG (2002) Alcohol and mitochondria: a dysfunctional relationship. Gastroenterology 122:2049–2063
Ilyas G, Zhao E, Liu K, Lin Y, Tesfa L, Tanaka KE, Czaja MJ (2016) Macrophage autophagy limits acute toxic liver injury in mice through down regulation of interleukin-1beta. J Hepatol 64(1):118–127
Kayagaki N, Stowe IB, Lee BL, O'Rourke K, Anderson K, Warming S, Cuellar T, Haley B, Roose-Girma M, Phung QT, Liu PS, Lill JR, Li H, Wu J, Kummerfeld S, Zhang J, Lee WP, Snipas SJ, Salvesen GS, Morris LX, Fitzgerald L, Zhang Y, Bertram EM, Goodnow CC, Dixit VM (2015) Caspase-11 cleaves gasdermin D for non-canonical inflammasome signalling. Nature 526:666–671
Kim KH, Kum YS, Park YY, Park JH, Kim SJ, Lee WR, Lee KG, Han SM, Park KK (2010) The protective effect of bee venom against ethanol-induced hepatic injury via regulation of the mitochondria-related apoptotic pathway. Basic Clin Pharmacol Toxicol 107:619–624
King AL, Swain TM, Mao Z, Udoh US, Oliva CR, Betancourt AM, Griguer CE, Crowe DR, Lesort M, Bailey SM (2014) Involvement of the mitochondrial permeability transition pore in chronic ethanol-mediated liver injury in mice. Am J Physiol Gastrointest Liver Physiol 306: G265–G277

Kohgo Y, Ikuta K, Ohtake T, Torimoto Y, Kato J (2007) Iron overload and cofactors with special reference to alcohol, hepatitis C virus infection and steatosis/insulin resistance. World J Gastroenterol 13:4699–4706

Kohgo Y, Ohtake T, Ikuta K, Suzuki Y, Torimoto Y, Kato J (2008) Dysregulation of systemic iron metabolism in alcoholic liver diseases. J Gastroenterol Hepatol 23(Suppl 1):S78–S81

Kroemer G, Marino G, Levine B (2010) Autophagy and the integrated stress response. Mol Cell 40:280–293

Lambert JC, Zhou Z, Kang YJ (2003) Suppression of Fas-mediated signaling pathway is involved in zinc inhibition of ethanol-induced liver apoptosis. Exp Biol Med (Maywood) 228:406–412

Lange LG, Sobel BE (1983) Mitochondrial dysfunction induced by fatty acid ethyl esters, myocardial metabolites of ethanol. J Clin Invest 72:724–731

Li H, Wang P, Sun Q, Ding WX, Yin XM, Sobol RW, Stolz DB, Yu J, Zhang L (2011) Following cytochrome c release, autophagy is inhibited during chemotherapy-induced apoptosis by caspase 8-mediated cleavage of Beclin 1. Cancer Res 71:3625–3634

Li P, Nijhawan D, Budihardjo I, Srinivasula SM, Ahmad M, Alnemri ES, Wang X (1997) Cytochrome c and dATP-dependent formation of Apaf-1/caspase-9 complex initiates an apoptotic protease cascade. Cell 91:479–489

Lin CW, Zhang H, Li M, Xiong X, Chen X, Dong XC, Yin XM (2013) Pharmacological promotion of autophagy alleviates steatosis and injury in alcoholic and non-alcoholic fatty liver conditions in mice. J Hepatol 58:993–999

Liu X, Zhang Z, Ruan J, Pan Y, Magupalli VG, Wu H, Lieberman J (2016) Inflammasome-activated gasdermin D causes pyroptosis by forming membrane pores. Nature 535:153–158

Liu Y, Levine B (2015) Autosis and autophagic cell death: the dark side of autophagy. Cell Death Differ 22:367–376

Liu Y, Shoji-Kawata S, Sumpter RM Jr, Wei Y, Ginet V, Zhang L, Posner B, Tran KA, Green DR, Xavier RJ, Shaw SY, Clarke PG, Puyal J, Levine B (2013) Autosis is a Na^+, K^+-ATPase-regulated form of cell death triggered by autophagy-inducing peptides, starvation, and hypoxia-ischemia. Proc Natl Acad Sci U S A 110:20364–20371

Lodder J, Denaes T, Chobert MN, Wan J, El-Benna J, Pawlotsky JM, Lotersztajn S, Teixeira-Clerc F (2015) Macrophage autophagy protects against liver fibrosis in mice. Autophagy 11:1280–1292

Lu Y, Cederbaum AI (2008) CYP2E1 and oxidative liver injury by alcohol. Free Radic Biol Med 44:723–738

Malhi H, Gores GJ, Lemasters JJ (2006) Apoptosis and necrosis in the liver: a tale of two deaths? Hepatology 43:S31–S44

Malhi H, Guicciardi ME, Gores GJ (2010) Hepatocyte death: a clear and present danger. Physiol Rev 90:1165–1194

Mandal P, Berger SB, Pillay S, Moriwaki K, Huang C, Guo H, Lich JD, Finger J, Kasparcova V, Votta B, Ouellette M, King BW, Wisnoski D, Lakdawala AS, DeMartino MP, Casillas LN, Haile PA, Sehon CA, Marquis RW, Upton J, Daley-Bauer LP, Roback L, Ramia N, Dovey CM, Carette JE, Chan FK, Bertin J, Gough PJ, Mocarski ES, Kaiser WJ (2014) RIP3 induces apoptosis independent of pronecrotic kinase activity. Mol Cell 56:481–495

Mansouri A, Gaou I, De Kerguenec C, Amsellem S, Haouzi D, Berson A, Moreau A, Feldmann G, Letteron P, Pessayre D, Fromenty B (1999) An alcoholic binge causes massive degradation of hepatic mitochondrial DNA in mice. Gastroenterology 117:181–190

Miao EA, Leaf IA, Treuting PM, Mao DP, Dors M, Sarkar A, Warren SE, Wewers MD, Aderem A (2010) Caspase-1-induced pyroptosis is an innate immune effector mechanism against intracellular bacteria. Nat Immunol 11:1136–1142

Nagy LE, Ding WX, Cresci G, Saikia P, Shah VH (2016) Linking pathogenic mechanisms of alcoholic liver disease with clinical phenotypes. Gastroenterology 150:1756–1768

Newton K, Dugger DL, Wickliffe KE, Kapoor N, de Almagro MC, Vucic D, Komuves L, Ferrando RE, French DM, Webster J, Roose-Girma M, Warming S, Dixit VM (2014) Activity of protein kinase RIPK3 determines whether cells die by necroptosis or apoptosis. Science 343: 1357–1360

Ni HM, Bockus A, Boggess N, Jaeschke H, Ding WX (2012) Activation of autophagy protects against acetaminophen-induced hepatotoxicity. Hepatology 55:222–232

Ni HM, Williams JA, Jaeschke H, Ding WX (2013) Zonated induction of autophagy and mitochondrial spheroids limits acetaminophen-induced necrosis in the liver. Redox Biol 1:427–432

Petrasek J, Bala S, Csak T, Lippai D, Kodys K, Menashy V, Barrieau M, Min SY, Kurt-Jones EA, Szabo G (2012) IL-1 receptor antagonist ameliorates inflammasome-dependent alcoholic steatohepatitis in mice. J Clin Invest 122:3476–3489

Ramachandran A, McGill MR, Xie Y, Ni HM, Ding WX, Jaeschke H (2013) Receptor interacting protein kinase 3 is a critical early mediator of acetaminophen-induced hepatocyte necrosis in mice. Hepatology 58:2099–2108

Roychowdhury S, Chiang DJ, Mandal P, McMullen MR, Liu X, Cohen JI, Pollard J, Feldstein AE, Nagy LE (2012) Inhibition of apoptosis protects mice from ethanol-mediated acceleration of early markers of CCl4 -induced fibrosis but not steatosis or inflammation. Alcohol Clin Exp Res 36:1139–1147

Roychowdhury S, McCullough RL, Sanz-Garcia C, Saikia P, Alkhouri N, Matloob A, Pollard KA, McMullen MR, Croniger CM, Nagy LE (2016) Receptor interacting protein 3 protects mice from high-fat diet-induced liver injury. Hepatology 64(5):1518–1533

Roychowdhury S, McMullen MR, Pisano SG, Liu X, Nagy LE (2013) Absence of receptor interacting protein kinase 3 prevents ethanol-induced liver injury. Hepatology 57:1773–1783

Sadrzadeh SM, Nanji AA, Price PL (1994) The oral iron chelator, 1,2-dimethyl-3-hydroxypyrid-4--one reduces hepatic-free iron, lipid peroxidation and fat accumulation in chronically ethanol-fed rats. J Pharmacol Exp Ther 269:632–636

Sakamaki J, Ryan KM (2016) Autophagy determines the path on the TRAIL to death. Dev Cell 37:291–293

Sborgi L, Ruhl S, Mulvihill E, Pipercevic J, Heilig R, Stahlberg H, Farady CJ, Muller DJ, Broz P, Hiller S (2016) GSDMD membrane pore formation constitutes the mechanism of pyroptotic cell death. EMBO J 35:1766–1778

Scorrano L, Korsmeyer SJ (2003) Mechanisms of cytochrome c release by proapoptotic BCL-2 family members. Biochem Biophys Res Commun 304:437–444

Sebastian BM, Roychowdhury S, Tang H, Hillian AD, Feldstein AE, Stahl GL, Takahashi K, Nagy LE (2011) Identification of a cytochrome P4502E1/Bid/C1q-dependent axis mediating inflammation in adipose tissue after chronic ethanol feeding to mice. J Biol Chem 286: 35989–35997

Shi J, Zhao Y, Wang K, Shi X, Wang Y, Huang H, Zhuang Y, Cai T, Wang F, Shao F (2015) Cleavage of GSDMD by inflammatory caspases determines pyroptotic cell death. Nature 526:660–665

Tenev T, Bianchi K, Darding M, Broemer M, Langlais C, Wallberg F, Zachariou A, Lopez J, MacFarlane M, Cain K, Meier P (2011) The ripoptosome, a signaling platform that assembles in response to genotoxic stress and loss of IAPs. Mol Cell 43:432–448

Thomes PG, Trambly CS, Fox HS, Tuma DJ, Donohue TM Jr (2015) Acute and chronic ethanol administration differentially modulate hepatic autophagy and transcription factor EB. Alcohol Clin Exp Res 39:2354–2363

Tsukamoto H, Horne W, Kamimura S, Niemela O, Parkkila S, Yla-Herttuala S, Brittenham GM (1995) Experimental liver cirrhosis induced by alcohol and iron. J Clin Invest 96:620–630

Vande Walle L, Lamkanfi M (2016) Pyroptosis. Curr Biol 26:R568–R572

Vanden Berghe T, Linkermann A, Jouan-Lanhouet S, Walczak H, Vandenabeele P (2014) Regulated necrosis: the expanding network of non-apoptotic cell death pathways. Nat Rev Mol Cell Biol 15:135–147

Vandenabeele P, Declercq W, Van Herreweghe F, Vanden Berghe T (2010) The role of the kinases RIP1 and RIP3 in TNF-induced necrosis. Sci Signal 3:re4

Verhagen AM, Ekert PG, Pakusch M, Silke J, Connolly LM, Reid GE, Moritz RL, Simpson RJ, Vaux DL (2000) Identification of DIABLO, a mammalian protein that promotes apoptosis by binding to and antagonizing IAP proteins. Cell 102:43–53

Vucic D, Dixit VM, Wertz IE (2011) Ubiquitylation in apoptosis: a post-translational modification at the edge of life and death. Nat Rev Mol Cell Biol 12:439–452

Wang S, Ni HM, Dorko K, Kumer SC, Schmitt TM, Nawabi A, Komatsu M, Huang H, Ding WX (2016) Increased hepatic receptor interacting protein kinase 3 expression due to impaired proteasomal functions contributes to alcohol-induced steatosis and liver injury. Oncotarget 7(14):17681–17698

Wang S, Pacher P, De Lisle RC, Huang H, Ding WX (2016b) A mechanistic review of cell death in alcohol-induced liver injury. Alcohol Clin Exp Res 40:1215–1223

Wei MC, Zong WX, Cheng EH, Lindsten T, Panoutsakopoulou V, Ross AJ, Roth KA, MacGregor GR, Thompson CB, Korsmeyer SJ (2001) Proapoptotic BAX and BAK: a requisite gateway to mitochondrial dysfunction and death. Science 292:727–730

Weinlich R, Green DR (2014) The two faces of receptor interacting protein kinase-1. Mol Cell 56:469–480

Werner J, Saghir M, Warshaw AL, Lewandrowski KB, Laposata M, Iozzo RV, Carter EA, Schatz RJ, Fernandez-Del Castillo C (2002) Alcoholic pancreatitis in rats: injury from nonoxidative metabolites of ethanol. Am J Physiol Gastrointest Liver Physiol 283:G65–G73

Williams JA, Ding WX (2015a) A mechanistic review of mitophagy and its role in protection against alcoholic liver disease. Biomolecules 5:2619–2642

Williams JA, Ding WX (2015b) Targeting Pink1-Parkin-mediated mitophagy for treating liver injury. Pharmacol Res 102:264–269

Williams JA, Manley S, Ding WX (2014) New advances in molecular mechanisms and emerging therapeutic targets in alcoholic liver diseases. World J Gastroenterol 20:12908–12933

Williams JA, Ni HM, Ding Y, Ding WX (2015) Parkin regulates mitophagy and mitochondrial function to protect against alcohol-induced liver injury and steatosis in mice. Am J Physiol Gastrointest Liver Physiol 309:G324–G340

Wu H, Bhopale KK, Ansari GA, Kaphalia BS (2008) Ethanol-induced cytotoxicity in rat pancreatic acinar AR42J cells: role of fatty acid ethyl esters. Alcohol Alcohol 43:1–8

Wu H, Cai P, Clemens DL, Jerrells TR, Ansari GA, Kaphalia BS (2006) Metabolic basis of ethanol-induced cytotoxicity in recombinant HepG2 cells: role of nonoxidative metabolism. Toxicol Appl Pharmacol 216:238–247

Xiong S, She H, Tsukamoto H (2004) Signaling role of iron in NF-kappa B activation in hepatic macrophages. Comp Hepatol 3(Suppl 1):S36

Xiong S, She H, Zhang AS, Wang J, Mkrtchyan H, Dynnyk A, Gordeuk VR, French SW, Enns CA, Tsukamoto H (2008) Hepatic macrophage iron aggravates experimental alcoholic steatohepatitis. Am J Physiol Gastrointest Liver Physiol 295:G512–G521

Yang WS, SriRamaratnam R, Welsch ME, Shimada K, Skouta R, Viswanathan VS, Cheah JH, Clemons PA, Shamji AF, Clish CB, Brown LM, Girotti AW, Cornish VW, Schreiber SL, Stockwell BR (2014) Regulation of ferroptotic cancer cell death by GPX4. Cell 156:317–331

Yang WS, Stockwell BR (2016) Ferroptosis: death by lipid peroxidation. Trends Cell Biol 26:165–176

Yin M, Wheeler MD, Kono H, Bradford BU, Gallucci RM, Luster MI, Thurman RG (1999a) Essential role of tumor necrosis factor alpha in alcohol-induced liver injury in mice. Gastroenterology 117:942–952

Yin XM, Ding WX (2003) Death receptor activation-induced hepatocyte apoptosis and liver injury. Curr Mol Med 3:491–508

Yin XM, Ding WX, Gao W (2008) Autophagy in the liver. Hepatology 47:1773–1785

Yin XM, Wang K, Gross A, Zhao Y, Zinkel S, Klocke B, Roth KA, Korsmeyer SJ (1999b) Bid-deficient mice are resistant to Fas-induced hepatocellular apoptosis. Nature 400:886–891

Yonekawa T, Gamez G, Kim J, Tan AC, Thorburn J, Gump J, Thorburn A, Morgan MJ (2015) RIP1 negatively regulates basal autophagic flux through TFEB to control sensitivity to apoptosis. EMBO Rep 16:700–708

Zakhari S (2006) Overview: how is alcohol metabolized by the body? Alcohol Res Health 29:245–254

Zelickson BR, Benavides GA, Johnson MS, Chacko BK, Venkatraman A, Landar A, Betancourt AM, Bailey SM, Darley-Usmar VM (2011) Nitric oxide and hypoxia exacerbate alcohol-induced mitochondrial dysfunction in hepatocytes. Biochim Biophys Acta 1807:1573–1582

Zelner I, Matlow JN, Natekar A, Koren G (2013) Synthesis of fatty acid ethyl esters in mammalian tissues after ethanol exposure: a systematic review of the literature. Drug Metab Rev 45:277–299

Zhang DW, Shao J, Lin J, Zhang N, Lu BJ, Lin SC, Dong MQ, Han J (2009) RIP3, an energy metabolism regulator that switches TNF-induced cell death from apoptosis to necrosis. Science 325:332–336

Zhong Z, Ramshesh VK, Rehman H, Liu Q, Theruvath TP, Krishnasamy Y, Lemasters JJ (2014) Acute ethanol causes hepatic mitochondrial depolarization in mice: role of ethanol metabolism. PLoS One 9:e91308

Zhou Z, Han V, Han J (2012) New components of the necroptotic pathway. Protein Cell 3:811–817

Zorov DB, Filburn CR, Klotz LO, Zweier JL, Sollott SJ (2000) Reactive oxygen species (ROS)-induced ROS release: a new phenomenon accompanying induction of the mitochondrial permeability transition in cardiac myocytes. J Exp Med 192:1001–1014

Chapter 7
Bile Acid-Induced Liver Injury in Cholestasis

Tiangang Li and John Y.L. Chiang

Abbreviations

UDCA	Ursodeoxycholic acid
CYP	Cytochrome p450
CA	Cholic acid
CDCA	Chenodeoxycholic acid
DCA	Deoxycholic acid
MCA	Muricholic acid
CYP7A1	Cholesterol 7α-hydroxylase
CYP8B1	Sterol 12α-hydroxylase
CYP27A1	Sterol 27-hydroxylase
CYP7B1	Oxysterol 7α-hydroxylase
BSH	Bile salt hydrolases
LCA	Lithocholic acid
BACS	Bile acyl-CoA synthetase
BACL	Bile acid-CoA ligase
BAAT	Bile acid-CoA:amino acid N-acyltransferase
NTCP	Na^+-dependent taurocholate transporter
OATP	Organic anion transporter
BSEP	Bile salt export pump
ABC	ATP-binding cassette transporter
MDR	Multi-drug resistant

T. Li (✉)
Department of Pharmacology, Toxicology and Therapeutics,
University of Kansas Medical Center, Kansas City, KS 66160, USA
e-mail: tli@kumc.edu

J.Y.L. Chiang
Department of Integrative Medical Sciences, Northeast Ohio Medical University,
Rootstown, OH 44272, USA

W.-X. Ding, X.-M. Yin (eds.), *Cellular Injury in Liver Diseases*, Cell Death
in Biology and Diseases, DOI 10.1007/978-3-319-53774-0_7

X.-M. Yin and Z. Dong (Series eds.), *Cell Death in Biology and Diseases*

MRP	MDR-related protein
OST	Organic solute transporter
ASBT	Apical sodium-dependent bile salt transporter
I-BABP	Intestinal bile acid-binding protein
ICP	Intrahepatic cholestasis of pregnancy
PFIC	Progressive familial intrahepatic cholestasis
TJP2	The junction protein 2 gene
PBC	Primary biliary cirrhosis
PSC	Primary sclerosing cholangitis
GCDCA	Glycochenodeoxycholic acid
JNK	c-Jun N-terminal kinase
PKC	Protein kinase C
EGFR	Epidermal growth factor receptor
ROS	Reactive oxygen species
MPT	Mitochondrial permeability transition pore
DAMP	Damage-associated molecular pattern molecules
ERK	Extracellular signal–regulated kinases
Egr-1	Early growth response factor 1
TNFα	Tumor necrosis factor α
IL-1β	Interleukin-1β
IL-6	Interleukin-6
ICAM	Intracellular adhesion molecule
FXR	Farnesoid X receptor
PXR	Pregnane X receptor
CAR	Constitutive androgen receptor
RXR	Retinoid X receptor
GPCR	G protein-coupled receptor
S1PR	Sphingosine-1-phosphate receptor
SHP	Small heterodimer partner
LRH-1	Liver receptor homolog-1
HNF4α	Hepatocyte nuclear factor 4α
FGF15	Fibroblast growth factor 15
FGFR4	Fibroblast growth factor receptor 4
SULT2A1	Sulfotransferase 2A1
UGT	UDP-glucuronosyltransferase
PKA	Protein kinase A
GLP-1	Glucagon-like peptide 1
NASH	Non-alcoholic steatohepatitis
SphK	Sphingosine kinase
norUDCA	nor-Ursodeoxycholic
PPARα	Peroxisome proliferator-activated receptor α
OCA	Obeticholic acid
GGT	γ-Glutamyltransferase

7.1 Introduction

Bile acids are synthesized from cholesterol only by the hepatocytes (Li and Chiang 2014; Russell and Setchell 1992). Bile acid synthesis accounts for about 50% of cholesterol turnover in humans (Chiang 2009). Bile acids are secreted across the apical membrane of the hepatocytes into the canaliculi and subsequently concentrated in the gallbladder. Biliary bile acid secretion plays an important role in generating bile flow and facilitating hepatobiliary secretion of various endogenous metabolites and xenobiotics (Trauner and Boyer 2003). Once released into the small intestine, bile acids act as physiological detergent molecules to facilitate the intestinal absorption of dietary lipids and fat-soluble vitamins. Majority of the bile acids in the intestine are reabsorbed and transported back to the liver via portal blood for re-secretion into the bile. Bile acids at high concentration can be cytotoxic. De novo bile acid synthesis in the liver is under a tight feedback inhibition by bile acids to maintain a relative constant bile acid pool size. Majority of bile acids are conjugated to amino acids mainly glycine and taurine, which increases the solubility of bile acids and decreases bile acid toxicity. In the bile, bile acids, together with phospholipids and cholesterol that are secreted by hepatocytes, form mixed micelles, which is critical in preventing free bile acid toxicity to the biliary epithelium. Disrupted or reduced bile flow out of the liver, which can be caused by both genetic and pathological factors, can lead to cholestasis, in which accumulation of bile acids in the liver causes hepatic inflammation and injury (Zollner and Trauner 2008). Cholestasis can result in liver failure and increased risk of liver cancer. The mechanisms of liver injury in cholestasis have been extensively studied in humans and in rodent models of cholestasis. In cholestasis, a number of protective mechanisms mediated by bile acid signaling are activated. Available therapies for human cholestasis are currently very limited. The hydrophilic bile acid ursodeoxycholic acid (UDCA) is the primary agent used to alleviate bile acid toxicity and liver injury in chronic human cholestasis such as primary biliary cirrhosis, while novel and effective therapies are still needed for cholestasis patients who do not respond adequately to UDCA treatment.

7.2 Bile Acid Biology

7.2.1 Bile Acid Composition

Bile acids are cholesterol derivatives. Cholesterol and cholic acid (CA) structures are shown in Fig. 7.1a. In humans, the bile acid pool consists of primary bile acid and secondary bile acid species. The primary bile acids are synthesized from cholesterol in the liver via multi-step reactions catalyzed by a number of cytochrome p450 (CYP) enzymes localized in the endoplasmic reticulum, mitochondria, cytosol, and peroxisomes (Li and Chiang 2014; Russell 2003). Once released into the

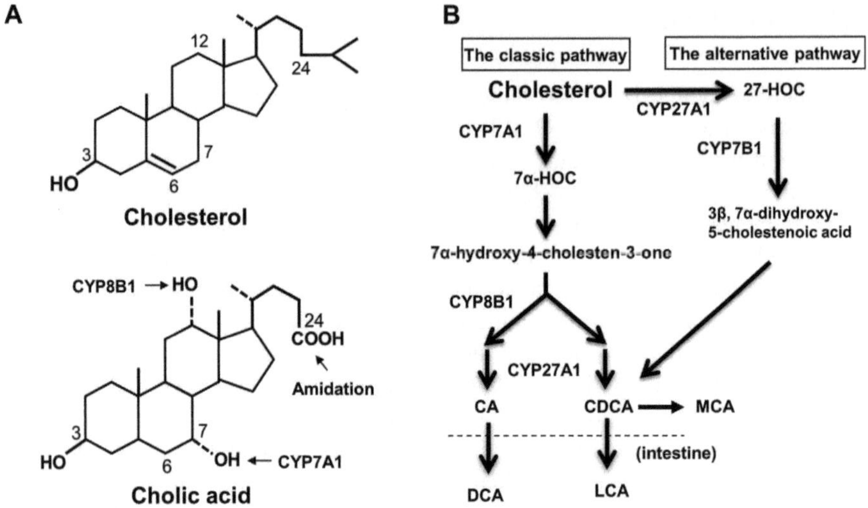

Fig. 7.1 Bile acid synthetic pathway. In the classic pathway, the rate-limiting enzyme cholesterol 7α-hydroxylase (CYP7A1) in the endoplasmic reticulum converts cholesterol into 7α-hydroxycholesterol (7α-HOC). The 7α-hydroxy-4 cholesten-3-one (C4) is a common precursor for CDCA and CA. The sterol 12α-hydroxylase (CYP8B1) catalyzes the C-12 hydroxylation and initiates the conversion of C4 to cholic acid (CA). The mitochondrial sterol 27-hydroxylase (CYP27A1) catalyzes the steroid side-chain oxidation in both CA and CDCA syntheses. In the alternative pathway, CYP27A1 catalyzes the first step to convert cholesterol to 27-hydroxycholesterol (27-HOC). In this pathway, C-7 hydroxylation is catalyzed by the oxysterol 7α- hydroxylase (CYP7B1). The alternative pathway only produces CDCA. In mice, majority of the CDCA is converted to muricholic acid (MCA). In the large intestine, some CA is converted to deoxycholic acid (DCA) and some CDCA is converted to lithocholic acid (LCA). Most bile acids are conjugated to glycine or taurine (amidation)

intestine, some primary bile acids will be further converted to secondary bile acids by gut bacterial enzymes. Human bile acid pool consists of two primary bile acid species: CA and chenodeoxycholic acid (CDCA) that make up approximately 80% of the total bile acid pool. The major secondary bile acid species in humans is deoxycholic acid (DCA), which accounts for about 20% of the total bile acids pool. Mice and rats have been used as major in vivo experimental models in research. In these rodents majority of CDCA is converted to muricholic acids (MCA) in the liver. In mice, α-MCA and β-MCA are the major MCA species that account for ~50% of the bile acid pool, while CDCA is usually less than 10% of the total bile acid pool (Li et al. 2012). MCAs have a hydroxyl group at the C-6 position. They are more hydrophilic and less toxic than their precursor. Due to these fundamental differences in bile acid composition and bile acid pool hydrophobicity between humans and mice, cautions need to be given when extrapolating findings from rodent models to humans in bile acid research.

7.2.2 Primary Bile Acid Synthesis in the Hepatocytes

Hepatic bile acid synthesis occurs through two pathways: the classic pathway and the alternative pathway, which are illustrated in Fig. 7.1. The classic bile acid synthetic pathway produces both CA and CDCA, and accounts for about 90% of total bile acid production in humans. The cholesterol 7α-hydroxylase (CYP7A1), a microsomal CYP enzyme, catalyzes the first and rate-limiting step to convert cholesterol to 7α-hydroxycholesterol (Chiang 2009; Myant and Mitropoulos 1977). The intermediate 7α-hydroxy-4-cholestene-3-one (C4) serves as the common precursor for CA and CDCA. Sterol 12α-hydroxylase (CYP8B1) catalyzes the hydroxylation at the C-12 position on C4, leading to the production of CA. The C4 that does not undergo 12-hydroxylation reaction is eventually converted to CDCA. The mitochondrial sterol 27-hydroxylase (CYP27A1) catalyzes the initial C-27 hydroxylation reaction in the sterol side chain oxidation of all bile acid species. In the alternative bile acid synthetic pathway, CYP27A1 catalyzes the first step to convert cholesterol to 27-hydroxycholesterol. In this pathway, the oxysterol 7α-hydroxylase (CYP7B1) catalyzes the hydroxylation at the C-7 position. The alternative pathway only produces CDCA. In addition to cholesterol, oxysterol intermediates formed in the peripheral tissues can enter the liver and feed into the alternative pathway for CDCA synthesis. The alternative pathway is thought to produce less than 10% of the total bile acids in humans under normal physiological conditions, but may account for about 50% of total bile acid production in some rodents (Li and Chiang 2014; Russell and Setchell 1992).

7.2.3 Secondary Bile Acid Synthesis in the Intestine

In the small and large intestine, bacterial bile salt hydrolases (BSH) de-conjugate some taurine- and glycine-conjugated bile acids to free bile acids (Ridlon et al. 2006). Then bacterial 7α-dehydroxylase activity converts CA to DCA and CDCA to either UDCA or lithocholic acid (LCA) (Fig. 7.2a) (Bjorkhem et al. 1989; Hylemon et al. 1991; Ridlon et al. 2006). UDCA can also be further converted to LCA. The secondary bile acids DCA and LCA are more hydrophobic and cytotoxic (Low-Beer and Nutter 1978; Marcus and Heaton 1988; Ridlon and Hylemon 2006). DCA is reabsorbed in the large intestine via passive absorption and transported back to the liver (Trauner and Boyer 2003). LCA can be sulfated and excreted in the bile (Alnouti 2009), or undergo hydroxylation (often at the C-6 position) and detoxification by CYP3A family enzymes in the liver and the intestine (Fig. 7.2a) (Araya and Wikvall 1999; Teixeira and Gil 1991). CYP3A can also convert some CDCA to MCA or hyocholic acid. Under normal conditions, only trace amount of LCA is found in the bile acid pool of humans and mice, and is mainly sulfated. Elevated LCA in cholestasis contributes to liver injury and inflammation (Staudinger et al. 2001). Phase-I bile acid metabolism is an important bile acid detoxification mechanism in cholestasis.

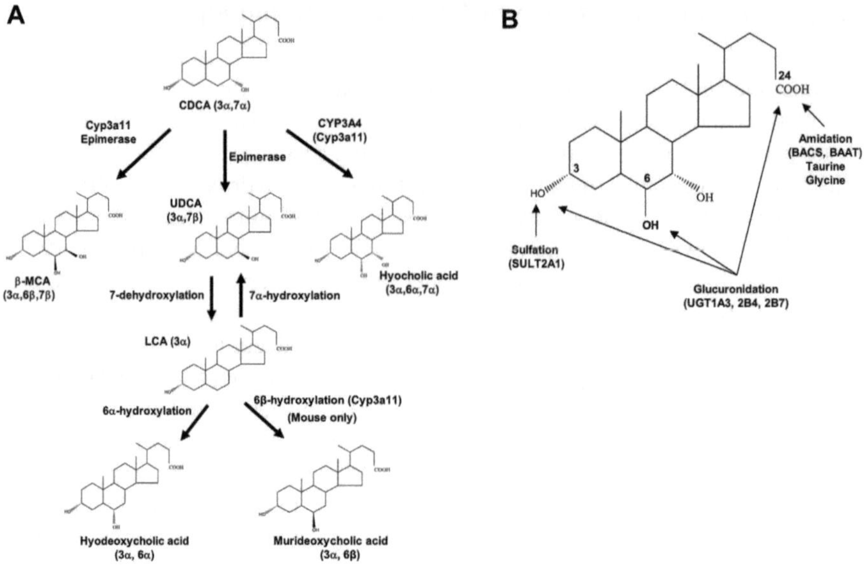

Fig. 7.2 Bile acid biotransformation. Phase-I and phase-II bile acid metabolisms are important bile acid detoxification mechanisms. (**a**). The synthesis of LCA from CDCA or UDCA involves 7α-dehydroxylation. LCA can be metabolized to hyodeoxycholic acid or murideoxycholic acid. LCA can also undergo 7α-hydroxylation to be converted back to UDCA. CYP3A enzymes play a key role in phase-I bile acid hydroxylation of LCA. CYP3A can also metabolize CDCA to MCA or hyocholic acid. (**b**). Under normal conditions, most of the bile acids are conjugated to amino acids taurine or glycine. Bile acids also undergo sulfation or glucuronidation mediated by SULTs or UGTs, respectively

7.2.4 Bile Acid Conjugation

Majority of the newly synthesized bile acids as well as de-conjugated bile acids returning to the liver from the intestine are conjugated to amino acid glycine or taurine on the side chain to form N-acyl amidates in humans, mice, and many other species (Shonsey et al. 2005). Amino acids are conjugated to bile acid molecules in a two-step enzymatic reaction (Fig. 7.2b). The bile acyl-CoA synthetase (BACS), also called bile acid-CoA ligase (BACL), first converts bile acids to an active intermediate CoA-thioester, which is then converted to amino acid conjugates by a bile acid-CoA:amino acid N-acyltransferase (BAAT) (Falany et al. 1994, 1997; Vessey et al. 1987; Wheeler et al. 1997). Human bile acid pool contains about three times more glycine-conjugated bile acids than taurine-conjugated bile acids, while mouse bile acid pool contains predominantly taurine-conjugated bile acids. Bile acid amidation increases bile acid aqueous solubility and decreases bile acid toxicity. Genetic defects in BACS and BAAT have been identified in pediatric human patients. These patients exhibited fat-soluble vitamin deficiency, growth retardation, and liver injury (Carlton et al. 2003; Chong et al. 2012; Setchell et al. 2013). Some bile acids also undergo sulfation and glucurondiation as phase-II bile acid bio-transformation (Alnouti 2009; Verreault et al. 2010), which also increases bile acid solubility for excretion (Fig. 7.2b). Bile acid sulfation is mediated by sulfotransferases (SULT). C-3 hydroxyl-group is a major site for bile acid sulfation. Small amount of LCA circulated to liver is immediately sulfated for biliary excretion. Glucuronidation of bile acids is mediated by UDP-glucuronosyltransferase (UGT) enzymes at the hydroxyl-group linked to C-3, C-6, or C-24 position. Bile acid sulfation and glucuronidation are minor bile acid conjugation pathways under healthy conditions, and may be upregulated during cholestasis conditions.

7.2.5 Enterohepatic Circulation of Bile Acids

Majority of the bile acids are absorbed in the distal intestine, transported to the liver via portal circulation, and re-secreted into the bile, which is referred to as the enterohepatic circulation. Cholecystokinin, a peptide hormone secreted by the epithelial cells in the duodenum upon food intake, stimulates gallbladder contraction and release of bile acids into the small intestine lumen (Otsuki 2000). Therefore, bile acid pool recycles a few times a day in the enterohepatic system, which is stimulated by meal intake. Absorption of bile acids in the intestine is highly efficient with only ~5% of total bile acids loss in the feces. As such, newly produced bile acids by de novo synthesis account for ~5% of the total bile acid pool, and a constant bile acid pool size is therefore maintained.

7.2.6 Hepatic and Intestine Bile Acid Transporters

Many specialized transporters are involved in mediating bile acid transport across the plasma membrane for uptake and secretion in the enterohepatic system (Meier 1995) (Fig. 7.3). Under normal conditions, bile acids returning to the liver from portal blood are efficiently taken up by the hepatocytes with an extraction rate of ~90%. The Na^+-dependent taurocholate transporter (NTCP, SLC10A1) is the major transporter that mediates conjugated bile acid transport across the basolateral membrane of the hepatocytes (Hagenbuch and Meier 1994; Hagenbuch et al. 1991; Meier and Stieger 2002). Deficient of NTCP resulted in elevated bile acid concentration in systemic circulation with no bile acid accumulation in the liver in both a human patient and *ntcp* knockout mice (Slijepcevic et al. 2015; Vaz et al. 2015). Several organic anion transporter (OATP2) isoforms mediate Na^+-independent uptake of mainly unconjugated bile acids into the hepatocytes (Trauner and Boyer 2003). NTCP has high substrate specificity for conjugated bile acids, while OATPs are also involved in the transport of other organic anions. Interestingly, recent studies revealed that NTCP also serves as a receptor for hepatitis B virus entry into hepatocytes (Ni et al. 2014; Yan et al. 2012). Hepatocyte secretion of bile acids into

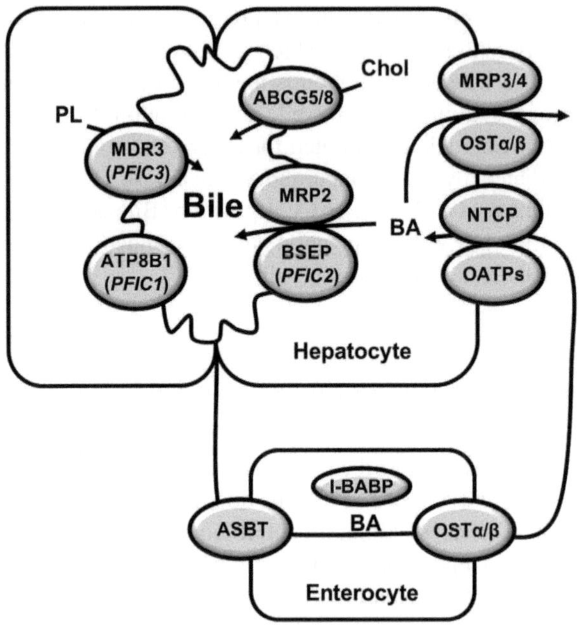

Fig. 7.3 Bile acid transport and enterohepatic bile acid circulation. Hepatocytes secrete bile acids into the bile via BSEP and MRP2. BSEP is the primary bile acid efflux transporter, while the multidrug resistance-associated protein-2 (MRP-2) can also secrete organic substrates including bile acids, bilirubin, and glutathione. ABCG5 and ABCG8 form heterodimers and transport cholesterol into the bile. Multidrug resistance-3 (MDR3) is responsible for phospholipid (PL) secretion. Cholesterol, bile acid, and phospholipid form mixed micelles in the bile. Genetic mutation of BSEP is linked to progressive familial intrahepatic cholestasis 1 (PFIC-1), and MDR3 mutation is linked to PFIC-3. ATP8B1 mutation is an aminophospholipid flippase and tis mutation is linked to PFIC-1. The Na^+-dependent taurocholate transporter (NTCP) is mainly responsible for Na^+-dependent uptake of conjugated bile acids. The organic anion transporters (OATPs) mediate the uptake of unconjugated bile acids. Organic solute transporters OSTα and OSTβ heterodimers, MRP3 and MRP4 secrete bile acids into the systemic circulation. In cholestasis, basolateral bile acid efflux is induced leading to subsequent renal bile acid excretion. In the intestine, the apical sodium-dependent bile acid transport (ASBT) mediates bile acid uptake into the enterocytes. Intracellular bile acids are bound to the intestinal bile acid-binding protein (I-BABP) and are transported to the basolateral membrane where bile acids are secreted into the portal circulation by the OSTα and OSTβ heterodimer

the bile against the concentration gradient is the rate-limiting step in bile formation (Boyer 2013). The bile salt export pump (BSEP, ABCB11) is the major bile acid transporter located at the canalicular membrane of the hepatocytes (Childs et al. 1995). Hepatocyte cholesterol secretion into the bile is mediated by the ATP-binding cassette transporters (ABCG5) and ABCG8 that form a functional heterodimer (Berge et al. 2000). Phosphatidylcholine secretion into the bile is mediated by the multi-drug resistant 3 (MDR3, ABCB4), which functions as canalicular phospholipid floppase (Smit et al. 1993). MDR-related protein 2 (MRP2) excretes conjugated bile acids, bilirubin, and other organic anions. Bile acids, cholesterol, and phospholipids form mixed micelles in the bile, which is critical in cholesterol solublization and also in reducing bile acid toxicity to the bile duct epithelium. As discussed later, loss-of-function mutations in either BSEP or ABCB4 are associated with cholestasis in humans.

Bile acids can be secreted across the basolateral membrane of the hepatocytes into the systemic circulation. This is not a significant process under normal physiology but can be up regulated during cholestasis (Ballatori et al. 2005; Boyer et al. 2006; Cui et al. 2009). The organic solute transporter α (OSTα) and OSTβ heterodimer, MRP3 and MRP4 are involved in basolateral bile acid efflux (Ballatori et al. 2005; Kullak-Ublick et al. 2000; Trauner and Boyer 2003). In the bile duct, unconjugated bile acids can passively diffuse into the cholangiocytes, and conjugated bile acids can enter the cholangiocytes through apical sodium-dependent bile salt transporter (ASBT) (Xia et al. 2006). Bile acids in cholangiocytes are secreted into the peribiliary plexus via the OSTα and OSTβ heterodimer and return to the hepatocytes for secretion into the bile or into the systemic circulation in cholestasis (Kullak-Ublick et al. 2000; Xia et al. 2006).

Bile acid transporters are highly expressed in the terminal ileum where majority of bile acids are absorbed. This is a highly efficient process and only about ~5% of the bile acids is loss via fecal excretion. The ASBT mediates bile acid uptake into the enterocytes (Shneider et al. 1995). Bile acids are carried by the intestinal bile acid-binding protein (I-BABP) in enterocytes to the basolateral membrane (Gong et al. 1994), where the OSTα and OSTβ heterodimer secretes bile acids into the portal circulation (Ballatori et al. 2005; Rao et al. 2008). This transport system is highly efficient in re-uptake of conjugated bile acids in the ileum. The unconjugated primary and secondary bile acids can be reabsorbed in the colon via passive diffusion. About 5% of the bile acid pool is lost daily in feces, and fecal bile acids are enriched with unconjugated bile acids compared to the endogenous bile acid pool composition.

7.3 Liver Injury in Cholestasis

7.3.1 Cholestasis

Cholestasis is a pathological condition where normal bile flow out of the liver is reduced or disrupted, leading to bile acid backup and intrahepatic accumulation of bile acids. Cholestasis could be resulted from genetic defects in canalicular transporters, mechanical obstruction of bile duct by gallstones or tumors, factors associated with pregnancy (intrahepatic cholestasis of pregnancy, ICP), autoimmune destruction of the bile ducts in and outside of the liver, drug-induced liver toxicity (Srivastava 2014; Zollner and Trauner 2008). In various cholestasis conditions, damage to the bile duct epithelium and elevated biliary pressure can cause bile duct rupture, exposure of hepatocytes to high concentrations of bile acids, inflammatory infiltration, and hepatocyte cell death. Hydrophobic bile acids are especially cytotoxic and cause hepatocyte cell death via various direct and indirect mechanisms. Chronic cholestasis causes liver fibrosis, cirrhosis, liver failure, or hepatocellular or cholangiocarcinomas.

Progressive familial intrahepatic cholestasis (PFIC) with chronic cholestasis and benign recurrent intrahepatic cholestasis (BRIC) with spontaneous resolving intermittent cholestasis are autosomal recessive diseases linked to mutations in *ATP8B1* (Type 1, PFIC1, also known as Byler disease), *BSEP* (Type 2, PFIC2), or *MDR3* (Type 3, PFIC3) (Srivastava 2014; Strautnieks et al. 1997) (Fig. 7.3). The *ATP8B1* gene encodes a P-type ATPase that functions as an aminophospholipid flippase that maintains membrane asymmetry by inward flipping of phosphatidylserine from the outer leaflet of the plasma membrane. Defective BSEP leads to intrahepatic bile acid accumulation and reduced biliary bile acids. MDR3 mediates canalicular phosphoatidylcholine secretion. Defective phospholipid secretion reduces mixed micelle formation, leading to "free" bile acid damage to the canalicular membrane and cholangiocytes. Congenital intrahepatic cholestasis due to defective canalicular transport is usually early onset and exhibits clinical features of pruritus and jaundice, and slow growth due to vitamin deficiency (Zollner and Trauner 2008). Genetic polymorphisms and heterozygote mutations of the *PFIC1*, *PFIC2*, and *PFIC3* genes can also affect susceptibility to acquired cholestasis such as intrahepatic cholestasis of pregnancy or drug-induced liver cholestasis (Lang et al. 2007; Noe et al. 2005; Pauli-Magnus et al. 2004). Dubin-Johnson syndrome has been linked to mutations in the *MRP2* gene that encodes the canalicular bilirubin transporter (Keitel et al. 2003). These patients have chronic hyperbilirubinemia, elevated bile acids and develop cholestasis. Neonatal cholestasis occurs in 1 of 2500 newborns and is even higher in premature newborns (Fischler and Lamireau 2014). Biliary atresia is a cholangiodestructive disease affecting both intra- and extra-hepatic biliary tract leading to cirrhosis and liver failure. Genetic defect in biliary tree development in infants causes decreased biliary bile flow and jaundice, liver cirrhosis, and death. Mutations in the junction protein 2 gene (*TJP2*) altered liver tight-junction structure and cause liver injury and progressive cholestasis (Sambrotta et al. 2014).

Primary biliary cholangitis (PBC) and primary sclerosing cholangitis (PSC) are two common types of slowly progressive chronic cholestasis diagnosed in adult humans (Kaplan and Gershwin 2005; Zein and Lindor 2010). The etiology of PBC and PSC is not fully clear but involves genetic, immune, and environmental factors. PBC is characterized by autoimmune destruction of small intrahepatic bile duct, leading to inflammation and impaired hepatic bile acid secretion and accumulation. PSC is an idiopathic liver disease characterized by chronic liver inflammation and progressive destruction of intrahepatic and extrahepatic bile ducts. PBC is predominantly diagnosed in females, while PSC is more prevalent in men.

7.3.2 Mechanism of Cell Death in Cholestasis

Although hepatocytes are routinely exposed to physiological concentration of bile acids returning from the portal circulation, exposing cultured primary human and mouse hepatocytes to supra-physiological levels of hydrophobic bile acids can directly cause hepatocyte cell death via necrosis. Cholestatic livers of rodent models and human patients exhibited histological characteristics of hepatocellular necrosis. Due to the detergent property of hydrophobic bile acids, hydrophobic bile acids at mM concentration range have been shown to damage plasma membrane lipid layers and alter membrane protein structure and function, and therefore disrupt plasma membrane integrity in cultured hepatocytes (Billington et al. 1980; Zhou et al. 2013). In addition, exposure of cultured hepatocytes to high concentrations of bile acids can cause necrosis, which is thought to depend on mitochondria dysfunction and ROS generation, leading to cellular ATP depletion, cell swelling, and lysis (Rodrigues et al. 1998b; Sokol et al. 2001; Spivey et al. 1993). Bile acids have also been shown to cause apoptosis in cultured hepatocytes, although recent experimental evidence suggests that necrosis is the major mechanism of cell death in mice after bile duct ligation (Woolbright and Jaeschke 2012). Treating mouse hepatocytes with 50 μM glycochenodeoxycholic acid (GCDCA) caused ligand-independent oligomerization of Fas receptor and caspase-8 activation, supporting the pro-apoptotic effect of hydrophobic bile acids (Faubion et al. 1999). In another study, taurolithocholate-3-sulfate was shown to trigger a rapid NADPH oxidase-dependent oxidative stress that initiates a signaling cascade leading to c-Jun N-terminal kinase (JNK), protein kinase C (PKC), and the epidermal growth factor receptor (EGFR) activation, which eventually caused Fas receptor phosphorylation and membrane translocation (Reinehr et al. 2005). This bile acid-induced signaling event resulted in mitochondrial permeability transition pore opening, cytochrome c release, caspase activation, and apoptosis. A study by Schoemaker et al. showed that NTCP-dependent GCDCA uptake into the hepatocytes was required for GCDCA-induced hepatocyte death, linking intracellularly accumulated bile acids to the initiation of cell death (Schoemaker et al. 2004). A number of studies have shown that bile acid can directly target mitochondria for reactive oxygen species (ROS) production and oxidative stress, which may be a key early event in bile acid-induced apoptosis. Direct exposure of isolated rodent and human mitochondria to hydrophobic bile acids impaired mitochondrial bioenergetics function and membrane potential and induced ROS generation (Krahenbuhl et al. 1994; Rolo et al. 2000; Sokol et al. 1993, 1995, 2005). In addition to the activation of extrinsic apoptotic pathway, bile acids have also been shown to directly cause mitochondrial membrane depolarization via ROS generation, mitochondrial permeability transition pore (MPT), mitochondrial Bax translocation, and cytochrome c release, and therefore activation of the intrinsic apoptotic pathway (Rodrigues et al. 1998a, 1999; Yerushalmi et al. 2001).

Direct toxicity of high concentrations of hydrophobic bile acids to hepatocytes has been demonstrated by numerous studies. However, cholestasis in vivo is associated with inflammation and immune cell infiltration, which plays more important roles in causing hepatocyte cell death in many forms of cholestasis (Woolbright and Jaeschke 2012). One possibility is that dead hepatocytes may release damage-associated molecular pattern molecules (DAMPs) that elicit a pro-inflammatory response in cholestatic livers. Furthermore, bile acids at pathological concentrations can act as pro-inflammatory molecules to induce cytokine and chemokine expressions in hepatocytes and macrophages (Zhang et al. 2012). Elevated pro-inflammatory cytokines and chemokines have been reported in human cholestasis (Neuman et al. 2002; Yamashiki et al. 1998). Bile acids are known to activate the MAPK signaling pathways including extracellular signal-regulated kinases 1/2 (ERK1/2) and JNK, which may be involved in bile acid induction of cytokine expression (Gupta et al. 2001, 2004; Miyake et al. 2000; Song et al. 2009). Through activation of the MAPK pathway, bile acids have been shown to induce early growth response factor 1 (Egr-1) in hepatocytes, which induced the production of cytokines, chemokines, and adhesion molecules that mediate the inflammatory response in cholestatic livers (Allen et al. 2010, 2011; Kim et al. 2006). These inflammatory responses as well as liver injury were significantly attenuated in *Egr-1* knockout mice after bile duct ligation (Kim et al. 2006). Neutrophil infiltration after initial hepatocyte injury has been shown to play important roles in cholestatic liver injury (Gujral et al. 2003, 2004). After bile duct ligation, there was extensive presence of neutrophils in the mouse liver parenchyma, while mice lacking CD18 or ICAM-1, which had significantly less neutrophil infiltration after bile duct ligation, also showed markedly attenuated liver injury (Gujral et al. 2003, 2004). In addition to hepatocytes, macrophages are a major source of cytokines including TNFα, IL-1β, and IL-6 in cholestasis. Cytokines produced by macrophages may mediate the hepatic pro-inflammatory response in cholestasis rather than causing direct hepatocyte cell death. IL-1 receptor deficient mice were partially protected against cholestatic liver injury after bile duct ligation (Sewnath et al. 2002). In contrast, studies found that depletion of Kupffer cells in the liver significantly increased liver injury and hepatocyte cell death in bile duct ligated mice, suggesting that Kupffer cells in general played a protective role in preventing liver injury and promoting cell survival and liver regeneration in cholestasis (Cheng et al. 2011; Gehring et al. 2006; Osawa et al. 2010; Roggin et al. 2000). Further studies conducted in mice deficient in pro-inflammatory cytokine signaling showed that bile duct ligation caused more severe hepatocellular necrosis in TNF receptor knockout mice, which was attributed to lack of TNFα-mediated Mrp3 induction and bile acid efflux (Bohan et al. 2003). Furthermore, mice lacking IL-6 exhibited significantly higher plasma transaminases and hepatocellular necrosis after bile duct ligation (Gehring et al. 2006). These studies collectively suggest that these pro-inflammatory cytokines produced by macrophages play a protective role after cholestatic liver injury.

7.4 Mechanisms of Bile Acid Detoxification in Cholestasis

A number of adaptive mechanisms are activated during cholestasis to reduce bile acid toxicity to the liver, including decreased de novo bile acid synthesis, increased phase I and phase II bile acid metabolism, and induction of basolateral bile acid secretion. Several ligand-activated nuclear receptors namely farnesoid X receptor (FXR), pregnane X receptor (PXR), and constitutive androstane receptor (CAR) play important roles by regulating the transcription of genes involved in these adaptive pathways. These nuclear receptors form heterodimers with retinoid X receptor

RXR and bind to their target gene promoters. Importantly, these nuclear receptors can be activated by bile acids or their derivatives in cholestasis, which thus tightly links bile acid sensing to the regulation of bile acid detoxification. In addition, bile acids can activate cell surface G protein couple receptors (GPCR) TGR5 and sphingosine-1-phosphate receptor (S1PR). These receptors have recently been implicated in the regulation of bile acid metabolism, inflammation, cell proliferation, and survival during cholestasis.

7.4.1 FXR

Hepatic bile acid synthesis is under negative feedback regulation by bile acids, which ensures a constant bile acid pool under normal physiology. In cholestasis, CYP7A1 expression may be repressed and hepatic de novo bile acid synthesis is reduced (Bertolotti et al. 2001; Schaap et al. 2009). The nuclear receptor FXR plays a central role in mediating bile acid feedback repression of bile acid synthetic genes in the liver. Hydrophobic bile acids including CDCA, CA, DCA, and LCA are endogenous FXR ligands, and among them CDCA is the most potent ligand with an EC_{50} = ~10 μmol/L (Makishima et al. 1999; Parks et al. 1999). Hydrophilic bile acids are not FXR ligands. FXR is expressed in both hepatocytes and in enterocytes, and activation of either hepatocyte FXR or intestine FXR leads to inhibition of *CYP7A1* gene transcription via two distinct signaling mechanisms. First, hepatocyte FXR can transcriptionally induce a nuclear receptor small heterodimer partner (SHP), which lacks a DNA binding domain and mainly acts as a co-repressor of other transcriptional factors via protein-protein interaction. Two nuclear receptors liver receptor homolog-1 (LRH-1) and hepatocyte nuclear factor 4α (HNF4α) bind to *CYP7A1* gene promoter and maintain the basal *CYP7A1* gene transcription (Goodwin et al. 2000; Lu et al. 2000). SHP has been shown to inhibit the transactivation activity of both HNF4α and LRH-1, leading to transcriptional inhibition of *CYP7A1*. Hepatocytes can efficiently secrete bile acids into the bile. In normal physiology, expansion of bile acid pool results in increased intestine bile acid content without significant intrahepatic bile acid accumulation (Li et al. 2010, 2012). Therefore, intestine bile acid sensing is very important in maintaining constant bile acid pool size. FXR activation in the intestine transcriptionally induces fibroblast growth factor 15 (FGF15) in mice. FGF15 can act as an endocrine hormone to activate the FGF receptor 4 (FGFR4) on hepatocytes. This leads to intracellular activation of ERK1/2 resulting in the transcriptional repression of *CYP7A1* gene (Inagaki et al. 2005; Goetz et al. 2007; Ito et al. 2005; Li et al. 2014; Lin et al. 2007; Wang et al. 2014). Human FGF19 shares ~51% amino acid sequence homology with mouse FGF15, and thus is the mouse FGF15 orthologue. FGF19 also acts via FGFR4 to repress human *CYP7A1* (Song et al. 2009). While mouse hepatocytes do not express FGF15, human hepatocytes express and secrete FGF19. Circulating FGF19 levels increased in human patients with obstructive cholestasis (Schaap et al. 2009), suggesting that hepatic bile acid accumulation in human cholestasis may induce both SHP and FGF19 to repress bile acid synthesis. It is worth mentioning that cholestasis is often accompanied with hepatic inflammation and elevation of cytokine expression. Numerous studies have demonstrated that pro-inflammatory cytokines repress *CYP7A1* gene transcription via activation of MAPK signaling pathways (Li and Chiang 2007; Li et al. 2006; Miyake et al. 2000). Hepatic FXR may be repressed in cholestasis (Alvarez et al. 2004; Chen et al. 2004). It is likely that pro-inflammatory signaling also plays a significant role in hepatic CYP7A1 down-regulation in chronic cholestasis.

Activation of FXR also reduces basolateral bile acid uptake via inhibition of NTCP (Denson et al. 2001), and promotes biliary bile acid secretion by inducing BSEP (Ananthanarayanan et al. 2001), MRP2 (Kast et al. 2002), and the phospholipid transporter ABCB4 (Liu et al. 2003). In cholestatic conditions, the basolateral transporters including OSTs and MRPs may be induced, resulting in direct bile acid efflux into the systemic blood circulation and elevated plasma bile acid concentration (Ballatori et al. 2005; Boyer et al. 2006; Cui et al. 2009). FXR induces the *OSTα* and *OSTβ* genes that are involved in basolateral bile acid secretion. In the intestine, FXR induces *OSTα* and *OSTβ* genes (Frankenberg et al. 2006) and represses *ASBT* gene (Chen et al. 2003; Neimark et al. 2004), and thus inhibits intestine bile acid absorption.

FXR mutations have been linked to ICP (Mullenbach et al. 2005; Painter et al. 2005; Van Mil et al. 2007). A recent study has identified mutations in the *FXR* gene in PFIC patients by whole-exome sequencing, and single nucleotide polymorphism study of two unrelated probands with severe cholestasis identified homozygous mutations in the *FXR* gene (Gomez-Ospina et al. 2016). The mutations caused altered DNA binding, non-responsive to ligand activation and non-detectable FXR.

7.4.2 PXR and CAR

The PXR is a xenobiotic sensor that is activated by a wide range of xenobiotics, endobiotics, and clinical drugs. PXR activation induces CYP3A and CYP2 family enzymes, conjugation enzymes, and transporters, and thus plays a key role in drug metabolism and drug-drug interaction (Kliewer et al. 2002). In cholestasis, PXR can also be activated by high levels of bile acids. Activation of PXR by bile acids or drugs has been shown to transcriptionally inhibit *CYP7A1* gene via its repressive effect on HNF4α (Bhalla et al. 2004; Li and Chiang 2005; Mason and Boyd 1978; Stahlberg 1995; Staudinger et al. 2001). In addition, PXR induces phase I bile acid metabolizing enzymes such as CYP3A and CYP2B that convert hydrophobic bile acids into more hydrophilic and less toxic metabolites via hydroxylation, which at the same time increases bile acid conjugation and secretion (Staudinger et al. 2001). In phase II and phase III bile acid metabolism, PXR also induces the bile acid conjugation enzyme SULT2A1 and UGT isoforms that play significant roles in detoxifying bile acids in cholestasis (Kliewer and Willson 2002). Consistently, mice lacking PXR were more susceptible to hepatotoxicity caused by LCA treatment or bile duct ligation (Staudinger et al. 2001, 2005), while pharmacological activation of PXR protected against liver injury in experimental cholestasis models (Stedman et al. 2005). The PXR agonist rifampicin has been used to treat pruritus in human cholestasis (Hofmann 2002).

The CAR is another key nuclear receptor in regulating drug and bile acid metabolizing genes in the liver (Stanley et al. 2006). Bile acids are not endogenous CAR ligands, but CAR can be activated by toxic metabolites in cholestatic livers. CAR also binds to the promoter of *CYP3A* and *CYP2B* genes and induces their expressions. CAR may play an important role in inducing sulfation of bile acids because *CAR* transgenic mice had increased levels of sulfated LCA and were resistant to LCA toxicity (Saini et al. 2004). Furthermore, CAR activation resulted in *CYP7A1* gene down-regulation in hepatocytes (Miao et al. 2006). A number of studies conducted in mice have demonstrated a protective role of CAR against cholestatic liver injury (Beilke et al. 2009; Guo et al. 2003; Saini et al. 2004; Stedman et al. 2005).

7.4.3 TGR5

TGR5 is a $G_{\alpha s}$ class of GPCR activated by bile acids (Kawamata et al. 2003; Maruyama et al. 2002). TGR5 activation stimulates adenylyl cyclase, which converts AMP to cAMP to activate protein kinase A (PKA). LCA and 3-keto-LCA are the most potent TGR5 ligands with an EC_{50} of less than 1 μM. Other hydrophobic bile acids such as DCA, CDCA, and CA activate TGR5 at EC_{50} of 1–10 μM range. In the liver, TGR5 is not expressed in hepatocytes, but is expressed in liver sinusoidal endothelial cells (Keitel et al. 2007), cholangiocytes, gallbladder epithelial cells, and Kupffer cells, but not in hepatocytes (Keitel et al. 2008). TGR5 is also expressed in the ileum and colon (Kawamata et al. 2003), where TGR5 regulates gut motility(Alemi et al. 2013b), barrier function and immune response (Cipriani et al. 2011; Yoneno et al. 2013), and glucagon-like peptide 1 (GLP-1) production (Katsuma et al. 2005; Thomas et al. 2009). Outside the enterohepatic system, TGR5 is expressed in white and brown adipose, spleen, kidney, pancreas, lung, and the central nervous system (Kawamata et al. 2003). Studies have suggested that TGR5 in these tissues can be targeted by pharmacological agents as potential therapeutics (Kumar et al. 2012, 2016; Wang et al. 2016; Watanabe et al. 2006). However, because these tissues are normally exposed to very low levels of bile acids, whether TGR5 mediates endogenous bile acid signaling in these tissues in physiological conditions is uncertain.

Experimental evidence supporting the involvement of TGR5 signaling in bile acid metabolism and cholestatic liver injury came from a few studies conducted in *Tgr5* knockout mice. These studies found that *Tgr5* knockout mice had reduced bile acid pool size (Maruyama et al. 2006), a more hydrophobic bile acid composition and showed more severe liver injury after bile duct ligation (Pean et al. 2013), suggesting a protective role of TGR5 in cholestasis. The reduced bile acid pool size and altered bile acid pool composition in *Tgr5* knockout mice cannot be fully accounted for by altered hepatic bile acid synthesis gene expression (Maruyama et al. 2006). The role of TGR5 in the regulation of bile acid transport in the enterohepatic systems remains to be determined. Cholangiocytes may be exposed to physiological and pathological concentrations of bile acids high enough to activate TGR5. In addition, cholangiocytes are subjected to direct toxicity of high concentrations of bile acids during cholestasis. TGR5 is localized to the primary cilium of cholangiocytes and exhibited proliferative and anti-apoptotic functions (Keitel et al. 2010; Reich et al. 2016). Cholangiocyte proliferation was significantly reduced in bile duct-ligated *Tgr5* KO mice, while TGR5 activation protected against cholangiocyte cell death, suggesting that cholangiocyte TGR5 is involved in protection against bile acid-induced bile duct injury. In addition, TGR5 promotes cholangiocyte proliferation and the development of cholangiocarcinoma in cholestasis (Reich et al. 2016).

Cholestasis is associated with hepatocyte cell death, cytokine production, and inflammatory infiltration. It was reported that $Tgr5^{-/-}$ mice had more severe liver necrosis and inflammation compared to wild-type mice (Wang et al. 2011). The TGR5-cAMP pathway inhibits LPS-induced cytokine production in Kupffer cells, which suggested that TGR5 in Kupffer may play a protective role against excessive cytokine production and liver injury in obstructive cholestasis (Keitel et al. 2008). Consistent with an anti-inflammatory role of TGR5 in macrophages, TGR5 activation is shown to be beneficial in other inflammation-associated disease models including atherosclerosis, non-alcoholic steatohepatitis (NASH), and colitis (Cipriani et al. 2011; Kawamata et al. 2003; Keitel et al. 2008; McMahan et al. 2013; Pean et al. 2013; Perino et al. 2014; Wang et al. 2011; Yoneno et al. 2013). Finally, recent studies showed that activation of TGR5 in sensory nerves by bile acids or their derivatives mediates itch associated with cholestasis (Alemi et al. 2013a; Lieu et al. 2014).

7.4.4 Sphingosine 1 Phosphate Receptor

By screening of bile acid-activated GPCR, Karimian et al. identified the sphingosine-1-phosphate receptor 2 (S1PR2) as a receptor for tauro- or glyco-conjugated hydrophobic and hydrophilic bile acids, including TCA, TDCA, TUDCA, GCA, GDCA, but not for unconjugated bile acids (Studer et al. 2012). Sphingosine can be produced via ceramide hydrolysis or via de novo synthesis from serine and palmitoyl-CoA. Sphingosine-1-phosphate (S1P) is a lipid mediator generated by intracellular lipid kinases sphingosine kinase 1 (SphK1) and sphingosine kinase 2 (SphK2) via sphingosine phosphorylation (Strub et al. 2010). S1P is transported outside of the cells and via autocrine/paracrine manners activates a number of SIPRs. Activation of S1PR2 by conjugated-bile acids leads to activation of ERK1/2 and AKT signaling pathways, and may mediate TUDCA regulation of cell survival pathways (Qiao et al. 2002; Schoemaker et al. 2004). In line with this, activation of SphK and S1P production has also been suggested to mediate the anti-apoptotic effects of TNFα on human hepatocytes (Osawa et al. 2001, 2005). Increased ceramide and sphingosine production is linked to obesity and NAFLD (Jiang et al. 2015a, 2015b; Qi et al. 2015). Conversion of ceramide to S1P may attenuate the effect of ceramide and high-fat diet induced insulin resistance (Bruce et al. 2012). S1P activation of S1PRs regulates inflammation, cell death, and insulin sensitivity in liver, muscle, pancreas, and immune cells (Adada et al. 2013). The role of bile acid activation of S1PR in modulating bile acid toxicity and hepatocyte injury requires further investigation.

7.5 Cholestasis Therapies

7.5.1 Ursodeoxycholic Acid

UDCA is a hydrophilic bile acid presented at very low levels (less than 5%) in human bile. UDCA is generally nontoxic to humans, and has been used for effective gallstone dissolution and also for digestive diseases in human patients for many years (Lioudaki et al. 2011). UDCA (Ursodiol™) has been approved by FDA for treating PBC, and has been shown to significantly improve liver tests and prolong the time needed for liver transplantation in these patients (Dyson et al. 2015). At recommended dose of 13–15 mg/kg/day, UDCA can be enriched to ~40% in the bile of patients (Dilger et al. 2012). UDCA does not activate bile acid receptors FXR or TGR5. Originally thought that UDCA can replace toxic hydrophobic bile acid and increase bile acid pool hydrophilicity, more recent evidence suggests that UDCA can activate intracellular signaling pathways such as PKC and MAPK and alleviate liver injury in cholestasis via several mechanisms. UDCA increases hepatobiliary secretion and promotes biliary $HCO3^-$ secretion that protects hepatocytes and cholangiocytes from hydrophobic bile acid insult (Beuers 2006; Hohenester et al. 2012; Prieto et al. 1999). In addition, studies also support that UDCA exhibits anti-inflammatory and pro-survival effects in cholestasis (Beuers 2006; Poupon 2012). Nor-ursodeoxycholic acid (norUDCA) is a side chain-shortened C_{23} homologue of UDCA (Yeh et al. 1997; Yoon et al. 1986). It has been shown that norUDCA is passively absorbed by cholangiocytes and undergoes cholehepatic shunting, and increases $HCO3^-$ secretion (Beuers 2006). So far, norUDCA has been shown to improve sclerosing cholangitis in the *Mdr2−/−* model of cholangiopathy (Halilbasic et al. 2009), and its potential benefits in human cholestasis will be tested in current and future clinical trials. About 40% of the PBC patients do not respond adequately to UDCA and still need novel effective therapies for disease management (Pares et al. 2006). UDCA is not effective in treating patient with PSC, and the efficacy of UDCA in other forms of cholestasis has not been demonstrated (European Association for the Study of the Liver 2009).

Fibrates, the peroxisome proliferator-activated receptor α (PPARα), have been used to treat patients with hypertriglyceridemia. Fibrate has been used as monotherapy for PBC patients, and in combination therapy with UDCA for PBC patients who do not adequately respond to UDCA (Ghonem et al. 2015). PPARα inhibits

CYP7A1, CYP8B1, and CYP27A1 to reduce bile acid synthesis. PPARα also induces MRP3, UGT2B4, and SULT2A1 for detoxification of bile acids. In clinical studies, bezafibrate reduced serum alkaline phosphatase and GTT in PBC patients who did not adequately respond to UDCA (Iwasaki et al. 2008). See detailed review of the use of fibrate drugs as adjunct therapy for cholestasis (Ghonem et al. 2015).

7.5.2 FXR Agonist

Obeticholic acid (OCA) is a potent FXR agonist developed by Intercept Pharmaceuticals Inc. OCA is a 6α-ethyl CDCA derivative that selectively activates FXR with a ~100-fold higher potency than CDCA (Pellicciari et al. 2002, 2004). Based on the understanding of FXR regulation of bile acid metabolism and inflammation, the potential benefit of OCA in treating cholestasis has been extensively investigated in both experimental animal models and in humans (Ali et al. 2015). OCA treatment showed effective protection in experimental cholestasis models (Fiorucci et al. 2005; Pellicciari et al. 2002). Recent clinical trials also showed that OCA significantly improved liver tests in patients with PBC (Hirschfield et al. 2015). In patients who did not respond adequately to UDCA therapy, OCA significantly reduced serum alkaline phosphatase, γ-glutamyltransferase (GGT), and alanine aminotransferase activities compared to placebo group (Hirschfield et al. 2015). OCA has recently been approved by FDA for treating PBC patients. In addition, OCA also improved NASH score in clinical trials, and is a promising therapy for fatty liver disease (Ali et al. 2015; Neuschwander-Tetri et al. 2015).

7.5.3 TGR5 Agonist and FXR/TGR5 Dual Agonist

INT-777 is a TGR5 selective agonist and has been shown to protect intestinal barrier and immune response to experimental colitis (Cipriani et al. 2011). TGR5 activation stimulates GLP-1 secretion from enteroendocrine L cell and stimulates insulin secretion from pancreatic β cells (Katsuma et al. 2005; Kumar et al. 2016). Potent TGR5 selective agonists are in the development for treating metabolic diseases (Sato et al. 2008; Tiwari and Maiti 2009). Activation of TGR5 protects cholangiocytes from bile acid toxicity in cholestasis and thus is a potential therapy for cholestasis (Cipriani et al. 2011; Pols et al. 2011). INT-767 is a dual FXR and TGR5 agonist (D'Amore et al. 2014; Rizzo et al. 2010). INT-767 has been shown to reduce liver injury in *Mdr2*$^{-/-}$ cholangiopathy mouse model (Baghdasaryan et al. 2011). INT-767 treatment to *db/db* mice improved hepatic steatosis and decreased inflammation (McMahan et al. 2013). However, TGR5 agonists may also promote proliferation and apoptosis and progression of cholangiocarcinoma (Reich et al. 2016). Pruritus is commonly associated with cholestasis and also with treatment with bile acid derivatives including both FXR and TGR5 agonists (Alemi et al. 2013a).

7.5.4 FGF19 Therapy

FGF19 is induced by FXR agonist to inhibit hepatic bile acid synthesis and thus is a potential therapeutic for cholestasis. An unwanted effect of FGF19 therapy is potential carcinogenesis because FGF19 is a growth factor that stimulates cell proliferation. An engineered FGF19 variant M70 mimics FGF19 in regulation of bile acid synthesis but is devoid of growth factor activity and does not cause hepatic tumorigenesis (Zhou et al. 2014). This FGF19 variant reduces CYP7A1 activity and protects mice from cholestatic liver injury, and also reduces serum 7α-hydroxy-4-cholesten-3-one, the serum marker for hepatic bile acid synthesis in human subjects (Luo et al. 2014). M70 reduces liver injury in *Mdr2*$^{-/-}$ sclerosing cholangitis mice (Zhou et al. 2016). Phase II clinical trial of M70 for primary sclerosing cholangitis patients is ongoing.

7.6 Conclusion

Bile acids are physiological detergent molecules that facilitate intestine fat and nutrient absorption, and are signaling molecules that regulate various cellular processes involved in metabolism, immune response, and cell growth. Bile acids are highly toxic and accumulation of bile acids in cholestasis leads to liver injury and inflammation. Both genetic defects and environmental factors can lead to cholestasis. During cholestasis, many protective mechanisms are activated to decrease bile acid synthesis and increase bile acid detoxification and excretion. Bile acid-activated receptors play important roles in mediating these protective mechanisms in the liver. Understanding of the mechanism of cell injury and the regulation of bile acid detoxification mechanisms in cholestasis have provided molecular basis for the development of novel therapeutic approaches for the treatment of cholestasis.

Acknowledgment This work is supported in part by NIH grant 1R01DK102487-01 (T.L.), the National Center for Research Resources (5P20RR021940-07), and the National Institute of General Medical Sciences (8 P20 GM103549-07) of the National Institutes of Health (T.L.); NIH grants DK58379 (J.C.) and DK44442 (J.C.).

References

Adada M, Canals D, Hannun YA, Obeid LM (2013) Sphingosine-1-phosphate receptor 2: a review. FEBS J 280:6354–6366

Alemi F, Kwon E, Poole DP, Lieu T, Lyo V, Cattaruzza F, Cevikbas F, Steinhoff M, Nassini R, Materazzi S, Guerrero-Alba R, Valdez-Morales E, Cottrell GS, Schoonjans K, Geppetti P, Vanner SJ, Bunnett NW, Corvera CU (2013a) The TGR5 receptor mediates bile acid-induced itch and analgesia. J Clin Invest 123:1513–1530

Alemi F, Poole DP, Chiu J, Schoonjans K, Cattaruzza F, Grider JR, Bunnett NW, Corvera CU (2013b) The receptor TGR5 mediates the prokinetic actions of intestinal bile acids and is required for normal defecation in mice. Gastroenterology 144:145–154

Ali AH, Carey EJ, Lindor KD (2015) Recent advances in the development of farnesoid X receptor agonists. Ann Transl Med 3:5

Allen K, Kim ND, Moon JO, Copple BL (2010) Upregulation of early growth response factor-1 by bile acids requires mitogen-activated protein kinase signaling. Toxicol Appl Pharmacol 243:63–67

Allen K, Jaeschke H, Copple BL (2011) Bile acids induce inflammatory genes in hepatocytes: a novel mechanism of inflammation during obstructive cholestasis. Am J Pathol 178:175–186

Alnouti Y (2009) Bile acid sulfation: a pathway of bile acid elimination and detoxification. Toxicol Sci 108:225–246

Alvarez L, Jara P, Sanchez-Sabate E, Hierro L, Larrauri J, Diaz MC, Camarena C, De la Vega A, Frauca E, Lopez-Collazo E, Lapunzina P (2004) Reduced hepatic expression of farnesoid X receptor in hereditary cholestasis associated to mutation in ATP8B1. Hum Mol Genet 13:2451–2460

Ananthanarayanan M, Balasubramanian N, Makishima M, Mangelsdorf DJ, Suchy FJ (2001) Human bile salt export pump promoter is transactivated by the farnesoid X receptor/bile acid receptor. J Biol Chem 276:28857–28865

Araya Z, Wikvall K (1999) 6alpha-hydroxylation of taurochenodeoxycholic acid and lithocholic acid by CYP3A4 in human liver microsomes. Biochim Biophys Acta 1438:47–54

Baghdasaryan A, Claudel T, Gumhold J, Silbert D, Adorini L, Roda A, Vecchiotti S, Gonzalez FJ, Schoonjans K, Strazzabosco M, Fickert P, Trauner M (2011) Dual farnesoid X receptor/TGR5 agonist INT-767 reduces liver injury in the Mdr2−/− (Abcb4−/−) mouse cholangiopathy model by promoting biliary HCO output. Hepatology 54:1303–1312

Ballatori N, Christian WV, Lee JY, Dawson PA, Soroka CJ, Boyer JL, Madejczyk MS, Li N (2005) OSTalpha-OSTbeta: a major basolateral bile acid and steroid transporter in human intestinal, renal, and biliary epithelia. Hepatology 42:1270–1279

Beilke LD, Aleksunes LM, Holland RD, Besselsen DG, Beger RD, Klaassen CD, Cherrington NJ (2009) Constitutive androstane receptor-mediated changes in bile acid composition contributes to hepatoprotection from lithocholic acid-induced liver injury in mice. Drug Metab Dispos 37:1035–1045

Berge KE, Tian H, Graf GA, Yu L, Grishin NV, Schultz J, Kwiterovich P, Shan B, Barnes R, Hobbs HH (2000) Accumulation of dietary cholesterol in sitosterolemia caused by mutations in adjacent ABC transporters. Science 290:1771–1775

Bertolotti M, Carulli L, Concari M, Martella P, Loria P, Tagliafico E, Ferrari S, Del Puppo M, Amati B, De Fabiani E, Crestani M, Amorotti C, Manenti A, Carubbi F, Pinetti A, Carulli N (2001) Suppression of bile acid synthesis, but not of hepatic cholesterol 7alpha-hydroxylase expression, by obstructive cholestasis in humans. Hepatology 34:234–242

Beuers U (2006) Drug insight: mechanisms and sites of action of ursodeoxycholic acid in cholestasis. Nat Clin Pract Gastroenterol Hepatol 3:318–328

Bhalla S, Ozalp C, Fang S, Xiang L and Kemper JK (2004) Ligand-activated PXR interferes with HNF-4 signaling by targeting a common coactivator PGC-1alpha: functional implications in hepatic cholesterol and glucose metabolism. J Biol Chem 279:45139–45147

Billington D, Evans CE, Godfrey PP, Coleman R (1980) Effects of bile salts on the plasma membranes of isolated rat hepatocytes. Biochem J 188:321–327

Bjorkhem I, Einarsson K, Melone P, Hylemon P (1989) Mechanism of intestinal formation of deoxycholic acid from cholic acid in humans: evidence for a 3-oxo-delta 4-steroid intermediate. J Lipid Res 30:1033–1039

Bohan A, Chen WS, Denson LA, Held MA, Boyer JL (2003) Tumor necrosis factor alpha-dependent up-regulation of Lrh-1 and Mrp3(Abcc3) reduces liver injury in obstructive cholestasis. J Biol Chem 278:36688–36698

Boyer JL (2013) Bile formation and secretion. Compr Physiol 3:1035–1078

Boyer JL, Trauner M, Mennone A, Soroka CJ, Cai SY, Moustafa T, Zollner G, Lee JY, Ballatori N (2006) Upregulation of a basolateral FXR-dependent bile acid efflux transporter OSTalpha-OSTbeta in cholestasis in humans and rodents. Am J Physiol Gastrointest Liver Physiol 290:G1124–G1130

Bruce CR, Risis S, Babb JR, Yang C, Kowalski GM, Selathurai A, Lee-Young RS, Weir JM, Yoshioka K, Takuwa Y, Meikle PJ, Pitson SM, Febbraio MA (2012) Overexpression of sphingosine kinase 1 prevents ceramide accumulation and ameliorates muscle insulin resistance in high-fat diet-fed mice. Diabetes 61:3148–3155

Carlton VE, Harris BZ, Puffenberger EG, Batta AK, Knisely AS, Robinson DL, Strauss KA, Shneider BL, Lim WA, Salen G, Morton DH, Bull LN (2003) Complex inheritance of familial hypercholanemia with associated mutations in TJP2 and BAAT. Nat Genet 34:91–96

Chen F, Ma L, Dawson PA, Sinal CJ, Sehayek E, Gonzalez FJ, Breslow J, Ananthanarayanan M, Shneider BL (2003) Liver receptor homologue-1 mediates species- and cell line-specific bile acid-dependent negative feedback regulation of the apical sodium-dependent bile acid transporter. J Biol Chem 278:19909–19916

Chen F, Ananthanarayanan M, Emre S, Neimark E, Bull LN, Knisely AS, Strautnieks SS, Thompson RJ, Magid MS, Gordon R, Balasubramanian N, Suchy FJ, Shneider BL (2004) Progressive familial intrahepatic cholestasis, type 1, is associated with decreased farnesoid X receptor activity. Gastroenterology 126:756–764

Cheng CW, Duwaerts CC, Rooijen N, Wintermeyer P, Mott S, Gregory SH (2011) NK cells suppress experimental cholestatic liver injury by an interleukin-6-mediated, Kupffer cell-dependent mechanism. J Hepatol 54:746–752

Chiang JY (2009) Bile acids: regulation of synthesis. J Lipid Res 50:1955–1966

Childs S, Yeh RL, Georges E, Ling V (1995) Identification of a sister gene to P-glycoprotein. Cancer Res 55:2029–2034

Chong CP, Mills PB, McClean P, Gissen P, Bruce C, Stahlschmidt J, Knisely AS, Clayton PT (2012) Bile acid-CoA ligase deficiency—a new inborn error of bile acid metabolism. J Inherit Metab Dis 35:521–530

Cipriani S, Mencarelli A, Chini MG, Distrutti E, Renga B, Bifulco G, Baldelli F, Donini A, Fiorucci S (2011) The bile acid receptor GPBAR-1 (TGR5) modulates integrity of intestinal barrier and immune response to experimental colitis. PLoS One 6:e25637

Cui YJ, Aleksunes LM, Tanaka Y, Goedken MJ, Klaassen CD (2009) Compensatory induction of liver efflux transporters in response to ANIT-induced liver injury is impaired in FXR-null mice. Toxicol Sci 110:47–60

D'Amore C, Di Leva FS, Sepe V, Renga B, Del Gaudio C, D'Auria MV, Zampella A, Fiorucci S, Limongelli V (2014) Design, synthesis, and biological evaluation of potent dual agonists of nuclear and membrane bile acid receptors. J Med Chem 57:937–954

Denson LA, Sturm E, Echevarria W, Zimmerman TL, Makishima M, Mangelsdorf DJ, Karpen SJ (2001) The orphan nuclear receptor, shp, mediates bile acid-induced inhibition of the rat bile acid transporter, ntcp. Gastroenterology 121:140–147

Dilger K, Hohenester S, Winkler-Budenhofer U, Bastiaansen BA, Schaap FG, Rust C, Beuers U (2012) Effect of ursodeoxycholic acid on bile acid profiles and intestinal detoxification machinery in primary biliary cirrhosis and health. J Hepatol 57:133–140

Dyson JK, Hirschfield GM, Adams DH, Beuers U, Mann DA, Lindor KD, Jones DE (2015) Novel therapeutic targets in primary biliary cirrhosis. Nat Rev Gastroenterol Hepatol 12(3):147–158

European Association for the Study of the Liver (2009) EASL clinical practice guidelines: management of cholestatic liver diseases. J Hepatol 51:237–267

Falany CN, Johnson MR, Barnes S, Diasio RB (1994) Glycine and taurine conjugation of bile acids by a single enzyme. Molecular cloning and expression of human liver bile acid CoA:amino acid N-acyltransferase. J Biol Chem 269:19375–19379

Falany CN, Fortinberry H, Leiter EH, Barnes S (1997) Cloning, expression, and chromosomal localization of mouse liver bile acid CoA:amino acid N-acyltransferase. J Lipid Res 38:1139–1148

Faubion WA, Guicciardi ME, Miyoshi H, Bronk SF, Roberts PJ, Svingen PA, Kaufmann SH, Gores GJ (1999) Toxic bile salts induce rodent hepatocyte apoptosis via direct activation of Fas. J Clin Invest 103:137–145

Fiorucci S, Clerici C, Antonelli E, Orlandi S, Goodwin B, Sadeghpour BM, Sabatino G, Russo G, Castellani D, Willson TM, Pruzanski M, Pellicciari R, Morelli A (2005) Protective effects of 6-ethyl chenodeoxycholic acid, a farnesoid X receptor ligand, in estrogen-induced cholestasis. J Pharmacol Exp Ther 313:604–612

Fischler B, Lamireau T (2014) Cholestasis in the newborn and infant. Clin Res Hepatol Gastroenterol 38:263–267

Frankenberg T, Rao A, Chen F, Haywood J, Shneider BL, Dawson PA (2006) Regulation of the mouse organic solute transporter alpha-beta, Ostalpha-Ostbeta, by bile acids. Am J Physiol Gastrointest Liver Physiol 290:G912–G922

Gehring S, Dickson EM, San Martin ME, van Rooijen N, Papa EF, Harty MW, Tracy TF Jr, Gregory SH (2006) Kupffer cells abrogate cholestatic liver injury in mice. Gastroenterology 130:810–822

Ghonem NS, Assis DN, Boyer JL (2015) Fibrates and cholestasis. Hepatology 62:635–643

Goetz R, Beenken A, Ibrahimi OA, Kalinina J, Olsen SK, Eliseenkova AV, Xu C, Neubert TA, Zhang F, Linhardt RJ, Yu X, White KE, Inagaki T, Kliewer SA, Yamamoto M, Kurosu H, Ogawa Y, Kuro-o M, Lanske B, Razzaque MS, Mohammadi M (2007) Molecular insights into the klotho-dependent, endocrine mode of action of fibroblast growth factor 19 subfamily members. Mol Cell Biol 27:3417–3428

Gomez-Ospina N, Potter CJ, Xiao R, Manickam K, Kim MS, Kim KH, Shneider BL, Picarsic JL, Jacobson TA, Zhang J, He W, Liu P, Knisely AS, Finegold MJ, Muzny DM, Boerwinkle E, Lupski JR, Plon SE, Gibbs RA, Eng CM, Yang Y, Washington GC, Porteus MH, Berquist WE, Kambham N, Singh RJ, Xia F, Enns GM, Moore DD (2016) Mutations in the nuclear bile acid receptor FXR cause progressive familial intrahepatic cholestasis. Nat Commun 7:10713

Gong YZ, Everett ET, Schwartz DA, Norris JS, Wilson FA (1994) Molecular cloning, tissue distribution, and expression of a 14-kDa bile acid-binding protein from rat ileal cytosol. Proc Natl Acad Sci U S A 91:4741–4745

Goodwin B, Jones SA, Price RR, Watson MA, McKee DD, Moore LB, Galardi C, Wilson JG, Lewis MC, Roth ME, Maloney PR, Willson TM, Kliewer SA (2000) A regulatory cascade of the nuclear receptors FXR, SHP-1, and LRH-1 represses bile acid biosynthesis. Mol Cell 6:517–526

Gujral JS, Farhood A, Bajt ML, Jaeschke H (2003) Neutrophils aggravate acute liver injury during obstructive cholestasis in bile duct-ligated mice. Hepatology 38:355–363

Gujral JS, Liu J, Farhood A, Hinson JA, Jaeschke H (2004) Functional importance of ICAM-1 in the mechanism of neutrophil-induced liver injury in bile duct-ligated mice. Am J Physiol Gastrointest Liver Physiol 286:G499–G507

Guo GL, Lambert G, Negishi M, Ward JM, Brewer HB Jr, Kliewer SA, Gonzalez FJ, Sinal CJ (2003) Complementary roles of farnesoid X receptor, pregnane X receptor, and constitutive androstane receptor in protection against bile acid toxicity. J Biol Chem 278:45062–45071

Gupta S, Stravitz RT, Dent P, Hylemon PB (2001) Down-regulation of cholesterol 7alpha-hydroxylase (CYP7A1) gene expression by bile acids in primary rat hepatocytes is mediated by the c-Jun N-terminal kinase pathway. J Biol Chem 276:15816–15822

Gupta S, Natarajan R, Payne SG, Studer EJ, Spiegel S, Dent P, Hylemon PB (2004) Deoxycholic acid activates the c-Jun N-terminal kinase pathway via FAS receptor activation in primary hepatocytes. Role of acidic sphingomyelinase-mediated ceramide generation in FAS receptor activation. J Biol Chem 279:5821–5828

Hagenbuch B, Meier PJ (1994) Molecular cloning, chromosomal localization, and functional characterization of a human liver Na+/bile acid cotransporter. J Clin Invest 93:1326–1331

Hagenbuch B, Stieger B, Foguet M, Lubbert H, Meier PJ (1991) Functional expression cloning and characterization of the hepatocyte Na+/bile acid cotransport system. Proc Natl Acad Sci U S A 88:10629–10633

Halilbasic E, Fiorotto R, Fickert P, Marschall HU, Moustafa T, Spirli C, Fuchsbichler A, Gumhold J, Silbert D, Zatloukal K, Langner C, Maitra U, Denk H, Hofmann AF, Strazzabosco M, Trauner M (2009) Side chain structure determines unique physiologic and therapeutic properties of norursodeoxycholic acid in Mdr2−/− mice. Hepatology 49:1972–1981

Hirschfield GM, Mason A, Luketic V, Lindor K, Gordon SC, Mayo M, Kowdley KV, Vincent C, Bodhenheimer HC Jr, Pares A, Trauner M, Marschall HU, Adorini L, Sciacca C, Beecher-Jones T, Castelloe E, Bohm O, Shapiro D (2015) Efficacy of obeticholic acid in patients with primary biliary cirrhosis and inadequate response to ursodeoxycholic acid. Gastroenterology 148(751–761):e758

Hofmann AF (2002) Rifampicin and treatment of cholestatic pruritus. Gut 51:756–757

Hohenester S, Wenniger LM, Paulusma CC, van Vliet SJ, Jefferson DM, Elferink RP, Beuers U (2012) A biliary HCO3- umbrella constitutes a protective mechanism against bile acid-induced injury in human cholangiocytes. Hepatology 55:173–183

Hylemon PB, Melone PD, Franklund CV, Lund E, Bjorkhem I (1991) Mechanism of intestinal 7 alpha-dehydroxylation of cholic acid: evidence that Allo-deoxycholic acid is an inducible side-product. J Lipid Res 32:89–96

Inagaki T, Choi M, Moschetta A, Peng L, Cummins CL, McDonald JG, Luo G, Jones SA, Goodwin B, Richardson JA, Gerard RD, Repa JJ, Mangelsdorf DJ, Kliewer SA (2005) Fibroblast growth factor 15 functions as an enterohepatic signal to regulate bile acid homeostasis. Cell Metab 2:217–225

Ito S, Fujimori T, Furuya A, Satoh J, Nabeshima Y, Nabeshima Y (2005) Impaired negative feedback suppression of bile acid synthesis in mice lacking betaKlotho. J Clin Invest 115:2202–2208

Iwasaki S, Ohira H, Nishiguchi S, Zeniya M, Kaneko S, Onji M, Ishibashi H, Sakaida I, Kuriyama S, Ichida T, Onishi S, Toda G, Study Group of Intractable Liver Diseases for Research on a Specific Disease HSRGMoHL and Welfare of J (2008) The efficacy of ursodeoxycholic acid and bezafibrate combination therapy for primary biliary cirrhosis: a prospective, multicenter study. Hepatol Res 38:557–564

Jiang C, Xie C, Li F, Zhang L, Nichols RG, Krausz KW, Cai J, Qi Y, Fang ZZ, Takahashi S, Tanaka N, Desai D, Amin SG, Albert I, Patterson AD, Gonzalez FJ (2015a) Intestinal farnesoid X receptor signaling promotes nonalcoholic fatty liver disease. J Clin Invest 125:386–402

Jiang C, Xie C, Lv Y, Li J, Krausz KW, Shi J, Brocker CN, Desai D, Amin SG, Bisson WH, Liu Y, Gavrilova O, Patterson AD, Gonzalez FJ (2015b) Intestine-selective farnesoid X receptor inhibition improves obesity-related metabolic dysfunction. Nat Commun 6:10166

Kaplan MM, Gershwin ME (2005) Primary biliary cirrhosis. N Engl J Med 353:1261–1273

Kast HR, Goodwin B, Tarr PT, Jones SA, Anisfeld AM, Stoltz CM, Tontonoz P, Kliewer S, Willson TM, Edwards PA (2002) Regulation of multidrug resistance-associated protein 2 (ABCC2) by the nuclear receptors pregnane X receptor, farnesoid X-activated receptor, and constitutive androstane receptor. J Biol Chem 277:2908–2915

Katsuma S, Hirasawa A, Tsujimoto G (2005) Bile acids promote glucagon-like peptide-1 secretion through TGR5 in a murine enteroendocrine cell line STC-1. Biochem Biophys Res Commun 329:386–390

Kawamata Y, Fujii R, Hosoya M, Harada M, Yoshida H, Miwa M, Fukusumi S, Habata Y, Itoh T, Shintani Y, Hinuma S, Fujisawa Y, Fujino M (2003) A G protein-coupled receptor responsive to bile acids. J Biol Chem 278:9435–9440

Keitel V, Nies AT, Brom M, Hummel-Eisenbeiss J, Spring H, Keppler D (2003) A common Dubin-Johnson syndrome mutation impairs protein maturation and transport activity of MRP2 (ABCC2). Am J Physiol Gastrointest Liver Physiol 284:G165–G174

Keitel V, Reinehr R, Gatsios P, Rupprecht C, Gorg B, Selbach O, Haussinger D, Kubitz R (2007) The G-protein coupled bile salt receptor TGR5 is expressed in liver sinusoidal endothelial cells. Hepatology 45:695–704

Keitel V, Donner M, Winandy S, Kubitz R, Haussinger D (2008) Expression and function of the bile acid receptor TGR5 in Kupffer cells. Biochem Biophys Res Commun 372:78–84

Keitel V, Ullmer C, Haussinger D (2010) The membrane-bound bile acid receptor TGR5 (Gpbar-1) is localized in the primary cilium of cholangiocytes. Biol Chem 391:785–789

Kim ND, Moon JO, Slitt AL, Copple BL (2006) Early growth response factor-1 is critical for cholestatic liver injury. Toxicol Sci 90:586–595

Kliewer SA, Willson TM (2002) Regulation of xenobiotic and bile acid metabolism by the nuclear pregnane X receptor. J Lipid Res 43:359–364

Kliewer SA, Goodwin B, Willson TM (2002) The nuclear pregnane X receptor: a key regulator of xenobiotic metabolism. Endocr Rev 23:687–702

Krahenbuhl S, Talos C, Fischer S, Reichen J (1994) Toxicity of bile acids on the electron transport chain of isolated rat liver mitochondria. Hepatology 19:471–479

Kullak-Ublick GA, Stieger B, Hagenbuch B, Meier PJ (2000) Hepatic transport of bile salts. Semin Liver Dis 20:273–292

Kumar DP, Rajagopal S, Mahavadi S, Mirshahi F, Grider JR, Murthy KS, Sanyal AJ (2012) Activation of transmembrane bile acid receptor TGR5 stimulates insulin secretion in pancreatic beta cells. Biochem Biophys Res Commun 427:600–605

Kumar DP, Asgharpour A, Mirshahi F, Park SH, Liu S, Imai Y, Nadler JL, Grider JR, Murthy KS, Sanyal AJ (2016) Activation of transmembrane bile acid receptor TGR5 modulates pancreatic islet alpha cells to promote glucose homeostasis. J Biol Chem 291:6626–6640

Lang C, Meier Y, Stieger B, Beuers U, Lang T, Kerb R, Kullak-Ublick GA, Meier PJ, Pauli-Magnus C (2007) Mutations and polymorphisms in the bile salt export pump and the multidrug resistance protein 3 associated with drug-induced liver injury. Pharmacogenet Genomics 17:47–60

Li T, Chiang JY (2005) Mechanism of rifampicin and pregnane X receptor inhibition of human cholesterol 7 alpha-hydroxylase gene transcription. Am J Physiol Gastrointest Liver Physiol 288:G74–G84

Li T, Chiang JY (2007) A novel role of transforming growth factor beta1 in transcriptional repression of human cholesterol 7alpha-hydroxylase gene. Gastroenterology 133:1660–1669

Li T, Chiang JY (2014) Bile acid signaling in metabolic disease and drug therapy. Pharmacol Rev 66:948–983

Li T, Jahan A, Chiang JY (2006) Bile acids and cytokines inhibit the human cholesterol 7 alpha-hydroxylase gene via the JNK/c-Jun pathway in human liver cells. Hepatology 43:1202–1210

Li T, Owsley E, Matozel M, Hsu P, Novak CM, Chiang JY (2010) Transgenic expression of cholesterol 7alpha-hydroxylase in the liver prevents high-fat diet-induced obesity and insulin resistance in mice. Hepatology 52:678–690

Li T, Francl JM, Boehme S, Ochoa A, Zhang Y, Klaassen CD, Erickson SK, Chiang JY (2012) Glucose and insulin induction of bile acid synthesis: mechanisms and implication in diabetes and obesity. J Biol Chem 287:1861–1873

Li S, Hsu DD, Li B, Luo X, Alderson N, Qiao L, Ma L, Zhu HH, He Z, Suino-Powell K, Ji K, Li J, Shao J, Xu HE, Li T, Feng GS (2014) Cytoplasmic tyrosine phosphatase Shp2 coordinates hepatic regulation of bile acid and FGF15/19 signaling to repress bile acid synthesis. Cell Metab 20:320–332

Lieu T, Jayaweera G, Zhao P, Poole DP, Jensen D, Grace M, McIntyre P, Bron R, Wilson YM, Krappitz M, Haerteis S, Korbmacher C, Steinhoff MS, Nassini R, Materazzi S, Geppetti P, Corvera CU, Bunnett NW (2014) The bile acid receptor TGR5 activates the TRPA1 channel to induce itch in mice. Gastroenterology 147:1417–1428

Lin BC, Wang M, Blackmore C, Desnoyers LR (2007) Liver-specific activities of FGF19 require Klotho beta. J Biol Chem 282:27277–27284

Lioudaki E, Ganotakis ES, Mikhailidis DP (2011) Lipid lowering drugs and gallstones: a therapeutic option? Curr Pharm Des 17:3622–3631

Liu Y, Binz J, Numerick MJ, Dennis S, Luo G, Desai B, MacKenzie KI, Mansfield TA, Kliewer SA, Goodwin B, Jones SA (2003) Hepatoprotection by the farnesoid X receptor agonist GW4064 in rat models of intra- and extrahepatic cholestasis. J Clin Invest 112:1678–1687

Low-Beer TS, Nutter S (1978) Colonic bacterial activity, biliary cholesterol saturation, and pathogenesis of gallstones. Lancet 2:1063–1065

Lu TT, Makishima M, Repa JJ, Schoonjans K, Kerr TA, Auwerx J, Mangelsdorf DJ (2000) Molecular basis for feedback regulation of bile acid synthesis by nuclear receptors. Mol Cell 6:507–515
Luo J, Ko B, Elliott M, Zhou M, Lindhout DA, Phung V, To C, Learned RM, Tian H, DePaoli AM, Ling L (2014) A nontumorigenic variant of FGF19 treats cholestatic liver diseases. Sci Transl Med **6**:247ra100
Makishima M, Okamoto AY, Repa JJ, Tu H, Learned RM, Luk A, Hull MV, Lustig KD, Mangelsdorf DJ, Shan B (1999) Identification of a nuclear receptor for bile acids. Science 284:1362–1365
Marcus SN, Heaton KW (1988) Deoxycholic acid and the pathogenesis of gall stones. Gut 29:522–533
Maruyama T, Miyamoto Y, Nakamura T, Tamai Y, Okada H, Sugiyama E, Itadani H, Tanaka K (2002) Identification of membrane-type receptor for bile acids (M-BAR). Biochem Biophys Res Commun 298:714–719
Maruyama T, Tanaka K, Suzuki J, Miyoshi H, Harada N, Nakamura T, Miyamoto Y, Kanatani A, Tamai Y (2006) Targeted disruption of G protein-coupled bile acid receptor 1 (Gpbar1/M-bar) in mice. J Endocrinol 191:197–205
Mason JI, Boyd GS (1978) The suppressive effect of the catatoxic steroid, pregnenolone-16alpha-carbonitrile, on liver microsomal cholesterol-7alpha-hydroxlyase. Steroids 31:849–854
McMahan RH, Wang XX, Cheng LL, Krisko T, Smith M, El Kasmi K, Pruzanski M, Adorini L, Golden-Mason L, Levi M, Rosen HR (2013) Bile acid receptor activation modulates hepatic monocyte activity and improves nonalcoholic fatty liver disease. J Biol Chem 288:11761–11770
Meier PJ (1995) Molecular mechanisms of hepatic bile salt transport from sinusoidal blood into bile. Am J Physiol 269:G801–G812
Meier PJ, Stieger B (2002) Bile salt transporters. Annu Rev Physiol 64:635–661
Miao J, Fang S, Bae Y, Kemper JK (2006) Functional inhibitory cross-talk between constitutive androstane receptor and hepatic nuclear factor-4 in hepatic lipid/glucose metabolism is mediated by competition for binding to the DR1 motif and to the common coactivators, GRIP-1 and PGC-1alpha. J Biol Chem 281:14537–14546
Miyake JH, Wang SL, Davis RA (2000) Bile acid induction of cytokine expression by macrophages correlates with repression of hepatic cholesterol 7alpha-hydroxylase. J Biol Chem 275:21805–21808
Mullenbach R, Bennett A, Tetlow N, Patel N, Hamilton G, Cheng F, Chambers J, Howard R, Taylor-Robinson SD, Williamson C (2005) ATP8B1 mutations in British cases with intrahepatic cholestasis of pregnancy. Gut 54:829–834
Myant NB, Mitropoulos KA (1977) Cholesterol 7 alpha-hydroxylase. J Lipid Res 18:135–153
Neimark E, Chen F, Li X, Shneider BL (2004) Bile acid-induced negative feedback regulation of the human ileal bile acid transporter. Hepatology 40:149–156
Neuman M, Angulo P, Malkiewicz I, Jorgensen R, Shear N, Dickson ER, Haber J, Katz G, Lindor K (2002) Tumor necrosis factor-alpha and transforming growth factor-beta reflect severity of liver damage in primary biliary cirrhosis. J Gastroenterol Hepatol 17:196–202
Neuschwander-Tetri BA, Loomba R, Sanyal AJ, Lavine JE, Van Natta ML, Abdelmalek MF, Chalasani N, Dasarathy S, Diehl AM, Hameed B, Kowdley KV, McCullough A, Terrault N, Clark JM, Tonascia J, Brunt EM, Kleiner DE, Doo E, Network NCR (2015) Farnesoid X nuclear receptor ligand obeticholic acid for non-cirrhotic, non-alcoholic steatohepatitis (FLINT): a multicentre, randomised, placebo-controlled trial. Lancet 385:956–965
Ni Y, Lempp FA, Mehrle S, Nkongolo S, Kaufman C, Falth M, Stindt J, Koniger C, Nassal M, Kubitz R, Sultmann H, Urban S (2014) Hepatitis B and D viruses exploit sodium taurocholate co-transporting polypeptide for species-specific entry into hepatocytes. Gastroenterology 146:1070–1083
Noe J, Kullak-Ublick GA, Jochum W, Stieger B, Kerb R, Haberl M, Mullhaupt B, Meier PJ, Pauli-Magnus C (2005) Impaired expression and function of the bile salt export pump due to three novel ABCB11 mutations in intrahepatic cholestasis. J Hepatol 43:536–543

Osawa Y, Banno Y, Nagaki M, Brenner DA, Naiki T, Nozawa Y, Nakashima S, Moriwaki H (2001) TNF-alpha-induced sphingosine 1-phosphate inhibits apoptosis through a phosphatidylinositol 3-kinase/Akt pathway in human hepatocytes. J Immunol 167:173–180
Osawa Y, Uchinami H, Bielawski J, Schwabe RF, Hannun YA, Brenner DA (2005) Roles for C16-ceramide and sphingosine 1-phosphate in regulating hepatocyte apoptosis in response to tumor necrosis factor-alpha. J Biol Chem 280:27879–27887
Osawa Y, Seki E, Adachi M, Suetsugu A, Ito H, Moriwaki H, Seishima M, Nagaki M (2010) Role of acid sphingomyelinase of Kupffer cells in cholestatic liver injury in mice. Hepatology 51:237–245
Otsuki M (2000) Pathophysiological role of cholecystokinin in humans. J Gastroenterol Hepatol 15(Suppl):D71–D83
Painter JN, Savander M, Ropponen A, Nupponen N, Riikonen S, Ylikorkala O, Lehesjoki AE, Aittomaki K (2005) Sequence variation in the ATP8B1 gene and intrahepatic cholestasis of pregnancy. Eur J Hum Genet 13:435–439
Pares A, Caballeria L, Rodes J (2006) Excellent long-term survival in patients with primary biliary cirrhosis and biochemical response to ursodeoxycholic acid. Gastroenterology 130:715–720
Parks DJ, Blanchard SG, Bledsoe RK, Chandra G, Consler TG, Kliewer SA, Stimmel JB, Willson TM, Zavacki AM, Moore DD, Lehmann JM (1999) Bile acids: natural ligands for an orphan nuclear receptor. Science 284:1365–1368
Pauli-Magnus C, Lang T, Meier Y, Zodan-Marin T, Jung D, Breymann C, Zimmermann R, Kenngott S, Beuers U, Reichel C, Kerb R, Penger A, Meier PJ, Kullak-Ublick GA (2004) Sequence analysis of bile salt export pump (ABCB11) and multidrug resistance p-glycoprotein 3 (ABCB4, MDR3) in patients with intrahepatic cholestasis of pregnancy. Pharmacogenetics 14:91–102
Pean N, Doignon I, Garcin I, Besnard A, Julien B, Liu B, Branchereau S, Spraul A, Guettier C, Humbert L, Schoonjans K, Rainteau D, Tordjmann T (2013) The receptor TGR5 protects the liver from bile acid overload during liver regeneration in mice. Hepatology 58:1451–1460
Pellicciari R, Fiorucci S, Camaioni E, Clerici C, Costantino G, Maloney PR, Morelli A, Parks DJ, Willson TM (2002) 6alpha-ethyl-chenodeoxycholic acid (6-ECDCA), a potent and selective FXR agonist endowed with anticholestatic activity. J Med Chem 45:3569–3572
Pellicciari R, Costantino G, Camaioni E, Sadeghpour BM, Entrena A, Willson TM, Fiorucci S, Clerici C, Gioiello A (2004) Bile acid derivatives as ligands of the farnesoid X receptor. Synthesis, evaluation, and structure-activity relationship of a series of body and side chain modified analogues of chenodeoxycholic acid. J Med Chem 47:4559–4569
Perino A, Pols TW, Nomura M, Stein S, Pellicciari R, Schoonjans K (2014) TGR5 reduces macrophage migration through mTOR-induced C/EBPbeta differential translation. J Clin Invest 124:5424–5436
Pols TW, Noriega LG, Nomura M, Auwerx J, Schoonjans K (2011) The bile acid membrane receptor TGR5: a valuable metabolic target. Dig Dis 29:37–44
Poupon R (2012) Ursodeoxycholic acid and bile-acid mimetics as therapeutic agents for cholestatic liver diseases: an overview of their mechanisms of action. Clin Res Hepatol Gastroenterol 36(Suppl 1):S3–12
Prieto J, Garcia N, Marti-Climent JM, Penuelas I, Richter JA, Medina JF (1999) Assessment of biliary bicarbonate secretion in humans by positron emission tomography. Gastroenterology 117:167–172
Qi Y, Jiang C, Cheng J, Krausz KW, Li T, Ferrell JM, Gonzalez FJ, Chiang JY (2015) Bile acid signaling in lipid metabolism: Metabolomic and lipidomic analysis of lipid and bile acid markers linked to anti-obesity and anti-diabetes in mice. Biochim Biophys Acta 1851:19–29
Qiao L, Yacoub A, Studer E, Gupta S, Pei XY, Grant S, Hylemon PB, Dent P (2002) Inhibition of the MAPK and PI3K pathways enhances UDCA-induced apoptosis in primary rodent hepatocytes. Hepatology 35:779–789
Rao A, Haywood J, Craddock AL, Belinsky MG, Kruh GD, Dawson PA (2008) The organic solute transporter alpha-beta, Ostalpha-Ostbeta, is essential for intestinal bile acid transport and homeostasis. Proc Natl Acad Sci U S A 105:3891–3896

Reich M, Deutschmann K, Sommerfeld A, Klindt C, Kluge S, Kubitz R, Ullmer C, Knoefel WT, Herebian D, Mayatepek E, Haussinger D, Keitel V (2016) TGR5 is essential for bile acid-dependent cholangiocyte proliferation in vivo and in vitro. Gut 65:487–501

Reinehr R, Becker S, Keitel V, Eberle A, Grether-Beck S, Haussinger D (2005) Bile salt-induced apoptosis involves NADPH oxidase isoform activation. Gastroenterology 129:2009–2031

Ridlon JM, Hylemon PB (2006) A potential role for resistant starch fermentation in modulating colonic bacterial metabolism and colon cancer risk. Cancer Biol Ther 5:273–274

Ridlon JM, Kang DJ, Hylemon PB (2006) Bile salt biotransformations by human intestinal bacteria. J Lipid Res 47:241–259

Rizzo G, Passeri D, De Franco F, Ciaccioli G, Donadio L, Orlandi S, Sadeghpour B, Wang XX, Jiang T, Levi M, Pruzanski M, Adorini L (2010) Functional characterization of the semisynthetic bile acid derivative INT-767, a dual farnesoid X receptor and TGR5 agonist. Mol Pharmacol 78:617–630

Rodrigues CM, Fan G, Ma X, Kren BT, Steer CJ (1998a) A novel role for ursodeoxycholic acid in inhibiting apoptosis by modulating mitochondrial membrane perturbation. J Clin Invest 101:2790–2799

Rodrigues CM, Fan G, Wong PY, Kren BT, Steer CJ (1998b) Ursodeoxycholic acid may inhibit deoxycholic acid-induced apoptosis by modulating mitochondrial transmembrane potential and reactive oxygen species production. Mol Med 4:165–178

Rodrigues CM, Ma X, Linehan-Stieers C, Fan G, Kren BT, Steer CJ (1999) Ursodeoxycholic acid prevents cytochrome c release in apoptosis by inhibiting mitochondrial membrane depolarization and channel formation. Cell Death Differ 6:842–854

Roggin KK, Papa EF, Kurkchubasche AG, Tracy TF Jr (2000) Kupffer cell inactivation delays repair in a rat model of reversible biliary obstruction. J Surg Res 90:166–173

Rolo AP, Oliveira PJ, Moreno AJ, Palmeira CM (2000) Bile acids affect liver mitochondrial bioenergetics: possible relevance for cholestasis therapy. Toxicol Sci 57:177–185

Russell DW (2003) The enzymes, regulation, and genetics of bile acid synthesis. Annu Rev Biochem 72:1370174

Russell DW, Setchell KD (1992) Bile acid biosynthesis. Biochemistry 31:4737–4749

Saini SP, Sonoda J, Xu L, Toma D, Uppal H, Mu Y, Ren S, Moore DD, Evans RM, Xie W (2004) A novel constitutive androstane receptor-mediated and CYP3A-independent pathway of bile acid detoxification. Mol Pharmacol 65:292–300

Sambrotta M, Strautnieks S, Papouli E, Rushton P, Clark BE, Parry DA, Logan CV, Newbury LJ, Kamath BM, Ling S, Grammatikopoulos T, Wagner BE, Magee JC, Sokol RJ, Mieli-Vergani G, University of Washington Center for Mendelian G, Smith JD, Johnson CA, McClean P, Simpson MA, Knisely AS, Bull LN and Thompson RJ (2014) Mutations in TJP2 cause progressive cholestatic liver disease. Nat Genet 46:326–328

Sato H, Macchiarulo A, Thomas C, Gioiello A, Une M, Hofmann AF, Saladin R, Schoonjans K, Pellicciari R, Auwerx J (2008) Novel potent and selective bile acid derivatives as TGR5 agonists: biological screening, structure-activity relationships, and molecular modeling studies. J Med Chem 51:1831–1841

Schaap FG, van der Gaag NA, Gouma DJ, Jansen PL (2009) High expression of the bile salt-homeostatic hormone fibroblast growth factor 19 in the liver of patients with extrahepatic cholestasis. Hepatology 49:1228–1235

Schoemaker MH, Conde de la Rosa L, Buist-Homan M, Vrenken TE, Havinga R, Poelstra K, Haisma HJ, Jansen PL, Moshage H (2004) Tauroursodeoxycholic acid protects rat hepatocytes from bile acid-induced apoptosis via activation of survival pathways. Hepatology 39:1563–1573

Setchell KD, Heubi JE, Shah S, Lavine JE, Suskind D, Al-Edreesi M, Potter C, Russell DW, O'Connell NC, Wolfe B, Jha P, Zhang W, Bove KE, Knisely AS, Hofmann AF, Rosenthal P, Bull LN (2013) Genetic defects in bile acid conjugation cause fat-soluble vitamin deficiency. Gastroenterology 144:945–955.e946. quiz e914–945

Sewnath ME, Van Der Poll T, Ten Kate FJ, Van Noorden CJ, Gouma DJ (2002) Interleukin-1 receptor type I gene-deficient bile duct-ligated mice are partially protected against endotoxin. Hepatology 35:149–158

Shneider BL, Dawson PA, Christie DM, Hardikar W, Wong MH, Suchy FJ (1995) Cloning and molecular characterization of the ontogeny of a rat ileal sodium-dependent bile acid transporter. J Clin Invest 95:745–754

Shonsey EM, Wheeler J, Johnson M, He D, Falany CN, Falany J, Barnes S (2005) Synthesis of bile acid coenzyme a thioesters in the amino acid conjugation of bile acids. Methods Enzymol 400:360–373

Slijepcevic D, Kaufman C, Wichers CG, Gilglioni EH, Lempp FA, Duijst S, de Waart DR, Oude Elferink RP, Mier W, Stieger B, Beuers U, Urban S, van de Graaf SF (2015) Impaired uptake of conjugated bile acids and hepatitis B virus preS1-binding in Na(+)-taurocholate cotransporting polypeptide knockout mice. Hepatology 62:207–219

Smit JJ, Schinkel AH, Oude Elferink RP, Groen AK, Wagenaar E, van Deemter L, Mol CA, Ottenhoff R, van der Lugt NM, van Roon MA et al (1993) Homozygous disruption of the murine mdr2 P-glycoprotein gene leads to a complete absence of phospholipid from bile and to liver disease. Cell 75:451–462

Sokol RJ, Devereaux M, Khandwala R, O'Brien K (1993) Evidence for involvement of oxygen free radicals in bile acid toxicity to isolated rat hepatocytes. Hepatology 17:869–881

Sokol RJ, Winklhofer-Roob BM, Devereaux MW, McKim JM Jr (1995) Generation of hydroperoxides in isolated rat hepatocytes and hepatic mitochondria exposed to hydrophobic bile acids. Gastroenterology 109:1249–1256

Sokol RJ, Straka MS, Dahl R, Devereaux MW, Yerushalmi B, Gumpricht E, Elkins N, Everson G (2001) Role of oxidant stress in the permeability transition induced in rat hepatic mitochondria by hydrophobic bile acids. Pediatr Res 49:519–531

Sokol RJ, Dahl R, Devereaux MW, Yerushalmi B, Kobak GE, Gumpricht E (2005) Human hepatic mitochondria generate reactive oxygen species and undergo the permeability transition in response to hydrophobic bile acids. J Pediatr Gastroenterol Nutr 41:235–243

Song KH, Li T, Owsley E, Strom S, Chiang JY (2009) Bile acids activate fibroblast growth factor 19 signaling in human hepatocytes to inhibit cholesterol 7alpha-hydroxylase gene expression. Hepatology 49:297–305

Spivey JR, Bronk SF, Gores GJ (1993) Glycochenodeoxycholate-induced lethal hepatocellular injury in rat hepatocytes. Role of ATP depletion and cytosolic free calcium. J Clin Invest 92:17–24

Srivastava A (2014) Progressive familial intrahepatic cholestasis. J Clin Exp Hepatol 4:25–36

Stahlberg D (1995) Effects of pregnenolone-16 alpha-carbonitrile on the metabolism of cholesterol in rat liver microsomes. Lipids 30:361–364

Stanley LA, Horsburgh BC, Ross J, Scheer N, Wolf CR (2006) PXR and CAR: nuclear receptors which play a pivotal role in drug disposition and chemical toxicity. Drug Metab Rev 38:515–597

Staudinger JL, Goodwin B, Jones SA, Hawkins-Brown D, MacKenzie KI, LaTour A, Liu Y, Klaassen CD, Brown KK, Reinhard J, Willson TM, Koller BH, Kliewer SA (2001) The nuclear receptor PXR is a lithocholic acid sensor that protects against liver toxicity. Proc Natl Acad Sci U S A 98:3369–3374

Stedman CA, Liddle C, Coulter SA, Sonoda J, Alvarez JG, Moore DD, Evans RM, Downes M (2005) Nuclear receptors constitutive androstane receptor and pregnane X receptor ameliorate cholestatic liver injury. Proc Natl Acad Sci U S A 102:2063–2068

Strautnieks SS, Kagalwalla AF, Tanner MS, Knisely AS, Bull L, Freimer N, Kocoshis SA, Gardiner RM, Thompson RJ (1997) Identification of a locus for progressive familial intrahepatic cholestasis PFIC2 on chromosome 2q24. Am J Hum Genet 61:630–633

Strub GM, Maceyka M, Hait NC, Milstien S, Spiegel S (2010) Extracellular and intracellular actions of sphingosine-1-phosphate. Adv Exp Med Biol 688:141–155

Studer E, Zhou X, Zhao R, Wang Y, Takabe K, Nagahashi M, Pandak WM, Dent P, Spiegel S, Shi R, Xu W, Liu X, Bohdan P, Zhang L, Zhou H, Hylemon PB (2012) Conjugated bile acids activate the sphingosine-1-phosphate receptor 2 in primary rodent hepatocytes. Hepatology 55:267–276

Teixeira J, Gil G (1991) Cloning, expression, and regulation of lithocholic acid 6 beta-hydroxylase. J Biol Chem 266:21030–21036

Thomas C, Gioiello A, Noriega L, Strehle A, Oury J, Rizzo G, Macchiarulo A, Yamamoto H, Mataki C, Pruzanski M, Pellicciari R, Auwerx J, Schoonjans K (2009) TGR5-mediated bile acid sensing controls glucose homeostasis. Cell Metab 10:167–177

Tiwari A, Maiti P (2009) TGR5: an emerging bile acid G-protein-coupled receptor target for the potential treatment of metabolic disorders. Drug Discov Today 14:523–530

Trauner M, Boyer JL (2003) Bile salt transporters: molecular characterization, function, and regulation. Physiol Rev 83:633–671

Van Mil SW, Milona A, Dixon PH, Mullenbach R, Geenes VL, Chambers J, Shevchuk V, Moore GE, Lammert F, Glantz AG, Mattsson LA, Whittaker J, Parker MG, White R, Williamson C (2007) Functional variants of the central bile acid sensor FXR identified in intrahepatic cholestasis of pregnancy. Gastroenterology 133:507–516

Vaz FM, Paulusma CC, Huidekoper H, de Ru M, Lim C, Koster J, Ho-Mok K, Bootsma AH, Groen AK, Schaap FG, Oude Elferink RP, Waterham HR and Wanders RJ (2015) Sodium taurocholate cotransporting polypeptide (SLC10A1) deficiency: conjugated hypercholanemia without a clear clinical phenotype. Hepatology 61(1):260-267

Verreault M, Kaeding J, Caron P, Trottier J, Grosse L, Houssin E, Paquet S, Perreault M, Barbier O (2010) Regulation of endobiotics glucuronidation by ligand-activated transcription factors: physiological function and therapeutic potential. Drug Metab Rev 42:110–122

Vessey DA, Benfatto AM, Kempner ES (1987) Bile acid: CoASH ligases from Guinea pig and porcine liver microsomes. Purification and characterization. J Biol Chem 262:5360–5365

Wang YD, Chen WD, Yu D, Forman BM, Huang W (2011) The G-protein-coupled bile acid receptor, Gpbar1 (TGR5), negatively regulates hepatic inflammatory response through antagonizing nuclear factor kappa light-chain enhancer of activated B cells (NF-kappaB) in mice. Hepatology 54:1421–1432

Wang C, Yang C, Chang JY, You P, Li Y, Jin C, Luo Y, Li X, McKeehan WL, Wang F (2014) Hepatocyte FRS2alpha is essential for the endocrine fibroblast growth factor to limit the amplitude of bile acid production induced by prandial activity. Curr Mol Med 14:703–711

Wang XX, Edelstein MH, Gafter U, Qiu L, Luo Y, Dobrinskikh E, Lucia S, Adorini L, D'Agati VD, Levi J, Rosenberg A, Kopp JB, Gius DR, Saleem MA, Levi M (2016) G protein-coupled bile acid receptor TGR5 activation inhibits kidney disease in obesity and diabetes. J Am Soc Nephrol 27:1362–1378

Watanabe M, Houten SM, Mataki C, Christoffolete MA, Kim BW, Sato H, Messaddeq N, Harney JW, Ezaki O, Kodama T, Schoonjans K, Bianco AC, Auwerx J (2006) Bile acids induce energy expenditure by promoting intracellular thyroid hormone activation. Nature 439:484–489

Wheeler JB, Shaw DR, Barnes S (1997) Purification and characterization of a rat liver bile acid coenzyme a ligase from rat liver microsomes. Arch Biochem Biophys 348:15–24

Woolbright BL, Jaeschke H (2012) Novel insight into mechanisms of cholestatic liver injury. World J Gastroenterol 18:4985–4993

Xia X, Francis H, Glaser S, Alpini G, LeSage G (2006) Bile acid interactions with cholangiocytes. World J Gastroenterol 12:3553–3563

Yamashiki M, Kosaka Y, Nishimura A, Watanabe S, Nomoto M, Ichida F (1998) Analysis of serum cytokine levels in primary biliary cirrhosis patients and healthy adults. J Clin Lab Anal 12:77–82

Yan H, Zhong G, Xu G, He W, Jing Z, Gao Z, Huang Y, Qi Y, Peng B, Wang H, Fu L, Song M, Chen P, Gao W, Ren B, Sun Y, Cai T, Feng X, Sui J, Li W (2012) Sodium taurocholate cotransporting polypeptide is a functional receptor for human hepatitis B and D virus. Elife 1:e00049

Yeh HZ, Schteingart CD, Hagey LR, Ton-Nu HT, Bolder U, Gavrilkina MA, Steinbach JH, Hofmann AF (1997) Effect of side chain length on biotransformation, hepatic transport, and choleretic properties of chenodeoxycholyl homologues in the rodent: studies with dinorchenodeoxycholic acid, norchenodeoxycholic acid, and chenodeoxycholic acid. Hepatology 26:374–385

Yerushalmi B, Dahl R, Devereaux MW, Gumpricht E, Sokol RJ (2001) Bile acid-induced rat hepatocyte apoptosis is inhibited by antioxidants and blockers of the mitochondrial permeability transition. Hepatology 33:616–626

Yoneno K, Hisamatsu T, Shimamura K, Kamada N, Ichikawa R, Kitazume MT, Mori M, Uo M, Namikawa Y, Matsuoka K, Sato T, Koganei K, Sugita A, Kanai T, Hibi T (2013) TGR5 signalling inhibits the production of pro-inflammatory cytokines by in vitro differentiated inflammatory and intestinal macrophages in Crohn's disease. Immunology 139:19–29

Yoon YB, Hagey LR, Hofmann AF, Gurantz D, Michelotti EL, Steinbach JH (1986) Effect of side-chain shortening on the physiologic properties of bile acids: hepatic transport and effect on biliary secretion of 23-nor-ursodeoxycholate in rodents. Gastroenterology 90:837–852

Zein CO, Lindor KD (2010) Latest and emerging therapies for primary biliary cirrhosis and primary sclerosing cholangitis. Curr Gastroenterol Rep 12:13–22

Zhang Y, Hong JY, Rockwell CE, Copple BL, Jaeschke H, Klaassen CD (2012) Effect of bile duct ligation on bile acid composition in mouse serum and liver. Liver Int 32:58–69

Zhou Y, Maxwell KN, Sezgin E, Lu M, Liang H, Hancock JF, Dial EJ, Lichtenberger LM, Levental I (2013) Bile acids modulate signaling by functional perturbation of plasma membrane domains. J Biol Chem 288:35660–35670

Zhou M, Wang X, Phung V, Lindhout DA, Mondal K, Hsu JY, Yang H, Humphrey M, Ding X, Arora T, Learned RM, DePaoli AM, Tian H, Ling L (2014) Separating Tumorigenicity from bile acid regulatory activity for endocrine hormone FGF19. Cancer Res 74:3306–3316

Zhou M, Learned RM, Rossi SJ, DePaoli AM, Tian H, Ling L (2016) Engineered fibroblast growth factor 19 reduces liver injury and resolves sclerosing cholangitis in Mdr2-deficient mice. Hepatology 63:914–929

Zollner G, Trauner M (2008) Mechanisms of cholestasis. Clin Liver Dis 12:1–26. vii

Chapter 8
Cell Death in Ischemia-Reperfusion-Induced Liver Injury

Julie Goswami, Heather Waring, Allan Tsung, and Hai Huang

Abbreviations

ASC	Apoptosis-associated speck-like protein-containing CARD
Atg	Autophagy
ATP	Adenosine triphosphate
Bax	Bcl-2-associated X
Bcl-2	B-cell lymphoma 2
BECN1	Beclin-1
Bid	BH3-interacting domain
CARS	Cysteinyl-tRNA-synthetase
c-FLIP	Cellular FLICE-like inhibitory protein
DAMP	Damage-associated molecular pattern
DC	Dendritic cell
FADD	Fas-associated death domain
GPX4	Glutathione peroxidase 4
GSH	Glutathione
GSSG	Oxidized glutathione
HMGB1	High-mobility group box protein 1
HSPB1	Heat shock protein beta 1
IRI	Ischemia-reperfusion injury
LSEC	Liver sinusoidal endothelial cell
MLKL	Mixed linkage kinase domain-like
MPT	Mitochondrial permeability transition
NASH	Non-alcoholic steatohepatitis
NET	Neutrophil extracellular trap

J. Goswami • H. Waring • H. Huang
Department of Surgery, University of Pittsburgh Medical Center, Pittsburgh, PA, USA

A. Tsung, M.D. (✉)
Department of Surgery, University of Pittsburgh Medical Center, Pittsburgh, PA, USA

Division of Hepatobiliary and Pancreatic Surgery, University of Pittsburgh,
3471 Fifth Avenue, Suite 300, Pittsburgh, PA 15213, USA
e-mail: tsunga@upmc.edu

W.-X. Ding, X.-M. Yin (eds.), *Cellular Injury in Liver Diseases*, Cell Death in Biology and Diseases, DOI 10.1007/978-3-319-53774-0_8

X.-M. Yin and Z. Dong (Series eds.), *Cell Death in Biology and Diseases*

NLRP3	Nucleotide-binding domain, leucine-rich repeat containing protein 3
NOX	Nitrogen oxide
NRF2	Nuclear respiratory factor 2
PAD4	Peptidyl-arginine-deaminase-4
PI3K	Phosphatidylinositol-3-kinase
PKB	Protein kinase B
PTEN	Phosphate and tensin homologue
RIPK	Receptor-interacting protein kinase
ROS	Reactive oxidant species
RSL	RAS-selective lethal
TFR1	Transferrin receptor 1
TLR	Toll-like receptor
TRADD	TNF-receptor-associated death domain
TRAIL	TNF-related apoptosis-inducing ligand
TUNEL	Terminal deoxynucleotidyl transferase-mediated deoxyuridine triphosphate nick-end labeling
VDAC	Voltage-dependent anion channel

8.1 Introduction

Hepatic ischemia-reperfusion injury (IRI) is an unavoidable consequence of transplantation, trauma, shock, and elective liver resection, where transient inflow occlusion or total vascular exclusion is utilized to reduce blood loss (Tsung et al. 2005). The initial phase of injury is due to direct cellular damage from ischemic insult. However, reperfusion of the liver has widespread effects through multiple mechanisms including production of reactive oxidant species (ROS), reactive nitrogen species, and excessive activation inflammatory pathways leading to cytokine release and migration of innate immune cells (Nastos et al. 2014). The end result is cell death that occurs through various mechanisms. In the setting of transplant, injury is mediated primarily through "cold" IRI in which cell death occurs primarily in sinusoidal endothelial cells (Zhai et al. 2013). In elective liver resection, "warm" IRI primarily leads to cell death of hepatocytes and results in significant enzyme release (Jaeschke and Woolbright 2012). In both settings, cell death typically does not manifest until the reperfusion phase. Additionally, an understanding of mechanisms of cell death after liver IRI is essential as the need for major liver surgery and transplantation increases with the increasing prevalence of hepatic parenchymal disease and associated malignancy (Ratziu et al. 2015).

Cell death occurs through a number of molecular mechanisms that can be classified in various ways and additionally affects various cell types. *Hepatocytes*, the major parenchymal cells of the liver, are subject to cell death through loss of calcium homeostasis, loss of cellular antioxidants, stimulation of catabolic enzymes, ATP depletion, and downstream mitochondrial dysfunction (Quesnelle et al. 2015).

Subsequently, dying hepatocytes release damage-associated molecular patterns (DAMPs) into the circulation, including histones and high-mobility group box protein (HMGB1), that trigger the systemic inflammatory response associated with IRI (Quesnelle et al. 2015). The liver also consists of non-parenchymal cells, including *Kupffer cells*, the resident macrophages of the liver, which generate cytokines (including IL-1, IL-6, TNF-α) leading to post-ischemic tissue damage and neutrophil recruitment (Peralta et al. 2013). *Liver sinusoidal endothelial cells (LSECs)* are additional important non-parenchymal cells present in the liver. A role for these LSECs is proposed in IRI and they are known to quickly suffer severe damage, but the mechanisms of LSEC damage after IRI have yet to be further elucidated (Caldwell-Kenkel et al. 1989; Peralta et al. 2013). A final type of non-parenchymal cell is the *hepatic stellate cell*, found in the perisunusoidal space, also known to suffer during IRI (Peralta et al. 2013). Innate immune cells, including *neutrophils*, additionally participate in the release of ROS, proteases, and degradative enzymes that lead to further cell death after IRI (Peralta et al. 2013). Additionally, monocytes, macrophages, dendritic cells (DCs), and natural killer cells are recruited to the site of injury and release cytokines, chemokines, lipid messengers, and ROS, which lead to hepatocyte death after IRI (Brenner et al. 2013). DCs, in particular, play a dual role in IRI, as blood-borne DCs can act as antigen-presenting cells and play a pro-inflammatory role after hepatic IRI, while liver-resident DCs have recently been shown to have a protective role in cold IRI associated with liver transplant (Zhang et al. 2013). Cell death mechanisms can be further distinguished by spontaneous vs. non-spontaneous cell death, the production of an inflammatory response or lack thereof, and biochemical and morphologic characteristics (Magna and Pisetsky 2016).

Until about a decade ago, cell death was viewed as simply *necrosis vs. apoptosis*, two distinct forms of cell death that occur as a consequence of hepatic IRI. The acute metabolic disruption associated with IRI leads to ATP depletion and oncotic necrosis. Meanwhile, apoptosis is an ATP-dependent form of programed cell death requiring specific stimuli, death ligand/receptor interactions, cascade of caspase activation, and ultimately leakage of cellular components (Majno and Joris 1995; Malhi et al. 2006). Necrosis affects large groups of cells while apoptosis generally affects individual cells that have been programmed to undergo this distinct form of cell death. Additionally, the two forms of cell death are morphologically and biochemically discrete. Necrosis occurs acutely after reperfusion and is characterized by cell swelling, karolysis, vacuolization, and enzyme release (Gujral et al. 2001; Malhi et al. 2006). Apoptosis, on the other hand, is characterized by cell shrinkage, chromatin condensation, DNA degradation, nuclear lobulation, and fragmentation, with ultimate fragmentation of cells into apoptotic bodies (Malhi et al. 2006).

However, necrosis and apoptosis occur in conjunction and do have shared features including mitochondrial permeability transition (MPT) and may in fact represent a spectrum of cell death rather than two distinct mechanisms or alternate pathways to cell death by the same mechanisms (Malhi et al. 2006). Recently, it has come to light that there are several other inter-related regulated forms of non-apoptotic cell death with significant inflammatory effects. These include necroptosis,

Table 8.1 Forms of cell death, morphology, and mechanisms

	Morphology	Main mechanism/ regulator	Associated inflammatory response?
Necrosis	Cell swelling Karyolysis Vacuolization	ATP depletion	Yes
Apoptosis	Nuclear condensation Apoptotic bodies	Caspase cascade	No (Rarely through secondary necrosis)
Necroptosis	Cell swelling Membrane rupture	RIPK3	Yes
Ferroptosis	Smaller mitochondria	Iron-dependent	Yes
Pyroptosis	Chromatin condensation	Caspase 1	Yes
Autophagy	Double-membraned vacuoles	PI3K ATG genes	Unknown
NETosis	Neutrophil release of chromatin fibers	Sepsis, sterile inflammation	Yes

autophagy, pyroptosis, and ferroptosis, and NETosis, as shown in Table 8.1. *Necroptosis*, morphologically similar to but biochemically distinct from necrosis, is triggered in the presence of caspase inhibition and mediated by receptor-interacting serine/threonine protein kinase 3 (RIPK3) (Linkermann and Green 2014). *Autophagy*, meanwhile, is a lysosome-mediated form of cell death used by both normal and damaged liver cells to remove long-lived or malformed proteins and organelles (Cursio et al. 2015). *Pyroptosis*, unlike apoptosis, is dependent specifically on casepase-1 and leads to significant release of pro-inflammatory cytokines (Tait et al. 2014). *Ferroptosis* is an oxidative iron-dependent form of cell death triggered by inactivation of cellular glutathione (GSH) (Cao and Dixon 2016). These interrelated forms of cell death are a growing area of interest and all participate in the widespread consequences seen after hepatic IRI.

8.2 Necrosis

One of the major effects of IRI is depletion of ATP as it cannot be produced by mitochondria. The result of this is significant morphological change from cytoskeletal alterations, including cell and mitochondrial swelling, dilation of the endoplasmic reticulum, and formation of plasma membrane outpouchings called "blebs" as shown in Fig. 8.1 (Lemasters et al. 1987). In short periods of ischemia, bleb formation can be rapidly reversed but after minutes to hours of ischemia, a metastable state develops with mitochondrial permealization, lysosome disruption, bleb growth and coalescence, cell swelling, and leakage of anionic (but not cationic) cell contents through opening of anionic death channels (Nieminen et al. 1988; Zahrebelski and Nieminen 1995; Nishimura and Lemasters 2001). Ultimately, bleb rupture and

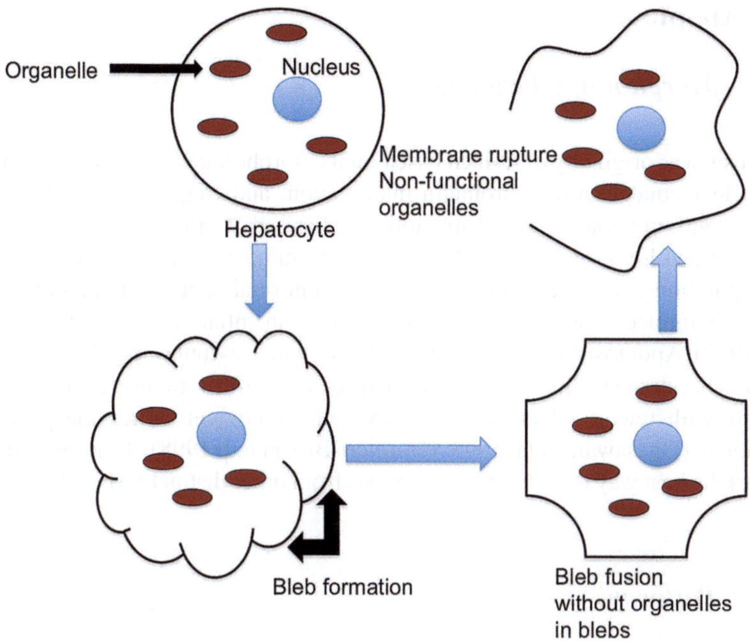

Fig. 8.1 Morphologic changes associated with necrosis after hepatic IRI

failure of the plasma membrane lead to cell death. Because cell contents are released, a significant inflammatory response is induced after cells die in this manner. Morphologic/histologic characteristics, reflective of membrane failure, include diffuse areas of cell swelling, vacuolation, karyolysis, cell content release, with significant inflammatory response (Jaeschke and Lemasters 2003).

8.2.1 Role in Cell Death and Hepatic IRI

Necrotic cells are thought to be the most prominent source of DAMPs, also known as alarmins, a diverse group of molecules that modulate sterile inflammation after IRI (Brenner et al. 2013). Among the inflammatory responders are macrophages, which act to resorb remnant necrotic tissue and expelled cell contents. This form of cell death, interchangeably termed necrosis, oncotic necrosis, or oncosis, is commonly thought to be responsible for 90% of cell death after IRI (Gujral et al. 2001). In several studies, the amino acid glycine has been shown to inhibit oncotic/necrotic cell death by blocking the final phase of cell swelling but does not act to restore metabolic derangements such as ATP depletion (Weinberg et al. 1987; Dickson et al. 1992; Marsh et al. 1993).

8.3 Apoptosis

8.3.1 Morphologic Features

Apoptosis was originally described based on its morphologic characteristics including nuclear condensation, chromatin margination, and fragmentation of both the nucleus and cytoplasm into "apoptotic bodies" that are ultimately degraded (Jaeschke and Lemasters 2003). Apoptosis, a form of cell death that affects select cell populations, was classically described with normal cellular organelles without release of intracellular contents or induction of an inflammatory response (Kerr et al. 1972). Apoptosis has since been shown to induce cytoplasmic changes including mitochondrial swelling, alterations to the endoplasmic reticulum, and secondary necrosis with release of proinflammatory cellular contents when the process is incomplete (Ogasawara et al. 1993; Camilleri-Broet et al. 1998). Thus, morphology remains the best way to distinguish apoptosis from other distinct forms of cell death.

8.3.2 Pathways

A variety of pathways lead to apoptotic cell death, including the "extrinsic pathway," the "intrinsic pathway" (mitochondrial mediated), and NF-κB signaling pathways, as shown in Fig. 8.2. The "extrinsic" pathway occurs due to ligands such as tumor necrosis factor TNF-α, Fas ligand, tumor necrosis factor-related apoptosis-inducing ligand (TRAIL) interact with receptors and adapter proteins including TNF-α receptor-associated death domain and Fas-associated death domain (FADD). This results in receptor oligomerization and proteolytic activation of pro-caspase-8 to caspase-8 (Jaeschke and Lemasters 2003). In the type 1 pathway, caspase-8 can directly activate pro-caspase-3 to induce apoptosis. However, hepatocytes require mitochondrial activation of the receptor signal through a type 2 pathway where caspase-8 cleaves BH3-interacting domain death agonist (Bid), a member of the B-cell lymphoma 2 (Bcl-2) family, to active fragment tBid which translocates to mitochondria and initiates caspase cleavage (Scaffidi et al. 1998; Yin 2000; Jaeschke and Lemasters 2003). The end result is the formation of the "apoptosome," a complex containing mitochondrial released cytochrome C, apoptosis-activating-factor-1, ATP, and pro-caspase-9, which then activates pro-caspase-3 and leads to the final morphologic changes of apoptosis (Korsmeyer et al. 2000; Wang 2001).

The type 2 pathway to apoptosis is faster and better regulated than the type 1 pathway but does require mitochondrial involvement and can be blocked by cyclosporin A (Hatano et al. 2000; Jaeschke and Lemasters 2003). Alternate pathways to apoptosis such as the "intrinsic" pathway are less completely understood and also lead to the activation of caspase 9 and caspase 3 through p53-dependent gene expression and translocation of pro-apoptotic Bcl-2-associated X (Bax) proteins (Jaeschke and Lemasters 2003). An additional pathway that leads to apoptosis is

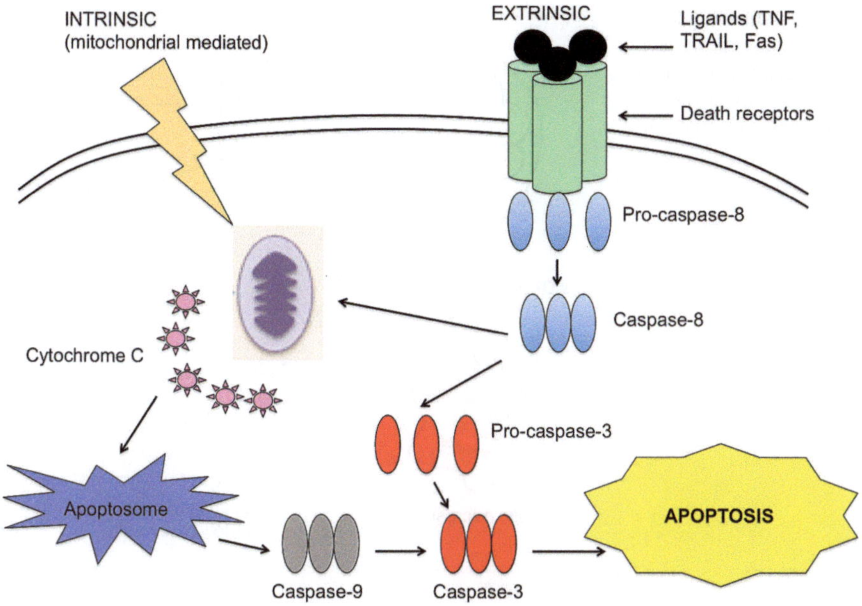

Fig. 8.2 Intrinsic and extrinsic pathways lead to apoptosis via activation of the caspase cascade

through TNF-α and Fas-mediated activation of TNF-α receptor-associated factor, which subsequently degrades inhibitor of nuclear factor κB (IκB) with the end result allowing NF-κB to induce apoptosis (Jaeschke and Lemasters 2003).

8.3.3 Role in Cell Death and Hepatic IRI

First described in 1996 in the setting of warm hepatic IRI, both sinusoidal endothelial cells and hepatocytes undergo apoptotic cell death (Sasaki et al. 1996). There is some controversy about the contribution of apoptosis to cell death after liver IRI. Observations using terminal deoxynucleotidyl transferase-mediated deoxyuridine triphosphate nick-end labeling (TUNEL) assay had suggested that up to 60–80% of susceptible sinusoidal endothelial cells and hepatocytes underwent apoptosis, potentially induced by Kupffer cell and platelet release of TNF-α, within 6 h of reperfusion (Kohli et al. 1999; Natori et al. 1999; Sindram et al. 2000; Sindram et al. 2001). However, as previously mentioned, apoptosis affects individual cells and so quantitatively apoptosis only comprises 2% of the liver cells at risk (Gujral et al. 2001). Therapeutically, pan-caspase inhibitors have been shown to have a modest protective effect on liver IRI by preventing Fas-mediated apoptosis (Cursio et al. 1999; Natori et al. 1999)

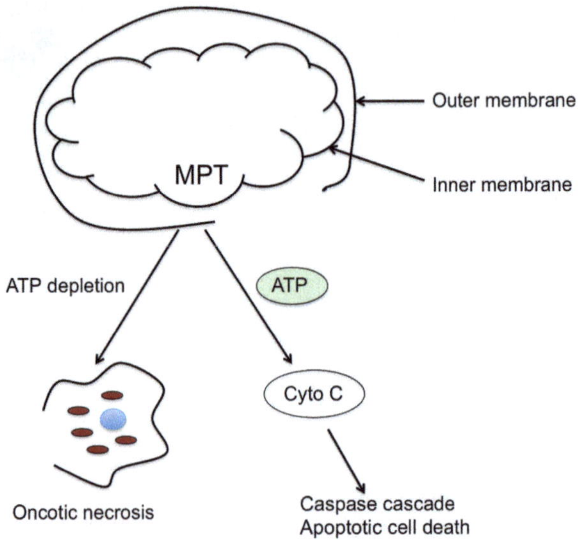

Fig. 8.3 Mitochondrial permeability transition (MPT) can lead to oncotic necrosis or apoptosis depending on the presence of ATP

8.4 Shared Features/Current Updates

An abundance of work done over the last decade has shown that mechanisms of necrosis and apoptosis are intricately intertwined. Mitochondrial permeability transition (MPT), depicted in Fig. 8.3, triggered by TNF-α and Fas-dependent signaling pathways in apoptosis, has been shown to also play a role in oncotic necrosis (Bernardi 1999; Jaeschke and Lemasters 2003). In particular, hepatocytes and endothelial cells are susceptible to pH-dependent necrotic cell death associated with reperfusion. Initial ischemic insult leads to acidosis that protects against necrotic cell death. However, upon reperfusion, restoration of pH leads to MPT and oncotic necrosis secondary to ATP depletion, without involvement of the caspase cascade (Malhi et al. 2006). While cell death in the early phases after reperfusion exhibits morphologic changes consistent with necrosis, i.e., cell swelling, karyolysis, vacuolization, enzyme release, the presence of ATP can act to switch from the necrosis to apoptosis pathway (Richter et al. 1996; Leist et al. 1997; Paxian et al. 2003). In the presence of a glycolytic substrate, such as fructose, ATP can be regenerated and leads to caspase-dependent apoptosis with ATP-mediated cytochrome C and apoptosome activation and resultant caspase-3 activation at a later time point, 8–12 h after reperfusion (Malhi et al. 2006). As these forms of cell death have been found to be inter-dependent, investigation has led to identification and description of additional cell death pathways including necroptosis, autophagy, pyroptosis, and ferroptosis.

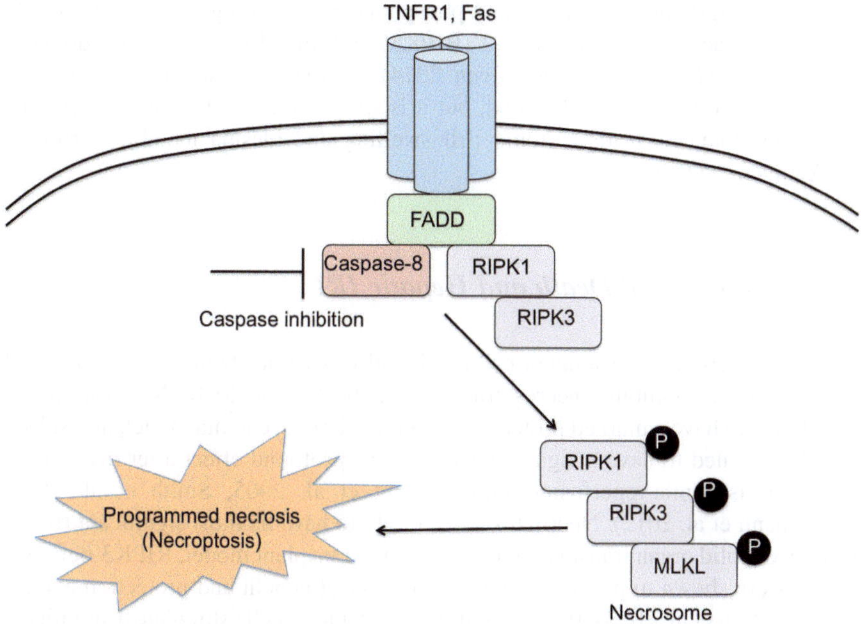

Fig. 8.4 Necroptosis is mediated by RIPK1 and RIPK3

8.5 Necroptosis

Necroptosis is a form of regulated active necrosis first described in 2005 that follows an intracellular signaling cascade dependent on receptor-interacting protein kinase 3 (RIPK3) (Teng et al. 2005; Linkermann and Green 2014). Necroptosis occurs when caspase-8 activation is inhibited and apoptosis cannot take place. It is hypothesized that this form of cell death developed as a method of "fail-safe" cell death when viruses/intracellular invaders interfere with caspase-8 activation (Linkermann and Green 2014). Morphologically, necroptosis does resemble necrosis but the pathways leading to necroptosis, shown in Fig. 8.4, are substantially different.

Necroptosis can be triggered by cytokines or agonists to the toll-like receptors (TLRs) in the setting of caspase inhibition. When caspase-8 is present, it can suppress necroptosis by association with Fas-associated death domain (FADD) and cellular FLICE-like inhibitory protein (c-FLIP) and also inactivates receptor-interacting protein kinases 1 and 3 (RIPK1 and RIPK3) by proteolytic cleavage (Vanlangenakker et al. 2011). However, when caspase-8 is inactivated or absent, RIPK1 and RIPK3 become phosphorylated instead of undergoing proteolytic cleavage (Vandenabeele et al. 2010; Zhao et al. 2015). Subsequently, death receptors and intracellular adaptor molecules, including FADD and TNF-receptor-associated death domain (TRADD), recruit RIPK1 and trigger assembly of the "necrosome" (Li et al. 2012; Linkermann and Green 2014). RIPK1 undergoes a series of ubiquitination, de-ubiquitination, and phosphorylation leading to association with receptor-interacting protein kinase 3 (RIPK3) and mixed linkage kinase domain-like (MLKL) (Linkermann and Green 2014). Further downstream mediators of necroptosis are less well understood, but it is speculated that disruption of plasma membrane channels leads to rapid cell swelling and plasma membrane rupture (Linkermann and Green 2014).

8.5.1 Role in Cell Death and Hepatic IRI

Interest in necroptosis as a major cell death pathway came about after discovery of a specific RIPK1 inhibitor, necrostatin (Vandenabeele et al. 2010). Necrostatin was first shown to have a marked protective effect in a brain ischemia model, and subsequently revealed to have a significant but not as profound effect after myocardial and renal ischemia-reperfusion injury (Teng et al. 2005; Smith et al. 2007; Linkermann et al. 2012). Necroptosis is thought to have its most significant role in the area of solid organ transplantation. In a renal transplant model, RIPK3-deficient donors were shown to provide a significant survival benefit and protection against rejection (Lau et al. 2013). It is thought that necroptotic cells stimulate a significant innate and adaptive immune response through the release of DAMPs and in turn promote the proinflammatory responses associated with IRI in solid organ transplant and ultimately rejection (Kaczmarek et al. 2013; Ladoire et al. 2014). Though several studies have been done focusing on necrostatin, there are several proposed points of intervention along the pathway for necroptosis including at the level of death receptors, RIPK1, RIPK3, MLKL, necrosome assembly, or downstream as-yet-incompletely understood pathways (Linkermann and Green 2014). Further elucidation of this pathway may lead to new therapeutic targets to prevent against the effects of IRI.

8.6 Ferroptosis

8.6.1 Discovery

Ferroptosis is a unique form of cell death triggered by the RAS-selective molecule erastin and dependent on intracellular iron (Dixon et al. 2012). Previously, RAS-selective lethal (RSL) compounds, including RSL3 and erastin, were identified and shown to be selectively lethal to RAS mutant tumor cells, important because RAS family GTPases are mutated in nearly one-third of cancers (Dolma et al. 2003; Yang and Stockwell 2008; Vigil et al. 2010). It was then observed that RSL-induced death, including erastin-induced cell death, lacks features of apoptosis such as cytochrome C release, caspase activation, and chromatin fragmentation, but is associated with ROS increase (Yagoda et al. 2007; Yang and Stockwell 2008). This led to the discovery of a new unique biochemical and morphological form of cell death that is prevented by iron chelation or genetic inhibition of iron uptake, now termed ferroptosis (Dixon et al. 2012).

8.6.2 Morphology and Mechanisms

Unlike other forms of cell death, the only distinct morphologic feature of ferroptosis is smaller mitochondria with increased membrane density and blunting of mitochondrial cristae (Yagoda et al. 2007; Dixon et al. 2012; Xie et al. 2016). Nuclear integrity is maintained without chromatin condensation (Friedmann Angeli et al. 2014). As mentioned above, ferroptosis is mediated by erastin that has two direct mitochondrial targets voltage-dependent anion channels 2 and 3 (VDAC 2/3), the activation of which leads to an iron-dependent accumulation of lethal lipid ROS (Yagoda et al. 2007; Dixon et al. 2012). In addition to VDAC2/3 activation, erastin also inhibits the cysteine/glutamate antiporter system x_c^- system leading to reduced glutathione (GSH) level with accelerated ROS accumulation (Dixon et al. 2012). The specific contribution of iron to ferroptosis is related to increased iron uptake and decreased iron storage that leads to iron overload. Excess iron then can produce ROS through the Fenton reaction; meanwhile, iron chelators have been shown to inhibit ferroptosis while exogenous iron enhances erastin-induced ferroptotic cell death (Louandre et al. 2013).

Besides the aforementioned VDAC 2/3 and erastin, there are some other known regulators of ferroptosis, as shown in Table 8.2. Transferrin receptor 1 (TFR1), responsible for importing ferric ion (Fe^{3+}), is known to be upregulated in cells sensitive to ferroptosis, while proteins from the iron storage complex are known to be downregulated in ferroptosis-sensitive cells (Yang and Stockwell 2008). The NOX protein family, responsible for oxygen reduction to superoxide, the ubiquitous tumor suppressor protein p53, and cysteinyl-tRNA synthetase (CARS) are additional positive regulators of ferroptosis (Dixon et al. 2012; Jiang et al. 2015; Hayano et al. 2016). As mentioned, system x_c^- is a negative regulator for ferroptosis. Additional negative regulators include GPX4, which converts GSH to oxidized glutathione (GSSG), heat shock protein beta 1 (HSPB1), and NRF2, with the major anti-ferroptosis role of NRF2 in hepatocellular carcinoma (Yang et al. 2014; Sun et al. 2015; Xie et al. 2016).

8.6.3 Role in Cell Death and Hepatic IRI

Ferroptosis is an intriguing target for modulating cell death and several inhibitors including antioxidants, iron chelators, ferostatin, liproxstatin, and zilueton (Xie et al. 2016). In particular, ferroptosis has been implicated in hepatic ischemia/

Table 8.2 Inducers, positive regulators, and negative regulators of ferroptosis

Role in ferroptosis	Compounds
Inducers	Erastin RSL3 and RSL5 Butionesulfoximine (BSO) Acetaminophen Small molecule ferroptosis-inducing agents (FINs) Lanperisone Slfasalasinc Sorfenib Artesunate
Positive regulators	VDAC2/3 Ras TFR1 NOX P53 Cysteinyl-tRNA synthetase (CARS)
Negative regulators	GPX4 System Xc^- HSPB1 NRF2

reperfusion injury in a murine model and is thought to contribute to the associated inflammatory response (Xie et al. 2016). Liproxstatin, which prevents ROS accumulation in glutathione peroxidase 4-deficient ($GPX4^{-/-}$) cells, has a protective effect with improved transaminase levels and reduced tissue injury seen after hepatic IRI (Friedmann Angeli et al. 2014). Like other non-apoptotic, non-necrotic forms of cell death that may play a role in hepatic IRI, downstream pathways of ferroptosis need to be elucidated to determine its exact contributions as well as the appropriate therapeutic targets.

8.7 Pyroptosis

Pyroptosis is a pro-inflammatory form of programmed lytic cell-death mediated by caspase-1, first described in macrophages in 1992 but shown to be distinct from other forms of cell death with the term pyroptosis coined in 2001 (Zychlinsky et al. 1992; Cookson and Brennan 2001; Miao et al. 2011). The key distinguishing feature of pyroptosis is that it occurs after caspase-1 activation and does not involve activation of the apoptotic caspases, as depicted in Fig. 8.5 (Miao et al. 2011). Morphologically, pyroptosis does lead to DNA damage with chromatin condensation, but unlike in apoptosis, the nucleus remains intact without karyorrhexis or fragmentation (Watson et al. 2000). Additionally, pyroptosis allows for cell membrane pore opening and thus leads to cell swelling, unlike apoptosis, which is associated with cell shrinkage (Fink and Cookson 2006). The rapid swelling leads to cell membrane protrusions, coalescence, rupture with release of cytosolic contents.

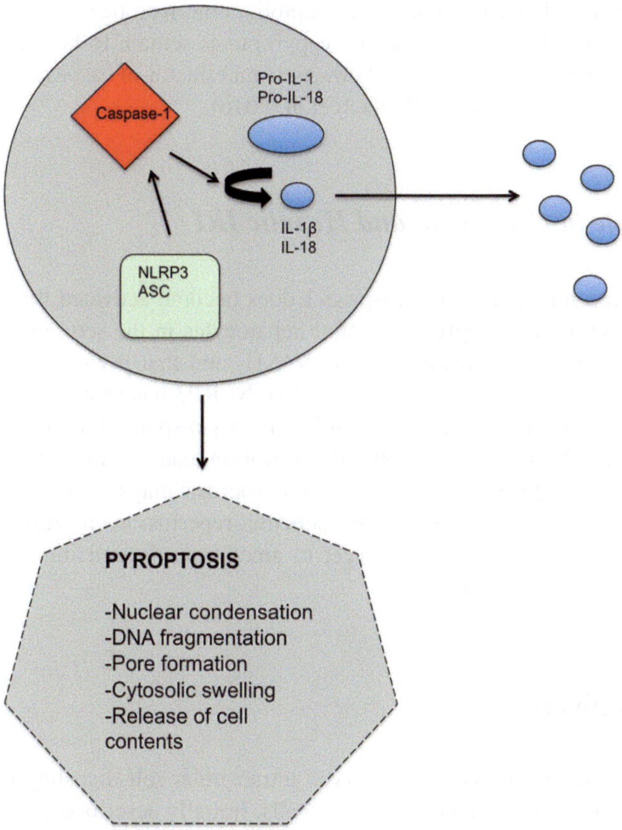

Fig. 8.5 Pyroptosis is mediated by caspase-1 and leads to release of cytokines with distinct morphologic changes

Of note, pyroptosis has been described primarily in macrophages and dendritic cells and can be activated by either intracellular pathways or TLR ligands (Edgeworth et al. 2002; Fink et al. 2008).

Pyroptosis works through the adapter protein apoptosis-associated speck-like protein containing CARD (ASC), which is recruited by inflammasomes including nucleotide-binding domain, leucine-rich repeat containing protein 3 (NLRP3), and then forms an ASC focus, also called the "pyroptosome" (Fernandes-Alnemri et al. 2007; Miao et al. 2011). The ASC focus recruits pro-caspase-1, activates it to caspase-1, and then causes proteolytic cleavage, allowing cleavage of pro-IL-1β and pro-IL-18 and release of IL-1β and IL-18 (Stehlik et al. 2003; Broz et al. 2010; Miao et al. 2011). In addition, pyroptosis exhibits significant pro-inflammatory effects through release of DAMPs including HMGB1 and IL-1α. It has recently been shown that pyroptosis, in particular, leads to the formation of an inflammatory form of HMGB1 through cysteine disulfide bonding, which can engage the TLR4 receptor and lead to release of additional cytokines such as TNF1 and IL-6

(Nystrom et al. 2013). Like other non-apoptotic, non-necrotic cell death pathways, the downstream effector mechanisms of pyroptosis remain to be elucidated. It is known that Gasdermin D is a shared downstream effector of caspase 1 and 2 but its role in pyroptosis is unknown (Wallach et al. 2016).

8.7.1 Role in Cell Death and Hepatic IRI

It has been recently shown that caspase-1 does become activated through NLRP3 inflammasome in both Kupffer cells and hepatocytes in the settings of both alcoholic and non-alcoholic steatohepatitis (NASH) and that pyroptosis can occur in hepatocytes in this setting (Wree et al. 2014). NLRP3 has recently been shown to play an important role in damage and inflammatory response after hepatic IRI, particularly in Kupffer cells, where NLRP3 activation leads to infiltration of immune cells (Huang et al. 2013). It is possible that pyroptosis plays a role in the release of both DAMPs and cytokines after liver ischemia-reperfusion and further investigation may reveal it to be a useful target to ameliorate the inflammatory response associated with hepatic IRI.

8.8 Autophagy

Autophagy is an evolutionarily conserved intracellular self-digesting pathway that is seen in both normal and damaged liver cells. Initially described as a cell survival mechanism for starvation, autophagy works through the formation of double-membrane vacuoles known as autophagosomes that transport long-lived or malformed proteins and damaged organelles to the lysosome for degradation (Klionsky and Emr 2000; Baehrecke 2005). Autophagy is carried out by the highly conserved *Atg* genes. The major regulators of autophagy include the class I and class III phosphatidylinositol 3-kinase (PI3K) signaling pathways (Petiot et al. 2000; Baehrecke 2005). Class I PI3K regulates autophagy in a negative manner by insulin-receptor-mediated activation with resultant activation of Akt/protein kinase B (PKB) (Baehrecke 2005). Meanwhile, phosphate and tensin homologue (PTEN), a known tumor suppressor, positively regulates autophagy by preventing activation of Akt/PKB (Baehrecke 2005). Class III PI3K positively regulates formation of the autophagosomal membrane. Autophagy is executed through five stages: initiation, nucleation, elongation, fusion, and degradation (Kim and Lee 2014; Go et al. 2015). After initiation and nucleation, ubiquitin-like pathways lead to expansion of the isolation membrane with regulation by multiple Atg proteins. Ultimately, the outer autophagosome membrane fuses with the lysosome and lysosomal hydrolases degrade the autophagosomal cargo (Baehrecke 2005).

8.8.1 *Role in Cell Death and Hepatic IRI*

In the liver specifically, autophagy allows for aminio acid and glucose production during starvation and also participates in hepatocellular lipid metabolism (Czaja et al. 2013). Much of the autophagy that occurs in the liver is selective autophagy of the mitochondria, i.e., mitophagy, in order to protect against mitochondrial generation of toxic ROS and pro-apoptotic proteins (Lemasters 2005). Overall, autophagy is a survival mechanism during stressful conditions such as metabolic stress, ischemia, and hypoxia (Mathew et al. 2007). It is however possible for autophagy to become a dysfunctional process and lead to cell death through excessive catabolism, cargo misrecognition, and activation of apoptotic pathways (Czaja et al. 2013). It is unclear exactly which mechanisms allow autophagy to transition from a cell survival pathway to promoting cell death; however, some studies in non-hepatic cells suggest that proteins involved in autophagy (such as ATG5) may directly interact with apoptotic and necrotic pathways, described above (Czaja et al. 2013).

During liver IRI, hepatocytes suffer from nutrient deprivation and acidosis that would stimulate autophagy in normal livers, IRI also leads to ATP depletion and decrease in certain proteins necessary for autophagy, especially ATG7 and BECN1, thought to be due to calcium-dependent proteases (Kim et al. 2008). The net result is that autophagy is insufficient to reduce the stress from reperfusion even with some temporary repolarization of mitochondria and ATP release. This effect is most profound in aged hepatocytes and may explain why aging livers are more susceptible to IRI, thought to be due to I/R-mediated depletion of ATG4B (Wang et al. 2011). Ultimately, developing strategies to enhance autophagy, particularly in the aging liver, may represent a protective strategy against the cell death encountered after hepatic IRI. Downstream effector pathways of autophagy as well as the role of autophagy in non-parenchymal hepatic cells remain to be elucidated.

8.9 Netosis

Neutrophils have long been known to play a role in liver IRI as the principal innate immune cells that lead to hepatic parenchymal damage and microvascular dysfunction (Jaeschke 2006). Neutrophils were discovered in the early 2000s to be able to form neutrophil extracellular traps (NETs) through the release of long chromatin fibers that are decorated with DAMPs including HMGB1 and histones (Brinkmann et al. 2004). NET formation was first described as a form of cell suicide with stepwise chromatin decondensation, nuclear swelling, nuclear spilling into cytoplasm, and then membrane perforation (Yipp and Kubes 2013). There is limited evidence that neutrophils may survive the process of forming NETs, though mechanisms through which this occurs are unknown. Only about 20–25% of neutrophils are thought to form NETs (Yipp and Kubes 2013).

This discovery and description of "NETosis" has led to the investigation of neutrophils as not just simple phagocytes but as sophisticated innate immune cells that can orchestrate inflammatory responses. NETs were initially described as a method of neutrophil cell death to allow for pathogen capture in the setting of septic shock. Recently, however, the role of NETs has been elucidated in settings of sterile inflammation, including hepatic ischemia reperfusion (Huang et al. 2015). NET formation is known to be increased with sterile liver injury. Additionally, targeting NETs with DNase (degrades extracellular DNA) or peptidyl-arginine-demainase 4 (PAD4) inhibitors (prevent chromatin decondensation) ameliorates the degree of liver injury seen after I/R (Huang et al. 2015). Thus, targeting this specific form of cell death in neutrophils may also represent a therapeutic potential in hepatic IRI.

8.10 Conclusions

Until about a decade ago, cell death was thought to be a dichotomy of apoptosis and necrosis. However, a large body of recent literature has revealed there to be many other forms of cell death with a high likelihood that these forms of cell death represent a continuum of interrelated mechanisms. Necroptosis, ferroptosis, pyroptosis, autophagy, and NETosis all may play a potential role in cell death in specific cell types and triggering the systemic inflammatory response after liver ischemia reperfusion injury. Further work is being done to determine exact downstream mechanisms of each of these forms of cell death. Ultimately, each may represent a potential therapeutic target against the widespread systemic effects of IRI.

References

Baehrecke EH (2005) Autophagy: dual roles in life and death? Nat Rev Mol Cell Biol 6(6):505–510

Bernardi P (1999) Mitochondrial transport of cations: channels, exchangers, and permeability transition. Physiol Rev 79(4):1127–1155

Brenner C, Galluzzi L, Kepp O, Kroemer G (2013) Decoding cell death signals in liver inflammation. J Hepatol 59(3):583–594

Brinkmann V, Reichard U, Goosmann C, Fauler B, Uhlemann Y, Weiss DS, Weinrauch Y, Zychlinsky A (2004) Neutrophil extracellular traps kill bacteria. Science 303(5663):1532–1535

Broz P, von Moltke J, Jones JW, Vance RE, Monack DM (2010) Differential requirement for Caspase-1 autoproteolysis in pathogen-induced cell death and cytokine processing. Cell Host Microbe 8(6):471–483

Caldwell-Kenkel JC, Currin RT, Tanaka Y, Thurman RG, Lemasters JJ (1989) Reperfusion injury to endothelial cells following cold ischemic storage of rat livers. Hepatology 10(3):292–299

Camilleri-Broet S, Vanderwerff H, Caldwell E, Hockenbery D (1998) Distinct alterations in mitochondrial mass and function characterize different models of apoptosis. Exp Cell Res 239(2):277–292

Cao JY, Dixon SJ (2016) Mechanisms of ferroptosis. Cell Mol Life Sci 73(11–12):2195–2209

Cookson BT, Brennan MA (2001) Pro-inflammatory programmed cell death. Trends Microbiol 9(3):113–114

Cursio R, Colosetti P, Gugenheim J (2015) Autophagy and liver ischemia-reperfusion injury. Biomed Res Int 2015:417590

Cursio R, Gugenheim J, Ricci JE, Crenesse D, Rostagno P, Maulon L, Saint-Paul MC, Ferrua B, Auberger AP (1999) A caspase inhibitor fully protects rats against lethal normothermic liver ischemia by inhibition of liver apoptosis. FASEB J 13(2):253–261

Czaja MJ, Ding WX, Donohue TM Jr, Friedman SL, Kim JS, Komatsu M, Lemasters JJ, Lemoine A, Lin JD, Ou JH, Perlmutter DH, Randall G, Ray RB, Tsung A, Yin XM (2013) Functions of autophagy in normal and diseased liver. Autophagy 9(8):1131–1158

Dickson RC, Bronk SF, Gores GJ (1992) Glycine cytoprotection during lethal hepatocellular injury from adenosine triphosphate depletion. Gastroenterology 102(6):2098–2107

Dixon SJ, Lemberg KM, Lamprecht MR, Skouta R, Zaitsev EM, Gleason CE, Patel DN, Bauer AJ, Cantley AM, Yang WS, Morrison B 3rd, Stockwell BR (2012) Ferroptosis: an iron-dependent form of nonapoptotic cell death. Cell 149(5):1060–1072

Dolma S, Lessnick SL, Hahn WC, Stockwell BR (2003) Identification of genotype-selective antitumor agents using synthetic lethal chemical screening in engineered human tumor cells. Cancer Cell 3(3):285–296

Edgeworth JD, Spencer J, Phalipon A, Griffin GE, Sansonetti PJ (2002) Cytotoxicity and interleukin-1beta processing following Shigella flexneri infection of human monocyte-derived dendritic cells. Eur J Immunol 32(5):1464–1471

Fernandes-Alnemri T, Wu J, Yu JW, Datta P, Miller B, Jankowski W, Rosenberg S, Zhang J, Alnemri ES (2007) The pyroptosome: a supramolecular assembly of ASC dimers mediating inflammatory cell death via caspase-1 activation. Cell Death Differ 14(9):1590–1604

Fink SL, Bergsbaken T, Cookson BT (2008) Anthrax lethal toxin and Salmonella elicit the common cell death pathway of caspase-1-dependent pyroptosis via distinct mechanisms. Proc Natl Acad Sci U S A 105(11):4312–4317

Fink SL, Cookson BT (2006) Caspase-1-dependent pore formation during pyroptosis leads to osmotic lysis of infected host macrophages. Cell Microbiol 8(11):1812–1825

Friedmann Angeli JP, Schneider M, Proneth B, Tyurina YY, Tyurin VA, Hammond VJ, Herbach N, Aichler M, Walch A, Eggenhofer E, Basavarajappa D, Radmark O, Kobayashi S, Seibt T, Beck H, Neff F, Esposito I, Wanke R, Forster H, Yefremova O, Heinrichmeyer M, Bornkamm GW, Geissler EK, Thomas SB, Stockwell BR, O'Donnell VB, Kagan VE, Schick JA, Conrad M (2014) Inactivation of the ferroptosis regulator Gpx4 triggers acute renal failure in mice. Nat Cell Biol 16(12):1180–1191

Go KL, Lee S, Zendejas I, Behrns KE, Kim JS (2015) Mitochondrial dysfunction and autophagy in hepatic ischemia/reperfusion injury. Biomed Res Int 2015:183469

Gujral JS, Bucci TJ, Farhood A, Jaeschke H (2001) Mechanism of cell death during warm hepatic ischemia-reperfusion in rats: apoptosis or necrosis? Hepatology 33(2):397–405

Hatano E, Bradham CA, Stark A, Iimuro Y, Lemasters JJ, Brenner DA (2000) The mitochondrial permeability transition augments Fas-induced apoptosis in mouse hepatocytes. J Biol Chem 275(16):11814–11823

Hayano M, Yang WS, Corn CK, Pagano NC, Stockwell BR (2016) Loss of cysteinyl-tRNA synthetase (CARS) induces the transsulfuration pathway and inhibits ferroptosis induced by cystine deprivation. Cell Death Differ 23(2):270–278

Huang H, Chen HW, Evankovich J, Yan W, Rosborough BR, Nace GW, Ding Q, Loughran P, Beer-Stolz D, Billiar TR, Esmon CT, Tsung A (2013) Histones activate the NLRP3 inflammasome in Kupffer cells during sterile inflammatory liver injury. J Immunol 191(5):2665–2679

Huang H, Tohme S, Al-Khafaji AB, Tai S, Loughran P, Chen L, Wang S, Kim J, Billiar T, Wang Y, Tsung A (2015) Damage-associated molecular pattern-activated neutrophil extracellular trap exacerbates sterile inflammatory liver injury. Hepatology 62(2):600–614

Jaeschke H (2006) Mechanisms of liver injury. II. Mechanisms of neutrophil-induced liver cell injury during hepatic ischemia-reperfusion and other acute inflammatory conditions. Am J Physiol Gastrointest Liver Physiol 290(6):G1083–G1088

Jaeschke H, Lemasters JJ (2003) Apoptosis versus oncotic necrosis in hepatic ischemia/reperfusion injury. Gastroenterology 125(4):1246–1257
Jaeschke H, Woolbright BL (2012) Current strategies to minimize hepatic ischemia-reperfusion injury by targeting reactive oxygen species. Transplant Rev (Orlando) 26(2):103–114
Jiang L, Kon N, Li T, Wang SJ, Su T, Hibshoosh H, Baer R, Gu W (2015) Ferroptosis as a p53-mediated activity during tumour suppression. Nature 520(7545):57–62
Kaczmarek A, Vandenabeele P, Krysko DV (2013) Necroptosis: the release of damage-associated molecular patterns and its physiological relevance. Immunity 38(2):209–223
Kerr JF, Wyllie AH, Currie AR (1972) Apoptosis: a basic biological phenomenon with wide-ranging implications in tissue kinetics. Br J Cancer 26(4):239–257
Kim JS, Nitta T, Mohuczy D, O'Malley KA, Moldawer LL, Dunn WA Jr, Behrns KE (2008) Impaired autophagy: a mechanism of mitochondrial dysfunction in anoxic rat hepatocytes. Hepatology 47(5):1725–1736
Kim KH, Lee MS (2014) Autophagy—a key player in cellular and body metabolism. Nat Rev Endocrinol 10(6):322–337
Klionsky DJ, Emr SD (2000) Autophagy as a regulated pathway of cellular degradation. Science 290(5497):1717–1721
Kohli V, Selzner M, Madden JF, Bentley RC, Clavien PA (1999) Endothelial cell and hepatocyte deaths occur by apoptosis after ischemia-reperfusion injury in the rat liver. Transplantation 67(8):1099–1105
Korsmeyer SJ, Wei MC, Saito M, Weiler S, Oh KJ, Schlesinger PH (2000) Pro-apoptotic cascade activates BID, which oligomerizes BAK or BAX into pores that result in the release of cytochrome c. Cell Death Differ 7(12):1166–1173
Ladoire S, Hannani D, Vetizou M, Locher C, Aymeric L, Apetoh L, Kepp O, Kroemer G, Ghiringhelli F, Zitvogel L (2014) Cell-death-associated molecular patterns as determinants of cancer immunogenicity. Antioxid Redox Signal 20(7):1098–1116
Lau A, Wang S, Jiang J, Haig A, Pavlosky A, Linkermann A, Zhang ZX, Jevnikar AM (2013) RIPK3-mediated necroptosis promotes donor kidney inflammatory injury and reduces allograft survival. Am J Transplant 13(11):2805–2818
Leist M, Single B, Castoldi AF, Kuhnle S, Nicotera P (1997) Intracellular adenosine triphosphate (ATP) concentration: a switch in the decision between apoptosis and necrosis. J Exp Med 185(8):1481–1486
Lemasters JJ (2005) Selective mitochondrial autophagy, or mitophagy, as a targeted defense against oxidative stress, mitochondrial dysfunction, and aging. Rejuvenation Res 8(1):3–5
Lemasters JJ, DiGuiseppi J, Nieminen AL, Herman B (1987) Blebbing, free Ca2+ and mitochondrial membrane potential preceding cell death in hepatocytes. Nature 325(6099):78–81
Li J, McQuade T, Siemer AB, Napetschnig J, Moriwaki K, Hsiao YS, Damko E, Moquin D, Walz T, McDermott A, Chan FK, Wu H (2012) The RIP1/RIP3 necrosome forms a functional amyloid signaling complex required for programmed necrosis. Cell 150(2):339–350
Linkermann A, Brasen JH, Himmerkus N, Liu S, Huber TB, Kunzendorf U, Krautwald S (2012) Rip1 (receptor-interacting protein kinase 1) mediates necroptosis and contributes to renal ischemia/reperfusion injury. Kidney Int 81(8):751–761
Linkermann A, Green DR (2014) Necroptosis. N Engl J Med 370(5):455–465
Louandre C, Ezzoukhry Z, Godin C, Barbare JC, Maziere JC, Chauffert B, Galmiche A (2013) Iron-dependent cell death of hepatocellular carcinoma cells exposed to sorafenib. Int J Cancer 133(7):1732–1742
Magna M, Pisetsky DS (2016) The Alarmin Properties of DNA and DNA-associated Nuclear Proteins. Clin Ther 38(5):1029–1041
Majno G, Joris I (1995) Apoptosis, oncosis, and necrosis. An overview of cell death. Am J Pathol 146(1):3–15
Malhi H, Gores GJ, Lemasters JJ (2006) Apoptosis and necrosis in the liver: a tale of two deaths? Hepatology 43(2 Suppl 1):S31–S44

Marsh DC, Vreugdenhil PK, Mack VE, Belzer FO, Southard JH (1993) Glycine protects hepatocytes from injury caused by anoxia, cold ischemia and mitochondrial inhibitors, but not injury caused by calcium ionophores or oxidative stress. Hepatology 17(1):91–98

Mathew R, Karantza-Wadsworth V, White E (2007) Role of autophagy in cancer. Nat Rev Cancer 7(12):961–967

Miao EA, Rajan JV, Aderem A (2011) Caspase-1-induced pyroptotic cell death. Immunol Rev 243(1):206–214

Nastos C, Kalimeris K, Papoutsidakis N, Tasoulis MK, Lykoudis PM, Theodoraki K, Nastou D, Smyrniotis V, Arkadopoulos N (2014) Global consequences of liver ischemia/reperfusion injury. Oxidative Med Cell Longev 2014:906965

Natori S, Selzner M, Valentino KL, Fritz LC, Srinivasan A, Clavien PA, Gores GJ (1999) Apoptosis of sinusoidal endothelial cells occurs during liver preservation injury by a caspase-dependent mechanism. Transplantation 68(1):89–96

Nieminen AL, Gores GJ, Wray BE, Tanaka Y, Herman B, Lemasters JJ (1988) Calcium dependence of bleb formation and cell death in hepatocytes. Cell Calcium 9(5–6):237–246

Nishimura Y, Lemasters JJ (2001) Glycine blocks opening of a death channel in cultured hepatic sinusoidal endothelial cells during chemical hypoxia. Cell Death Differ 8(8):850–858

Nystrom S, Antoine DJ, Lundback P, Lock JG, Nita AF, Hogstrand K, Grandien A, Erlandsson-Harris H, Andersson U, Applequist SE (2013) TLR activation regulates damage-associated molecular pattern isoforms released during pyroptosis. EMBO J 32(1):86–99

Ogasawara J, Watanabe-Fukunaga R, Adachi M, Matsuzawa A, Kasugai T, Kitamura Y, Itoh N, Suda T, Nagata S (1993) Lethal effect of the anti-Fas antibody in mice. Nature 364(6440):806–809

Paxian M, Bauer I, Rensing H, Jaeschke H, Mautes AE, Kolb SA, Wolf B, Stockhausen A, Jeblick S, Bauer M (2003) Recovery of hepatocellular ATP and "pericentral apoptosis" after hemorrhage and resuscitation. FASEB J 17(9):993–1002

Peralta C, Jimenez-Castro MB, Gracia-Sancho J (2013) Hepatic ischemia and reperfusion injury: effects on the liver sinusoidal milieu. J Hepatol 59(5):1094–1106

Petiot A, Ogier-Denis E, Blommaart EF, Meijer AJ, Codogno P (2000) Distinct classes of phosphatidylinositol 3′-kinases are involved in signaling pathways that control macroautophagy in HT-29 cells. J Biol Chem 275(2):992–998

Quesnelle KM, Bystrom PV, Toledo-Pereyra LH (2015) Molecular responses to ischemia and reperfusion in the liver. Arch Toxicol 89(5):651–657

Ratziu V, Goodman Z, Sanyal A (2015) Current efforts and trends in the treatment of NASH. J Hepatol 62(1 Suppl):S65–S75

Richter C, Schweizer M, Cossarizza A, Franceschi C (1996) Control of apoptosis by the cellular ATP level. FEBS Lett 378(2):107–110

Sasaki H, Matsuno T, Tanaka N, Orita K (1996) Activation of apoptosis during the reperfusion phase after rat liver ischemia. Transplant Proc 28(3):1908–1909

Scaffidi C, Fulda S, Srinivasan A, Friesen C, Li F, Tomaselli KJ, Debatin KM, Krammer PH, Peter ME (1998) Two CD95 (APO-1/Fas) signaling pathways. EMBO J 17(6):1675–1687

Sindram D, Porte RJ, Hoffman MR, Bentley RC, Clavien PA (2000) Platelets induce sinusoidal endothelial cell apoptosis upon reperfusion of the cold ischemic rat liver. Gastroenterology 118(1):183–191

Sindram D, Porte RJ, Hoffman MR, Bentley RC, Clavien PA (2001) Synergism between platelets and leukocytes in inducing endothelial cell apoptosis in the cold ischemic rat liver: a Kupffer cell-mediated injury. FASEB J 15(7):1230–1232

Smith CC, Davidson SM, Lim SY, Simpkin JC, Hothersall JS, Yellon DM (2007) Necrostatin: a potentially novel cardioprotective agent? Cardiovasc Drugs Ther 21(4):227–233

Stehlik C, Lee SH, Dorfleutner A, Stassinopoulos A, Sagara J, Reed JC (2003) Apoptosis-associated speck-like protein containing a caspase recruitment domain is a regulator of procaspase-1 activation. J Immunol 171(11):6154–6163

Sun X, Ou Z, Xie M, Kang R, Fan Y, Niu X, Wang H, Cao L, Tang D (2015) HSPB1 as a novel regulator of ferroptotic cancer cell death. Oncogene 34(45):5617–5625

Tait SW, Ichim G, Green DR (2014) Die another way—non-apoptotic mechanisms of cell death. J Cell Sci 127(Pt 10):2135–2144

Teng X, Degterev A, Jagtap P, Xing X, Choi S, Denu R, Yuan J, Cuny GD (2005) Structure-activity relationship study of novel necroptosis inhibitors. Bioorg Med Chem Lett 15(22):5039–5044

Tsung A, Sahai R, Tanaka H, Nakao A, Fink MP, Lotze MT, Yang H, Li J, Tracey KJ, Geller DA, Billiar TR (2005) The nuclear factor HMGB1 mediates hepatic injury after murine liver ischemia-reperfusion. J Exp Med 201(7):1135–1143

Vandenabeele P, Galluzzi L, Vanden Berghe T, Kroemer G (2010) Molecular mechanisms of necroptosis: an ordered cellular explosion. Nat Rev Mol Cell Biol 11(10):700–714

Vanlangenakker N, Bertrand MJ, Bogaert P, Vandenabeele P, Vanden Berghe T (2011) TNF-induced necroptosis in L929 cells is tightly regulated by multiple TNFR1 complex I and II members. Cell Death Dis 2:e230

Vigil D, Cherfils J, Rossman KL, Der CJ (2010) Ras superfamily GEFs and GAPs: validated and tractable targets for cancer therapy? Nat Rev Cancer 10(12):842–857

Wallach D, Kang TB, Dillon CP, Green DR (2016) Programmed necrosis in inflammation: toward identification of the effector molecules. Science 352(6281):aaf2154

Wang JH, Ahn IS, Fischer TD, Byeon JI, Dunn WA Jr, Behrns KE, Leeuwenburgh C, Kim JS (2011) Autophagy suppresses age-dependent ischemia and reperfusion injury in livers of mice. Gastroenterology 141(6):2188–2199. e2186

Wang X (2001) The expanding role of mitochondria in apoptosis. Genes Dev 15(22):2922–2933

Watson PR, Gautier AV, Paulin SM, Bland AP, Jones PW, Wallis TS (2000) Salmonella enterica serovars Typhimurium and Dublin can lyse macrophages by a mechanism distinct from apoptosis. Infect Immun 68(6):3744–3747

Weinberg JM, Davis JA, Abarzua M, Rajan T (1987) Cytoprotective effects of glycine and glutathione against hypoxic injury to renal tubules. J Clin Invest 80(5):1446–1454

Wree A, Eguchi A, McGeough MD, Pena CA, Johnson CD, Canbay A, Hoffman HM, Feldstein AE (2014) NLRP3 inflammasome activation results in hepatocyte pyroptosis, liver inflammation, and fibrosis in mice. Hepatology 59(3):898–910

Xie Y, Hou W, Song X, Yu Y, Huang J, Sun X, Kang R, Tang D (2016) Ferroptosis: process and function. Cell Death Differ 23(3):369–379

Yagoda N, von Rechenberg M, Zaganjor E, Bauer AJ, Yang WS, Fridman DJ, Wolpaw AJ, Smukste I, Peltier JM, Boniface JJ, Smith R, Lessnick SL, Sahasrabudhe S, Stockwell BR (2007) RAS-RAF-MEK-dependent oxidative cell death involving voltage-dependent anion channels. Nature 447(7146):864–868

Yang WS, SriRamaratnam R, Welsch ME, Shimada K, Skouta R, Viswanathan VS, Cheah JH, Clemons PA, Shamji AF, Clish CB, Brown LM, Girotti AW, Cornish VW, Schreiber SL, Stockwell BR (2014) Regulation of ferroptotic cancer cell death by GPX4. Cell 156(1–2):317–331

Yang WS, Stockwell BR (2008) Synthetic lethal screening identifies compounds activating iron-dependent, nonapoptotic cell death in oncogenic-RAS-harboring cancer cells. Chem Biol 15(3):234–245

Yin XM (2000) Bid, a critical mediator for apoptosis induced by the activation of Fas/TNF-R1 death receptors in hepatocytes. J Mol Med (Berl) 78(4):203–211

Yipp BG, Kubes P (2013) NETosis: how vital is it? Blood 122(16):2784–2794

Zahrebelski G, Nieminen AL, al-Ghoul K, Qian T, Herman B, Lemasters JJ (1995) Progression of subcellular changes during chemical hypoxia to cultured rat hepatocytes: a laser scanning confocal microscopic study. Hepatology 21(5):1361–1372

Zhai Y, Petrowsky H, Hong JC, Busuttil RW, Kupiec-Weglinski JW (2013) Ischaemia-reperfusion injury in liver transplantation—from bench to bedside. Nat Rev Gastroenterol Hepatol 10(2):79–89

Zhang M, Ueki S, Kimura S, Yoshida O, Castellaneta A, Ozaki KS, Demetris AJ, Ross M, Vodovotz Y, Thomson AW, Stolz D B, Geller DA, Murase N (2013) Roles of dendritic cells in murine hepatic warm and liver transplantation-induced cold ischemia/reperfusion injury. Hepatology 57(4):1585–1596

Zhao H, Jaffer T, Eguchi S, Wang Z, Linkermann A, Ma D (2015) Role of necroptosis in the pathogenesis of solid organ injury. Cell Death Dis 6:e1975

Zychlinsky A, Prevost MC, Sansonetti PJ (1992) Shigella flexneri induces apoptosis in infected macrophages. Nature 358(6382):167–169

Chapter 9
Autophagy in Liver Homeostasis

Bilon Khambu, Nazmul Huda, Jun Zhou, Shengmin Yan, and Xiao-Ming Yin

Abbreviation

AFLD	Alcoholic fatty liver disease
AMPK	AMP-activated protein kinase
ASH	Alcoholic steatohepatitis
ATF4	Activating transcription factor 4
ATG	Autophagy-related
ATZ	Alpha-1-antitrypsin Z
AUP1	Ancient ubiquitous protein 1
BA	Bile acid
BAR	Bile acid-activated nuclear receptor
CARM1	Coactivator-associated arginine methyl transferase
CBZ	Carbamazepine
CMA	Chaperonin-Mediated Autophagy
CREB	cAMP response element-binding protein
DDR	DNA damage response
DFCP1	Zinc finger FYVE domain-containing protein1
DR1	Direct repeat 1
Dyn2	Dynamin 2
ER	Endoplasmic reticulum

B. Khambu • N. Huda • S. Yan • X.-M. Yin (✉)
Department of Pathology and Laboratory Medicine, Indiana University School of Medicine, Indianapolis, IN 46202, USA
e-mail: xmyin@iu.edu

J. Zhou
Department of Pathology and Laboratory Medicine, Indiana University School of Medicine, Indianapolis, IN 46202, USA

Center of Minimally Invasive Surgery, Xiangya Second Hospital, Central South University, Changsha, Hunan 410011, China

W.-X. Ding, X.-M. Yin (eds.), *Cellular Injury in Liver Diseases*, Cell Death in Biology and Diseases, DOI 10.1007/978-3-319-53774-0_9

X.-M. Yin and Z. Dong (Series eds.), *Cell Death in Biology and Diseases*

ERAD	ER-associated degradation
ERQC	Endoplasmic reticulum quality control
FA	Fatty acid
FGF21	Fibroblast growth factor 21
FIP200	FAK Family Kinase-Interacting Protein of 200 kDa
FOXO3a	Fork head box O 3a
HIF2α	Hypoxia inducible factor 2α
HSL	Hormone-sensitive lipase
IR	Ionizing radiation
KFERQ	Lysine-phenylalanine-Glutamic acid-Arginine-Glutamine
LAMP-2A	Lysosome-associated membrane protein type-2A
LD	Lipid droplet
MDB	Mallory–Denk bodies
MGL	Monoacylglycerol lipase
mTORC1	Mammalian target of rapamycin complex 1
NAFLD	Non-alcoholic fatty liver disease
NASH	Non-alcoholic steatohepatitis
NBR1	Neighbor of BRCA1 gene 1
NDP52	Nuclear domain 10 protein 52 kDa
NPC1L1	Niemann-Pick-Type C1-Like1
OPTN	Optineurin
PE	Phosphatidylethanolamine
PLIN	Perilipin
PMP	Peroxisome membrane proteins
ROS	Reactive oxygen species
SKP2	S-phase kinase-associated protein 2
TFEB	Transcription factor EB
TSC2	Tuberous sclerosis 2
Ube2g2	E2 ubiquitin conjugases
ULK1	Unc-51 Like Autophagy-Activating Kinase 1
UVRAG	UV Radiation Resistance-Associated Gene

9.1 Introduction

Autophagy (from the Greek, "auto" oneself, "phagy" to eat) is an evolutionarily conserved lysosome-mediated degradative process. Its basic function is to break down and to recycle macromolecules, damaged or dysfunctional cellular organelles, and to reduce cellular waste or superfluous intracellular components (Stolz et al. 2014). Because of the liver's multi-dimensional functions, the autophagic turnover of different hepatocellular components is important for the maintenance of the hepatic homeostasis (Czaja et al. 2013).

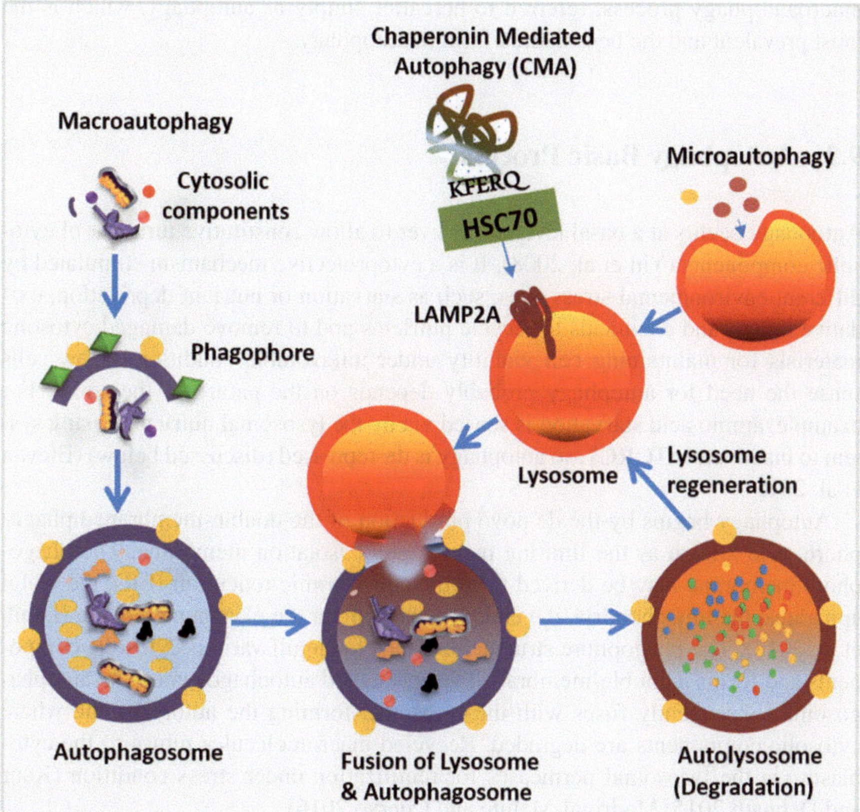

Fig. 9.1 Schematics of three types of autophagy

Autophagy is broadly classified into three types: macroautophagy, microautophagy, chaperonin mediate autophagy (CMA) (Fig. 9.1). The presence of these different autophagy pathways provides functional diversity in cells. Macroautophagy is the best characterized autophagy pathway where an expanding membrane called phagophore sequesters cytosolic components and forms double-membraned vesicles called autophagosomes. Autophagosomes fuse with lysosomes to break down the engulfed cytosolic components (Mizushima and Komatsu 2011). Microautophagy involves an invagination of lysosomal membrane to trap cytosolic proteins, forming vesicles that pinch off into the lysosomal lumen and are degraded by the lysosomal proteases (Li et al. 2012). In chaperone-mediated autophagy (CMA), targeted proteins are transported by chaperones, such as heat shock-cognate protein of 70 kDa (Hsc70), to the lysosomal lumen crossing the lysosomal membrane. Hsc70 recognizes the KFERQ (Lysine-phenylalanine-Glutamic acid-Arginine-Glutamine)-like pentapeptide motif present in the amino acid sequences of all CMA targets. CMA substrates interact with lysosome-associated membrane protein type-2A (LAMP-2A) at the lysosome surface and are subsequently internalized for the lysosomal degradation (Cuervo and Wong 2014). In this chapter, we will discuss in detail the macroautophagy process, referred to hereafter simply as autophagy, which is the most prevalent and the best studied form of autophagy.

9.2 Autophagy Basic Process

Autophagy occurs at a basal level in the liver to allow constitutive turnover of cytosolic components (Yin et al. 2008). It is a cytoprotective mechanism stimulated by different environmental stress cues, such as starvation or nutrient deprivation, oxidative injury, and chemicals to recycle nutrients and to remove damaged cytosolic materials for maintaining cell viability under unfavorable conditions. How cells sense the need for autophagy probably depends on the nature of the stress. For example, amino acid starvation is sensed via by the lysosomal nutrient sensing system to inactivate mTORC1, so autophagy is de-repressed (discussed below) (Efeyan et al. 2015).

Autophagy begins by the de novo production of the double-membraned phagophore, also known as the limiting membrane or isolation membrane, The phagophore membrane may be derived from the endoplasmic reticulum (ER), the Golgi apparatus, the mitochondria, the endocytic system, or the plasma membrane (Lamb et al. 2013). The phagophore structure expands to engulf various cytosolic components and forms a double-membraned vesicle called autophagosome. The autophagosome subsequently fuses with the lysosome, forming the autolysosome where cytosolic components are degraded. Recycled macromolecules return to the cytoplasm via the lysosomal permeases for reutilization under stress condition (Kaur and Debnath 2015; Madrigal-Matute and Cuervo 2016).

Autophagy induction is driven by a concerted action of a suite of proteins designated as ATG or "autophagy-related" proteins. So far 35 different ATG proteins have been characterized (Nakatogawa et al. 2009). They work together in a concerted manner to induce autophagy. ATG proteins could be generally listed in six functional groups that cooperate to perform key processes in autophagosome formation.

1. ULK1 kinase complex: comprised of ULK1, FIP200, Atg13L, and Atg101.
2. VPS34 kinase complex:VPS34, VPS15, Beclin-1 and Atg14, or UVRAG.
3. PtdIns(3)P proteins and Zinc finger FYVE domain-containing protein1(DFCP1).
4. Atg5-Atg12 ubiquitin-like conjugation system: Atg12, Atg5, Atg16.
5. LC3-phosphatidylethanolamine (PE) conjugation system: Atg8 or its homologues such as LC3.
6. Atg9a.

One of the earliest detectable events in autophagy initiation is the formation of ULK1 punctae. ULK1 punctae formation lies upstream of the recruitment of other ATG proteins (Lamb et al. 2013). Activated ULK protein complex recruits the VPS kinase complex to the phagophore, and phosphorylates the class III PI3K together with the ULK1 complex, generating a local pool of phosphatidylinositol 3-phosphate.

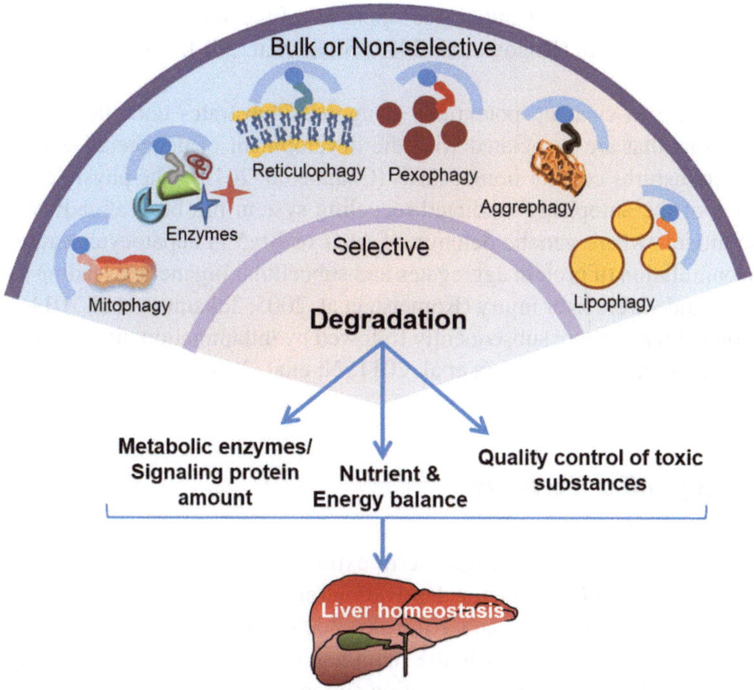

Fig. 9.2 Selective versus non-selective autophagy and their role in liver homeostasis

This is followed by the local recruitment of oligomers of Atg12-conjugated Atg5 in complex with Atg16L. The formation of the Atg12-Atg5-Atg16 complex is mediated by Atg7 and Atg11. The Atg12-Atg5-Atg16 complex is recruited to the autophagosome membrane. This complex facilitates the Atg8/LC3 lipidation via the conjugation to PE on the autophagosomal membrane. The latter event causes expansion and enclosure of the autophagosomal membrane. Recent studies have also found that deacetylation of a nuclear pool of LC3 and translocation into the cytosol occurs during starvation-induced autophagy to meet the increased demand for autophagosome biogenesis (Huang et al. 2015).

Autophagy can perform a bulk cargo degradation (nonselective) or selective cargo degradation depending on cellular stimulus (Fig. 9.2). The multi-origin nature of membranes contributing to autophagosome formation in mammalian cells has been proposed (Lamb et al. 2013). It is possible that the selectivity of the autophagy process is governed by the organelle that give rise to the membrane. Various intracellular substrates such as protein aggregates (proteophagy), damaged mitochondria (mitophagy), peroxisomes (pexophagy), lipid droplets (lipophagy), ferritin (ferritinophagy), and even intracellular microorganisms (xenophagy) could be targeted for degradation (Stolz et al. 2014; Okamoto 2014). For selective cargo degradation, autophagy *receptors*, such as p62/SQSTM1, NBR1 (neighbor of BRCA1 gene 1), NDP52 (nuclear domain 10 protein 52 kDa), and OPTN (Optineurin), bind to the cargo (often ubiquitinated) and key components of the autophagy machinery, especially the LC3 protein (Rogov et al. 2014; Stolz et al. 2014; Johansen and Lamark 2011).

The failure to clear up potentially dangerous substrates underlies the cellular dysfunctions that are associated with the development of degenerative condition, and hence disturbs cellular homeostasis (Czaja et al. 2013). The physiological significance of the autophagy-mediated recycling system has been already shown in mouse models where genetic deletion of Atg7 or Atg5 in hepatocytes causes massive accumulation of protein aggregates and subcellular organelles, leading to hepatomegaly and severe liver injury (Komatsu et al. 2005; Takamura et al. 2011). These pathological features are subsequently followed by inflammation, fibrosis, cirrhosis, and tumor formation (Takamura et al. 2011; Ni et al. 2014).

9.2.1 Signaling Pathway

Autophagy is upregulated in response to extra- or intracellular stress and signals, such as starvation, growth factor deprivation, endoplasmic reticulum (ER) stress, and pathogen infection (He and Klionsky 2009). One of the well-characterized autophagy signaling pathways is the mammalian target of rapamycin complex 1 (mTORC1) pathway. mTORC1 activation suppresses autophagy, whereas suppression of mTORC1 activates autophagy (Kim and Guan 2015; Efeyan et al. 2015). mTORC1 is the central node of the nutrient response system, which senses the amino acid levels and also the growth factor signaling (Efeyan et al. 2015). Being a kinase complex, it regulates the phosphorylation status of ULK1 complex-phosphorylating ULK1 and Atg13 to suppress the autophagy induction function of the ULK1 complex. Nutrient starvation or other stress signals inactivate mTORC1 complex and hence dephosphorylate ULK1 complex, which is activated to recruit other autophagy proteins and induces autophagy (Efeyan et al. 2015; Kim and Guan 2015). A low nutrient level also increases the cellular AMP/ATP ratio, activating AMP-activated protein kinase (AMPK), an energy-sensing molecule, which phosphorylates tuberous sclerosis 2 (TSC2) and the mTORC1 component Raptor, leading to inactivation of mTORC1 and subsequent activation of the ULK1 complex. AMPK can also phosphorylate ULK1 to activate it. Formation of autophagosome is the hallmark of autophagy induction. Lysosomal degradation of cellular components releases the nutrients, including amino acids, which can reactivate mTORC1 and thereby attenuate the ULK1-dependent autophagy (Kim and Guan 2015). This regulatory feedback inhibition of autophagy prevents excessive autophagy that may lead to apoptosis. The reactivation of mTORC1 also activates the lysosome biogenesis by a program called "autophagic lysosome reformation" to reset lysosome networks after prolonged autophagy (Chen and Yu 2013; Yu et al. 2010). In contrary to conventional idea, mTORC1-independent pathway for autophagy induction has been recently reported for cells deprived of glucose or cultured in glutamine-containing medium (Eng et al. 2010; Cheong et al. 2011). Interestingly, during glucose depletion, induction of LC3-lipidation does not require ULK1/ULK2 in this model, likely due to the accumulation of ammonia associated with amino acid catabolism in the cell (Cheong et al. 2011; Eng et al. 2010).

9.2.2 Regulation of Autophagy

Autophagy is a cellular protective mechanism to combat environmental stressors such as starvation, hypoxia, heat stress, or ROS accumulation (Schwartz et al. 1992; Huang et al. 2011). Different key modulators such as hormones (insulin or glucagon), energy status, nutrient levels, or environmental stressors regulate autophagy. Considering its key role in degradative process, autophagy is tightly regulated at different levels—transcriptional, post-transcriptional, or translation level—to maintain cellular homeostasis (Feng et al. 2015). Transcriptional regulation involves different transcriptional factors affecting autophagy genes involved in the various steps of autophagy (Feng et al. 2015). Recently, researchers have started to identify transcription factors that modulate the magnitude and duration of autophagy through the control of the expression of different autophagy genes. These transcription factors are E2F family genes, p53, STAT3, FoxO1/3, TFEB, FXR, PPARα, PPAR β/δ/√, HIF1α, GATA, and ATF4 (Feng et al. 2015).

Below we will discuss the best characterized transcription factors with regard to their role in autophagy process and in liver metabolism and homeostasis.

Transcription factor EB (TFEB) has been recently shown to coordinate autophagosome formation by driving expression of autophagy genes to sustain this process (Settembre et al. 2013a). TFEB-targeted autophagy genes include Atg4, Atg9, LC3, p62, UVRAG, Wili1. Corroborating this increase in autophagy gene expression, degradation of long-lived proteins is augmented by TFEB (Settembre et al. 2011). Liver-directed TFEB expression corrects hepatic disease by increasing SERPINA1 polymer degradation via enhanced autophagy flux (Pastore et al. 2013). Besides increasing expression of autophagy genes involved in the sequestration, membrane closure, and fusion with the lysosome, TFEB induces lysosome biogenesis, thus favoring autophagy degradation (Settembre et al. 2011).

TFEB is a target of mTORC1 (Settembre et al. 2013b). mTORC1 regulates the subcellular localization and activity of TFEB. Phosphorylation of TFEB at Ser211 by mTORC1 on lysosomal surface retains TFEB in the cytosol via the interaction with the members of the YWHA (14-3-3) family of proteins (Settembre et al. 2013b). mTORC1 inhibition dephosphorylates TFEB, which dissociates TFEB from the YWHA proteins, followed by a rapid translocation to the nucleus to initiate the downstream transcription (Settembre et al. 2013b; Martina et al. 2012). Nuclear localization and activity of TFEB is also regulated by phosphorylation by the extracellular signal-regulated kinase-2 (Erk2), whose activity is tuned by the levels of extracellular nutrients (Settembre et al. 2011).

During nutrient starvation, AMPK phosphorylates Forkhead box O 3a (FOXO3a) in the nucleus, which in turn transcriptionally represses S-phase kinase-associated protein 2 (SKP2) (Shin et al. 2016). This repression leads to increased levels of Coactivator-associated arginine methyltransferase (CARM1) protein and subsequent increases in histone H3 Arg17 dimethylation. Genome-wide analyses reveal that CARM1 exerts transcriptional co-activator function on autophagy-related and lysosomal genes through TFEB. Thus, CARM1-dependent histone arginine methylation is a crucial nuclear event in autophagy, and represents a new signaling axis of AMPK–SKP2–CARM1 in the regulation of autophagy induction after nutrient starvation (Shin et al. 2016).

Other transcription factors affecting autophagy genes include the Fork head (FOXO) family of transcription factors. FOXO has been reported to regulate genes especially in the core machinery, currently known to enhance autophagic flux by direct stimulation of autophagy-related genes (Sengupta et al. 2011). The role of this factor in autophagy is well characterized in murine models of muscular atrophy, an age-related condition (Mammucari et al. 2007; Zhao et al. 2007). Finally, FXR and PPARα are transcription factors recently reported to modulate autophagy at transcriptional level and impact the liver metabolism (described below).

Besides transcriptional regulation, autophagy is regulated both at the post transcriptional and at the posttranslational level. mRNA stabilization of key autophagy genes and temporary maintenance of degradative pool of key autophagy genes such as ULK1, Atg12, and LC3 have been reported (Khambu et al. 2011). Other posttranslational regulatory processes such as phosphorylation, ubiquitination, acetylation, epigenetic regulation, and protein-protein interaction have all been well reported to regulate autophagy process (Feng et al. 2015).

9.3 Role of Autophagy in Metabolic Homeostasis

Autophagy is tightly linked to nutrients and energy balance. Even though it has been classically viewed as a "protein-centric" catabolic process, recent studies have suggested that autophagy can mobilize diverse cellular energy stores, such as lipids and carbohydrates, to replenish various metabolites during normal and stressed conditions (Mizushima and Klionsky 2007). Moreover, it could target many dysfunctional or superfluous organelles and toxic protein aggregates and hence assist in maintaining liver over all homeostasis (Fig. 9.2). Due to a high biosynthetic activity and a role in protein turnover and carbohydrate storage, hepatocytes are particularly dependent on basal autophagy for their normal physiological function. Two important features of autophagy suite this dynamic homeostatic function-constant operation for an intracellular quality control and rapid up-regulation of autophagy flux in response to a wide range of stimuli, including (but not limited to) nutritional, metabolic, oxidative, pathogenic, proteotoxic stimuli. Broadly, autophagy appears to play role in liver homeostasis in three ways:

1. *Nutrient recycle*—Autophagy provides essential biomolecules such as amino acids, fatty acids, and carbohydrates for cellular energy generation (recycling) by mobilization and hydrolysis of proteins, lipid, or glycogen.
2. *Quality control of organelles and toxic aggregates*—Autophagy regulates the number, quality, and dynamics (functionality) of mitochondria, ER, and other organelles by selective targeting process such as mitophagy and ER-phagy.
3. *Altering the level of metabolic enzymes and signaling protein*—Different enzymes involved in metabolic pathways such as glycolysis are targeted for cellular degradation by autophagy (Schneider et al. 2014). Moreover, multifunctional signaling hub protein such as p62/SQSTM1 is also targeted by autophagy process (Katsuragi et al. 2015).

Recently, tissue-specific intracellular degradative role of autophagy has been shown to be communicated to other organs by the release of specific mitokine signaling protein, suggesting the regulatory role of autophagy in coordinating the inter-organ metabolism. For example, fibroblast growth factor 21 (FGF21) is a mitokine induced during stress response due to defective mitophagy. Activating transcription factor 4 (ATF4) promotes FGF21 expression in autophagy-defective tissue. FGF21 released from mitophagy-defective tissue acts on neighboring tissue to modulate their local metabolism (Kim et al. 2013). In the following section, we will discuss how autophagy regulates liver metabolism in terms of particular metabolites or cytoplasmic components.

9.3.1 Autophagy in Lipid Homeostasis

The liver actively participates in the lipid metabolism together with the adipose tissue, the muscle, and the brain. Fatty acid (FA) uptake, de novo synthesis, β-oxidation of FAs, cholesterol synthesis and biotransformation into bile acid, lipoprotein uptake and secretion are all part of the lipid metabolism engaged by hepatocytes. A key component subjected to metabolic regulation is lipid droplet (LD), the storage form of lipids.

Hepatocytes can store excessive lipids within LD under nutrient sufficient condition. It is the second largest repository of stored lipid in the body after adipose tissue. LD synthesis occurs in the endoplasmic reticulum. It contains a lipid and a protein component. LDs were initially considered "inert" lipid deposits-physiologically inactive particles. It is now considered to be dynamic in nature, and defined equivalently to an intracellular organelle. LDs are active due to dynamic interactions with other organelles and to the buffering capacity to misfolded proteins (Cole et al. 2002) or infectious agents (Filipe and McLauchlan 2015). Its size ranges from 0.1 to 10 μm. It contains lipid components (neutral lipid esters, triglycerides, and cholesteryl esters) (Khor et al. 2013), surrounded by a phospholipid monolayer embedded with LD coat proteins that are classified into a family of perilipin (PLIN) (Kimmel et al. 2010). When nutrients are scarce, hepatocytes rapidly deplete their energy reserves, including LDs, in order to meet their basic energetic demand. LDs degradation is basically mediated by (1) *cytoplasmic lipases*-LD-associated cytosolic lipases, such as adipose triacylglycerol lipase (ATGL alias PNPLA2, desnutrin, FP17548, TTS-2.2) or hormone-sensitive lipase (HSL) and monoacylglycerol lipase (MGL), directly dock to LD surfaces for hydrolysis of triglycerides into FAs and glycerol), or by (2) *autophagy* that sequesters the LDs and delivers them to lysosomes. However, considering the relative paucity of cytosolic lipases in hepatocytes in comparison with those in adipocytes, it is reasonable to assume that lipophagy is the key to LD catabolism.

Recent studies have revealed an important role for autophagy in LD breakdown by lipophagy or macrolipophagy. Pharmacological inhibition with 3-methyladenine or by using RNA interference against Atg7 or Atg5 increased the hepatocellular triglyceride level, increase in number and size of hepatic lipid droplets (Singh et al. 2009). Ablation of liver-specific autophagy also leads to excessive hepatic lipid accumulation and the development of fatty liver (Amir and Czaja 2011). Lipophagy is a ubiquitous pathway for in-bulk mobilization of cellular lipid droplets where LDs are engulfed by autophagosomes and, after lysosome fusion, degraded by lysosomal acid lipases. How the lipid droplets are selectively targeted and the molecular entities (receptor/s) involved are still largely unknown. Segregation of different sizes of lipid droplets for sequestration by autophagosomes is also unknown. Studies in mouse hepatocytes have shown a clear colocalization of LDs and the LC3 autophagy markers. Electron microscopic micrograph analysis also shows clearly engulfed LDs inside the double membranous autophagosome membranes (Singh et al. 2009) and enrichment of autophagy proteins such as LC3B, Atg7, Atg5, and Beclin-1 on the LDs surface has also been found (Kaushik and Cuervo 2015). What signals the recruitment of ATG proteins associated with the recognition process to the lipid cargo is not clear.

Given the differences in LD size, it is likely that autophagosome assembly occurs at the surface of the LDs, the autophagosome sequesters portions of larger lipid droplets (piecemeal) or possibly consumes smaller lipid droplets (complete) for lysosomal delivery. The removal of perilipin proteinic barrier (perilipin-2 and perilipin-3) from the lipid droplet surfaces may occur prior to the docking of autophagy proteins and cytosolic lipases. This facilitates lipolysis by cytosolic lipases and assembly of autophagosomes (Kaushik and Cuervo 2015).

Given the lipid-associated ability of Atg14 and Atg3, these two proteins have been postulated to be the critical determinants for recruitment and expansion of the autophagic machinery at the LD surface (Fan et al. 2011; Nath et al. 2014). It is possible that similar to other organelles targeting (see below), the LD-associated proteins may be ubiquitinated and hence selectively targeted for its lysosomal delivery via p62/SQSTM1 adaptor protein. A direct molecular link between LDs and the cellular ubiquitinylation machinery has been recently shown by Spandl et al. (2011). Recruitment and integration of ancient ubiquitous protein 1 (AUP1) binds to E2 ubiquitin conjugases (Ube2g2) by its G2BR domain. Formation of AUP1-Ube2g2 molecular complex to the lipid droplets has been postulated as a potential mechanism for labeling of the lipid droplets components (Spandl et al. 2011). Moreover, a recent study has shown that active Rab7 "priming" of LDs is a prerequisite for its breakdown (Schroeder et al. 2015). Rab7 is a regulator of the transport and maturation of the late endocytic vesicles, but can also normally associate with LDs. Another study showed that dynamin 2 (Dyn2), a large GTPase involved in membrane scission, favored the degradation of hepatic LDs by promoting the maturation of the autophagic compartment and maintaining lysosomal homeostasis (Schulze et al. 2013).

Besides the catabolic role of autophagy in regulating LDs, Shibata and colleagues (Shibata et al. 2009) found that autophagy proteins such as LC3 conjugation system are required for LD formation. In an in vivo model of autophagy deficiency starvation-induced lipid accumulation failed to occur (Shibata et al. 2009). Interestingly, LDs have been reported to be a potential lipid contributing source to the biogenesis of autophagosomes and thus aid in autophagic initiation. PNPLA5, a neutral lipase present in the LD, was found to be involved in mobilization of the lipid precursors from LDs (Dupont et al. 2014).

Recently, activation of autophagy and catabolism of cellular lipid droplets has been shown to be regulated by a complex transcriptional regulatory network. TFEB (described above) is also shown to be involved in lipid catabolism (Settembre et al. 2013a) by increasing the expression of PGC1α and PPARα and hence enhancing the fatty acid β-oxidation. PGC1α is the co-activator of the nuclear receptor PPARα and one of the major cellular regulators of lipid catabolism (PGC1α-PPARα-lipophagy axis). PGC1α and PPARα are well known for their regulation of the mitochondrial and peroxisomal enzymes to affect the final catabolic steps of free FA and other intermediate lipid products. Together with TFEB orthologue HLH-30, MXL-3 was also found to modulate lysosomal acidic lipase LIPL-1 and LIPL-3-dependent lipid droplets breakdown through lipophagy in *Caenorrhabditis elegans* model (O'Rourke and Ruvkun 2013). FXR-CREB axis (described below) promotes lipophagy in the fasted state via direct transcriptional activation of TFEB and autophagy genes by recruiting the coactivator CRTC2. In contrast nutrient availability suppresses this effect (Lee et al. 2014).

Abnormal accumulation of lipid droplets is considered the manifestation of fatty liver diseases, making liver susceptible to further development of other chronic liver pathologies such as steatohepatitis, fibrosis, cirrhosis, and tumor generation. Autophagy is induced following acute lipidogenic stimulus. Chronic lipid stimulation, however, compromises autophagy function by affecting autophagy flux (autophagosome-lysosome fusion). It has been postulated that high-fat diet alters the lipid contents and hence autophagic vesicular fusion (Koga et al. 2010). Evidence has been accumulated that autophagy can counteract excess lipid buildup. Pharmacological reduction of hepatic steatosis in alcoholic fatty liver disease (AFLD) and non-alcoholic fatty liver disease (NAFLD) by carbamazepine (CBZ) and rapamycin, both of which are autophagy inducers, was observed and was most likely related to the enhancement of autophagy degradation of lipids (Lin et al. 2013).

9.3.2 Autophagy and Nuclear Receptors

The nuclear receptor FXR or BAR (Bile acid-activated nuclear receptor) has recently been shown to be a repressor of autophagy in the liver besides regulating bile acid (BA), lipid, and glucose metabolism (Li and Chiang 2014; Evans and Mangelsdorf 2014). In fed state (postprandial period), bile acids-activated FXR can modulate the autophagy network in the liver independent of the mTORC1 pathway (Seok et al. 2014; Lee et al. 2014). FXR directly binds to the direct repeat 1 (DR1) (promoter) sites of autophagy genes, such as LC3, Atg4, Atg7, Atg10, Wipi1, Dfcp1, Ulk1, LAMPs, and p62. FXR competes with PPARα for binding to shared sites in the promoter regions of these genes and represses their expression during fed condition. Interestingly, PPARα activation during fasting can bind to these promoters of autophagy genes and promote the production of energy from the degradation of hepatic FA. Similarly, FXR-mediated regulation of autophagy genes expression was observed by Seok et al. (2014). By contrast, in this study, they proposed FXR activation blocks autophagy by inhibiting the transcriptional activity of a fasting transcriptional activator cAMP response element-binding protein (CREB), a protein that promotes the expression of several autophagy genes through TFEB expression. FXR inhibits the formation of a functional interaction complex between the CREB and its coactivator protein CRTC2 in inhibiting lipophagy in the liver of fed mice (FXR-CREB-CRTC2 pathway).

While FXR is known to suppress autophagy in response to nutrient signals, whether normal autophagy function is important for modulating FXR nuclear receptor signaling and maintaining normal BA metabolism is unknown. Considering the physiological importance of the FXR and the autophagy pathway, autophagy may in turn also regulate the FXR signaling pathway. Decreased autophagy could impact BA metabolism or FXR indirectly via deregulation of cholesterol metabolism. Autophagy delivers intracellularly stored cholesterol (LDs) into the lysosome where cholesterol ester is hydrolyzed for subsequent cellular distribution, efflux and BA synthesis. In this perspective, decreased autophagy may prevent the availability of free cholesterol in hepatocytes for the biosynthesis of BA. BA is biosynthesized especially in the hepatocytes from cholesterol.

BA are amphiphatic water soluble biomolecules participating in body cholesterol disposal as well as generation of bile flow and biliary lipid secretion. Its synthesis and transport is an important function of hepatocytes. Intracellular accumulation of BA is toxic to cells due to its inherent detergent nature and membrane-disruption properties. So intracellular BA level is stringently regulated and maintained within a narrow concentration range by FXR mediated transcriptional regulation of genes involved in its synthesis and transport (Ananthanarayanan et al. 2001). In vivo knock out of FXR in hepatocytes elevated the intracellular BA, which in turn caused impairment of autophagic flux. Similar impairment in autophagic flux was observed when BA was added to HepG2 cells. BA impairs autophagosome-lysosomal fusion in hepatocytes without affecting key autophagy genes expression (Manley et al. 2014). BA treatment decreased the co-localization of LC3 protein with the Rab7. All the evidence suggests a tight connection of intracellular BA and autophagy process.

9.3.3 Autophagy and Carbohydrate Metabolism

An accumulating pile of evidence now suggests there is an intimate connection between glucose metabolism and autophagy. In conditions of glucose deprivation, AMPK is activated by the decrease of cellular ATP levels, which then leads to suppression of mTORC1 and induction of autophagy (Kim and Lee 2014). Meanwhile, AMPK also positively regulates VPS34 activity and autophagy via directly phosphorylates ULK1 and Beclin-1 (Kim and Lee 2014). In addition to the AMPK pathway, glucose might also regulate autophagy by activation of FOXO1 or Rag-dependent activation of mTORC1 (Kim and Lee 2014; Salih and Brunet 2008).

As an evolutionarily conserved process of cellular degradation process, autophagy has been suggested to play a critical role in the maintenance of energy homeostasis (Kim and Lee 2014). The role of autophagy in hepatic glucose metabolism has also been investigated with effects on gluconeogenesis and glycogen storage (Kim and Lee 2014). In liver-specific Atg7-deficient mice, the absence of the autophagic amino acid release has been suggested to affect gluconeogenesis during starvation (Ezaki et al. 2011). Another work also reported that Atg7 was necessary for statin-induced gluconeogenesis in the liver (Wang et al. 2015). However, a similar study using tissue-specific Atg5-deficient mice indicated that Atg5-related autophagy had no effect on gluconeogenesis but was essential for ketogenesis during starvation (Takagi et al. 2016). These contradictive studies imply differential roles of ATG genes in glucose metabolism, which still need to be further investigated. In addition to gluconeogenesis, autophagy has been suggested to play a role in glycogen breakdown in newborn hepatocytes of the rat as well as in the skeletal muscles of *Drosophila melanogaster* (Kotoulas et al. 2006; Zirin et al. 2013). Nonetheless, liver-specific Atg7- or VPS34-deficient mice showed a decrease or no change in hepatic glycogen content in other studies (Komatsu et al. 2005; Jaber et al. 2012; Ezaki et al. 2011). Obviously, the differential effects of autophagy on glycogen metabolism still need to be further clarified.

9.3.4 Protein Quality Control (Aggrephagy)

Intracellular proteins are generally classified into short-lived proteins (half-life, 10–20 min) and long-lived proteins (Mizushima and Klionsky 2007). Although short-lived proteins are less than 1% of the total proteins in hepatocytes, due to their rapid turnover, they still contribute to about one-third of total protein degradation (Mizushima and Klionsky 2007). The rate of protein degradation has been reported with a basal rate at 1.5–4.5%/h under starvation in perfused rat liver (Mizushima and Klionsky 2007).

During nutritional stress in most organisms, the carbohydrate store supplies nutrients at the initial phase of fasting. Then the proteins are broken down to provide substrates for gluconeogenesis (Mizushima and Klionsky 2007). After these stores are exhausted during starvation, cells employ autophagy to reutilize existing macromolecules. Therefore, autophagy plays an important role in the response of organisms to nutritional stress (Mizushima and Klionsky 2007). Under starvation conditions, wild-type mice decrease about 30% hepatic proteins after fasting for 24 h, which, however, becomes insignificant in Atg7 deletion mice (Komatsu et al. 2005; Mizushima and Klionsky 2007), indicating that autophagy may play a critical role in hepatic protein degradation under starvation.

Other than a starvation response, protein turnover via autophagy also has been connected to various hepatic pathophysiological conditions, and pharmacological modulation of autophagy may be a potential therapy for these liver diseases (Puri and Chandra 2014). For example, alpha-1-antitrypsin (AT) is expressed in hepatocytes and secreted into the blood as an inhibitor for neutrophil proteases (Puri and Chandra 2014). In alpha-1-antitrypsin deficiency a mutation in the AT gene leads to the mutant protein that cannot be degraded by the proteasome, resulting in an accumulation of the AT protein in the hepatocytes endoplasmic reticulum (Puri and Chandra 2014). In addition to the decrease of AT in the blood, AT deficiency also induces hepatic fibrosis and carcinogenesis in the liver due to the toxicity of the mutant alpha-1-antitrypsin Z (ATZ) accumulation (Hidvegi et al. 2010). Accumulating evidences have shown that autophagy forms an important pathway for the clearance of misfolded mutant proteins. Autophagy-enhancing drug CBZ has been suggested to decrease the hepatic load of ATZ as well as hepatic fibrosis in a mouse model with AT deficiency related liver disease (Hidvegi et al. 2010). Another study also observed a decrease of ATZ in human primary hepatocytes using Ezetimibe, which activates autophagy through the inhibition of Niemann-Pick-Type C1-Like1 (NPC1L1) (Yamamura et al. 2014). Autophagy also has been suggested to be a potential therapy in another rare inherited disease called fibrinogen storage disease, which is characterized by hyperfibrinogenemia and accumulation of fibrinogen aggregates in the liver (Puri and Chandra 2014; Puls et al. 2013).

Inclusion bodies can be observed in various neuronal, muscular, and other human diseases. Several molecular constituents including p62, chaperons, and proteasome subunits are common in different types of inclusion bodies (Strnad et al. 2008). One of the best studied inclusion bodies, Mallory–Denk bodies (MDB), are one of characteristic morphologic features of alcoholic steatohepatitis (ASH), but also could be observed in various other chronic liver diseases including non-alcoholic steatohepatitis (NASH) (Zatloukal et al. 2007). The formation of MDB may include protein misfolding, chaperone alterations, disproportional protein expression with keratin 8 level higher than keratin 18 level, and subsequent keratin 8 crosslinking by transglutaminase. Among these factors, one crucial event for MDB formation is p62 (Strnad et al. 2008). Although the function of MDBs in related diseases is still unclear, MDBs are important histological and potential progression markers in various liver diseases (Zatloukal et al. 2007). Previous study reported that further activation of autophagy by rapamycin could significantly decrease the amount of inclusion bodies in proteasome inhibitor-treated keratin 8 transgenic mice, in which MDB-like inclusion bodies occurred in the liver (Harada et al. 2008). The study suggests autophagy may involve in the formation of cytoplasmic inclusions, and also provides a potential therapy via modulation of autophagy for diseases that may result from the accumulation of inclusion bodies.

9.3.5 *Organelle Quality Control*

Cellular viability depends on the quality and integrity of intracellular organelles. Cellular homeostasis and adaptation of organisms to adjustable environments necessitate critical maintenance of organelle quality and quantity. Several lines of evidence suggest that the quality control of organelles is achieved via autophagy, the process is known as organellophagy via bulk or selective autophagy (Okamoto 2014). Below, we describe briefly the different forms of organellophagy and their roles in quality control and hence in overall liver homeostasis. Quality control of protein aggregates (aggrephagy) has been described in the previous section.

9.3.5.1 Mitophagy

Hepatocytes are rich in mitochondria (about 800/hepatocyte). Being metabolically active cells, hepatocytes generate a high amount of reducing equivalents such as NADH and FADH2 through mitochondrial nutrient oxidation. Generation of ATP from these reducing equivalents generates reactive oxygen species (ROS). Under normal conditions, about 1–2% of mitochondria oxygen consumption results in ROS production (Boveris and Chance 1973). Excessive ROS generation can morphologically and functionally impact the mitochondria by oxidizing lipid, proteins, and nucleic acids (particularly mtDNA). Defective and unhealthy mitochondria can boost cellular oxidative stress, which may induce cell death through apoptosis. Removal of these defective mitochondria by autophagy is vital for mitochondrial quality control and hence for the hepatocyte homeostasis. Emerging studies demonstrate that mitophagy contributes to mitochondrial quality and quantity control. This selective removal of the dysfunctional mitochondria needs to be highly regulated and requires coordinated functions of mitochondrial and cytosolic proteins to spare the healthy mitochondria and other organelles. The key question is how an autophagosome selectively engulf mitochondria and how that process is triggered without the activation of cell-wide autophagy? Recent studies highlight the PINK1/Parkin pathway of mitophagy as an important regulator of mitochondrial homeostasis (Eiyama and Okamoto 2015; Wei et al. 2015). The recruitment of Parkin to defective mitochondria requires the activity of PINK1 (Youle and Narendra 2011). As a matter of fact, overexpression of PINK1 or induced accumulation of PINK1 on mitochondria is adequate to induce Parkin translocation and mitophagy in the absence of mitochondrial uncoupling. In addition, NIX has been described to promote Parkin translocation and Parkin-mediated mitophagy in mouse embryonic fibroblasts (Ding et al. 2010). This pathway is orchestrated by an intricate array of proteins that are frequently being revised and enhanced. For example, recent studies revealed that PINK1 is stabilized in the mitochondria in reaction to reduced membrane potential, recruits Parkin, which is able to ubiquitinate VDAC, TOMs, Fis1, Mfn1 & 2, and MIRO, to induce mitophagy (Zhang 2013). Defects in mitophagy may lead to Parkinson's disease. PINK1 and Parkin are reported to be mutated in autosomal-recessive Parkinson's disease (Youle and Narendra 2011). More research needs to be done to answer: how mitophagy is controlled to titrate mitochondrial numbers and how mitochondria are attached to the autophagosomal recognition.

9.3.5.2 Reticulophagy

Reticulophagy or ER-phagy (autophagy of the endoplasmic reticulum) is a process that delivers defective or excessive ER to the lysosome for degradation. Role of autophagy in degrading excessive ER is evidenced by an increased level of hepatic ER in liver-specific Atg7 or Atg5-deficient mice (Komatsu et al. 2005; Yang et al. 2016). The liver is responsible for the selective uptake, concentration, metabolism, and excretion of the majority of drugs and chemicals that are introduced into the body. These xenobiotic substances are processed by a variety of soluble or membrane-bound enzymes, which are especially associated with hepatocyte ER. So exposure of xenobiotic generally causes proliferation of ER in hepatocytes. Autophagy regulates ER degradation in hepatocytes (Yang et al. 2016). Interestingly, it appears that p62 acts like an adaptor for ER-phagy in the liver. ER-phagy induced by ER stress was recently categorized as micro-ER-phagy, which does not necessitate the use of macroautophagy modules (Schuck et al. 2014; Khaminets et al. 2015).

The endoplasmic reticulum quality control (ERQC) system shuttles misfolded proteins for degradation by the proteasome through the well-defined ER-associated degradation (ERAD) pathway. Autophagy is known to participate in ERQC as ER stress-induced autophagy complements ERAD (Ding and Yin 2008). Recently, the macro-ER-phagy pathway and its role in ERQC have been reported (Lipatova and Segev 2015). Unlike the micro-ER-phagy pathway, the macro-ER-phagy pathway requires core autophagy-specific proteins, and Ypt1, a Rab GTPase in the yeast (Lipatova et al. 2013; Lipatova and Segev 2014). Importantly, Atg9 plays a role in the assembly of ER-to-autophagy membranes. Other core Atgs and Ypt1 assemble the Atg-protein complex on autophagosome membrane to form the pre-autophagosomal structure (Lipatova and Segev 2015). Beyond exploring how autophagy of the ER is regulated, researchers are working to understand why there are different autophagy receptors for perinuclear and cytoplasmic ER and searching for perinuclear autophagy receptors in mammalian cells.

9.3.5.3 Pexophagy

Peroxisomes play important roles for proper cell function in all eukaryotic organisms. They participate in the metabolism of lipids and production and degradation of hydrogen peroxide. The quality control of peroxisomes has been carefully examined since many physiological complications and diseases are associated with the peroxisomal deficiencies or dysfunctions. There is significant involvement of the autophagy machinery in peroxisome degradation (pexophagy) in mammalian cells (Nordgren et al. 2013).

Peroxisomes need to be created and degraded in a rapid and constant turnover as the half-life of it is short (1–2 days). The autophagy–lysosome pathway is one of the main pathways for the degradation of peroxisomes. In the yeast, both macro- and micro-autophagy are involved in peroxisome removal. Peroxisomes are associated with oxidative reactions including the β-oxidation of branched and long-chain FAs. ROS generated from the reactions are associated with autophagy induction. For example, mTORC1 is inhibited by ROS-mediated ATM kinase signaling.

Several Pex proteins are involved in macropexophagy. The elimination of Pex3 from the peroxisomal membrane is mediated by macropexophagy, followed by its degradation by the proteasome. When Pex3 fails to be removed, for example, in a mutant lacking VPS34, peroxisomes are not sequestered by autophagy and stay steadily in the cytosol (Manjithaya et al. 2010). Other than Pex3, the presence of Pex14 at the peroxisomal membrane is required for the recognition of the organelle by the macropexophagy machinery (Neuhaus et al. 2014). The ubiquitinylated Pex5 recruits the autophagy adapter protein p62 through engaging the autophagic core machinery (Subramani 2015). Recently, hypoxia inducible factor 2α (HIF2α), an oncoprotein, has been reported to induce hepatocyte pexophagy (Walter et al. 2014). It is speculated that HIF2α increases expression of autophagy adaptors, such as NBR1 and p62/SQSTM1, and induces clustering and subsequent degradation of peroxisomes. HIF2α may also induce an E3 ubiquitin ligase that mediates the ubiquitination of particular peroxisome membrane proteins (PMP), enhancing the NBR1 accumulation on peroxisome, which in turn serves as a platform for the recruitment of p62/SQSTM1 for pexophagy (Walter et al. 2014). Even though a number of outstanding research works have been reported toward the mechanism of pexophagy, understanding the functions of the proteins shared by the different modes of pexophagy is still much needed.

9.3.6 DNA Damage and Genomic Quality

Both endogenous and exogenous factors create DNA lesions in cells, which activates DNA damage response (DDR) and DNA repair mechanisms. One of the important endogenous factors is ROS, which is generated in normal cellular process, e.g., respiration in mitochondria.

Using the DNA-damaging agents and ionizing radiation (IR), several reports demonstrated a connection between DNA damage and autophagy. Many types of DNA damage are able to trigger PARP-1. DNA damage induced by doxorubicin has been recently shown to be paralleled by autophagy. Interestingly, doxorubicin-induced autophagy was experimentally perceived to be governed by PARP-1 (Munoz-Gamez et al. 2009). The p53 molecule is a critical protein in DDR. Autophagy and DDR are connected by two central factors, p53 and mTORC1. On one hand, p53 can either activate AMPK, which disables mTORC1, thus activating autophagy, or stimulate the transcriptional expression of damage-regulated autophagy modulator, a lysosomal protein that enables the autophagic process. Importantly, p53 activation also has effects in the expression of PTEN, an inhibitor of PI3K/Akt signaling pathway, which suppresses mTORC1 and then activates autophagy. On the other hand, compared to nuclear p53, cytoplasmic p53 acts as a repressor for autophagy. p53 can interact with Atg7 and regulate the transcription of cell cycle inhibitor p21 (Lee et al. 2012). Therefore, depending on the cellular context and the degree of DNA damage, autophagy can have a protective or cytotoxic role in cellular viability.

Both ATM and FOXO3a have been proven to control autophagy (Kroemer et al. 2010). Genotoxic and oxidative agents cause DNA damage and ATM activation in cells, causing suppression of mTORC1 signaling via AMPK activation. The crosstalk between autophagy and DDR and its function in regulating cell fate is an important topic to be explored. Knowing the detailed and complicated relationship between DDR and autophagy will have insightful impacts on several fields of medical interest, such as cancer, aging, and neurodegeneration. The current evidence connecting DNA damage to autophagy suggests that both are involved in the normal physiology and pathological conditions. More detailed and careful studies of the pathways connecting DDR to autophagy would likely result in therapeutic approaches.

9.4 Conclusion

Normal liver function requires basal autophagy for maintaining homeostasis. Autophagy process resides in the middle of cellular catabolic and anabolic process by directly or indirectly engaging in the metabolism of lipids, carbohydrates, proteins, and other metabolites. All of these multidimensional functions of autophagy are mediated either through its bulk or selective targeting and catabolic function. Further understanding of the selective targeting mechanisms and how autophagy mobilizes various nutrient pools toward specific anabolic function is urgently needed. Only with this information will it be possible to therapeutically modulate autophagy to harness its benefit for treatment of different autophagy-associated metabolic liver diseases.

References

Amir M, Czaja MJ (2011) Autophagy in nonalcoholic steatohepatitis. Expert Rev Gastroenterol Hepatol 5(2):159–166

Ananthanarayanan M, Balasubramanian N, Makishima M, Mangelsdorf DJ, Suchy FJ (2001) Human bile salt export pump promoter is transactivated by the farnesoid X receptor/bile acid receptor. J Biol Chem 276(31):28857–28865

Boveris A, Chance B (1973) The mitochondrial generation of hydrogen peroxide. General properties and effect of hyperbaric oxygen. Biochem J 134(3):707–716

Chen Y, Yu L (2013) Autophagic lysosome reformation. Exp Cell Res 319(2):142–146

Cheong H, Lindsten T, Wu J, Lu C, Thompson CB (2011) Ammonia-induced autophagy is independent of ULK1/ULK2 kinases. Proc Natl Acad Sci U S A 108(27):11121–11126

Cole NB, Murphy DD, Grider T, Rueter S, Brasaemle D, Nussbaum RL (2002) Lipid droplet binding and oligomerization properties of the Parkinson's disease protein alpha-synuclein. J Biol Chem 277(8):6344–6352

Cuervo AM, Wong E (2014) Chaperone-mediated autophagy: roles in disease and aging. Cell Res 24(1):92–104

Czaja MJ, Ding WX, Donohue TM Jr, Friedman SL, Kim JS, Komatsu M, Lemasters JJ, Lemoine A, Lin JD, Ou JH, Perlmutter DH, Randall G, Ray RB, Tsung A, Yin XM (2013) Functions of autophagy in normal and diseased liver. Autophagy 9(8):1131–1158

Ding WX, Yin XM (2008) Sorting, recognition and activation of the misfolded protein degradation pathways through macroautophagy and the proteasome. Autophagy 4(2):141–150

Ding WX, Ni HM, Li M, Liao Y, Chen X, Stolz DB, Dorn GW, Yin XM (2010) Nix is critical to two distinct phases of mitophagy, reactive oxygen species-mediated autophagy induction and Parkin-ubiquitin-p62-mediated mitochondrial priming. J Biol Chem 285(36):27879–27890

Dupont N, Chauhan S, Arko-Mensah J, Castillo EF, Masedunskas A, Weigert R, Robenek H, Proikas-Cezanne T, Deretic V (2014) Neutral lipid stores and lipase PNPLA5 contribute to autophagosome biogenesis. Curr Biol 24(6):609–620

Efeyan A, Comb WC, Sabatini DM (2015) Nutrient-sensing mechanisms and pathways. Nature 517(7534):302–310

Eiyama A, Okamoto K (2015) PINK1/Parkin-mediated mitophagy in mammalian cells. Curr Opin Cell Biol 33:95–101

Eng CH, Yu K, Lucas J, White E, Abraham RT (2010) Ammonia derived from glutaminolysis is a diffusible regulator of autophagy. Sci Signal 3(119):ra31

Evans RM, Mangelsdorf DJ (2014) Nuclear Receptors, RXR, and the Big Bang. Cell 157(1):255–266

Ezaki J, Matsumoto N, Takeda-Ezaki M, Komatsu M, Takahashi K, Hiraoka Y, Taka H, Fujimura T, Takehana K, Yoshida M, Iwata J, Tanida I, Furuya N, Zheng DM, Tada N, Tanaka K, Kominami E, Ueno T (2011) Liver autophagy contributes to the maintenance of blood glucose and amino acid levels. Autophagy 7(7):727–736

Fan W, Nassiri A, Zhong Q (2011) Autophagosome targeting and membrane curvature sensing by Barkor/Atg14(L). Proc Natl Acad Sci U S A 108(19):7769–7774

Feng Y, Yao Z, Klionsky DJ (2015) How to control self-digestion: transcriptional, post-transcriptional, and post-translational regulation of autophagy. Trends Cell Biol 25(6):354–363

Filipe A, McLauchlan J (2015) Hepatitis C virus and lipid droplets: finding a niche. Trends Mol Med 21(1):34–42

Harada M, Hanada S, Toivola DM, Ghori N, Omary MB (2008) Autophagy activation by rapamycin eliminates mouse Mallory-Denk bodies and blocks their proteasome inhibitor-mediated formation. Hepatology 47(6):2026–2035

He C, Klionsky DJ (2009) Regulation mechanisms and signaling pathways of autophagy. Annu Rev Genet 43:67–93

Hidvegi T, Ewing M, Hale P, Dippold C, Beckett C, Kemp C, Maurice N, Mukherjee A, Goldbach C, Watkins S, Michalopoulos G, Perlmutter DH (2010) An autophagy-enhancing drug promotes degradation of mutant alpha1-antitrypsin Z and reduces hepatic fibrosis. Science 329(5988):229–232

Huang J, Lam GY, Brumell JH (2011) Autophagy signaling through reactive oxygen species. Antioxid Redox Signal 14(11):2215–2231

Huang R, Xu Y, Wan W, Shou X, Qian J, You Z, Liu B, Chang C, Zhou T, Lippincott-Schwartz J, Liu W (2015) Deacetylation of nuclear LC3 drives autophagy initiation under starvation. Mol Cell 57(3):456–466

Jaber N, Dou Z, Chen JS, Catanzaro J, Jiang YP, Ballou LM, Selinger E, Ouyang X, Lin RZ, Zhang J, Zong WX (2012) Class III PI3K Vps34 plays an essential role in autophagy and in heart and liver function. Proc Natl Acad Sci U S A 109(6):2003–2008

Johansen T, Lamark T (2011) Selective autophagy mediated by autophagic adapter proteins. Autophagy 7(3):279–296

Katsuragi Y, Ichimura Y, Komatsu M (2015) p62/SQSTM1 functions as a signaling hub and an autophagy adaptor. FEBS J 282(24):4672–4678
Kaur J, Debnath J (2015) Autophagy at the crossroads of catabolism and anabolism. Nat Rev Mol Cell Biol 16(8):461–472
Kaushik S, Cuervo AM (2015) Degradation of lipid droplet-associated proteins by chaperone-mediated autophagy facilitates lipolysis. Nat Cell Biol 17(6):759–770
Khambu B, Uesugi M, Kawazoe Y (2011) Translational repression stabilizes messenger RNA of autophagy-related genes. Genes Cells 16(8):857–867
Khaminets A, Heinrich T, Mari M, Grumati P, Huebner AK, Akutsu M, Liebmann L, Stolz A, Nietzsche S, Koch N, Mauthe M, Katona I, Qualmann B, Weis J, Reggiori F, Kurth I, Hubner CA, Dikic I (2015) Regulation of endoplasmic reticulum turnover by selective autophagy. Nature 522(7556):354–358
Khor VK, Shen WJ, Kraemer FB (2013) Lipid droplet metabolism. Curr Opin Clin Nutr Metab Care 16(6):632–637
Kim YC, Guan KL (2015) mTOR: a pharmacologic target for autophagy regulation. J Clin Invest 125(1):25–32
Kim KH, Lee MS (2014) Autophagy—a key player in cellular and body metabolism. Nat Rev Endocrinol 10(6):322–337
Kim KH, Jeong YT, Oh H, Kim SH, Cho JM, Kim YN, Kim SS, Kim DH, Hur KY, Kim HK, Ko T, Han J, Kim HL, Kim J, Back SH, Komatsu M, Chen H, Chan DC, Konishi M, Itoh N, Choi CS, Lee MS (2013) Autophagy deficiency leads to protection from obesity and insulin resistance by inducing Fgf21 as a mitokine. Nat Med 19(1):83–92
Kimmel AR, Brasaemle DL, McAndrews-Hill M, Sztalryd C, Londos C (2010) Adoption of PERILIPIN as a unifying nomenclature for the mammalian PAT-family of intracellular lipid storage droplet proteins. J Lipid Res 51(3):468–471
Koga H, Kaushik S, Cuervo AM (2010) Altered lipid content inhibits autophagic vesicular fusion. FASEB J 24(8):3052–3065
Komatsu M, Waguri S, Ueno T, Iwata J, Murata S, Tanida I, Ezaki J, Mizushima N, Ohsumi Y, Uchiyama Y, Kominami E, Tanaka K, Chiba T (2005) Impairment of starvation-induced and constitutive autophagy in Atg7-deficient mice. J Cell Biol 169(3):425–434
Kotoulas OB, Kalamidas SA, Kondomerkos DJ (2006) Glycogen autophagy in glucose homeostasis. Pathol Res Pract 202(9):631–638
Kroemer G, Marino G, Levine B (2010) Autophagy and the integrated stress response. Mol Cell 40(2):280–293
Lamb CA, Yoshimori T, Tooze SA (2013) The autophagosome: origins unknown, biogenesis complex. Nat Rev Mol Cell Biol 14(12):759–774
Lee IH, Kawai Y, Fergusson MM, Rovira II, Bishop AJ, Motoyama N, Cao L, Finkel T (2012) Atg7 modulates p53 activity to regulate cell cycle and survival during metabolic stress. Science 336(6078):225–228
Lee JM, Wagner M, Xiao R, Kim KH, Feng D, Lazar MA, Moore DD (2014) Nutrient-sensing nuclear receptors coordinate autophagy. Nature 516(7529):112–115
Li T, Chiang JY (2014) Bile acid signaling in metabolic disease and drug therapy. Pharmacol Rev 66(4):948–983
Li WW, Li J, Bao JK (2012) Microautophagy: lesser-known self-eating. Cell Mol Life Sci 69(7):1125–1136
Lin CW, Zhang H, Li M, Xiong X, Chen X, Chen X, Dong XC, Yin XM (2013) Pharmacological promotion of autophagy alleviates steatosis and injury in alcoholic and non-alcoholic fatty liver conditions in mice. J Hepatol 58(5):993–999
Lipatova Z, Segev N (2014) Ypt/Rab GTPases regulate two intersections of the secretory and the endosomal/lysosomal pathways. Cell Logist 4(3):e954870
Lipatova Z, Segev N (2015) A role for macro-ER-phagy in ER quality control. PLoS Genet 11(7):e1005390
Lipatova Z, Shah AH, Kim JJ, Mulholland JW, Segev N (2013) Regulation of ER-phagy by a Ypt/Rab GTPase module. Mol Biol Cell 24(19):3133–3144

Madrigal-Matute J, Cuervo AM (2016) Regulation of liver metabolism by autophagy. Gastroenterology 150(2):328–339
Mammucari C, Milan G, Romanello V, Masiero E, Rudolf R, Del Piccolo P, Burden SJ, Di Lisi R, Sandri C, Zhao J, Goldberg AL, Schiaffino S, Sandri M (2007) FoxO3 controls autophagy in skeletal muscle in vivo. Cell Metab 6(6):458–471
Manjithaya R, Nazarko TY, Farre JC, Subramani S (2010) Molecular mechanism and physiological role of pexophagy. FEBS Lett 584(7):1367–1373
Manley S, Ni HM, Kong B, Apte U, Guo G, Ding WX (2014) Suppression of autophagic flux by bile acids in hepatocytes. Toxicol Sci 137(2):478–490
Martina JA, Chen Y, Gucek M, Puertollano R (2012) MTORC1 functions as a transcriptional regulator of autophagy by preventing nuclear transport of TFEB. Autophagy 8(6):903–914
Mizushima N, Klionsky DJ (2007) Protein turnover via autophagy: implications for metabolism. Annu Rev Nutr 27:19–40
Mizushima N, Komatsu M (2011) Autophagy: renovation of cells and tissues. Cell 147(4):728–741
Munoz-Gamez JA, Rodriguez-Vargas JM, Quiles-Perez R, Aguilar-Quesada R, Martin-Oliva D, de Murcia G, Menissier de Murcia J, Almendros A, Ruiz de Almodovar M, Oliver FJ (2009) PARP-1 is involved in autophagy induced by DNA damage. Autophagy 5(1):61–74
Nakatogawa H, Suzuki K, Kamada Y, Ohsumi Y (2009) Dynamics and diversity in autophagy mechanisms: lessons from yeast. Nat Rev Mol Cell Biol 10(7):458–467
Nath S, Dancourt J, Shteyn V, Puente G, Fong WM, Nag S, Bewersdorf J, Yamamoto A, Antonny B, Melia TJ (2014) Lipidation of the LC3/GABARAP family of autophagy proteins relies on a membrane-curvature-sensing domain in Atg3. Nat Cell Biol 16(5):415–424
Neuhaus A, Kooshapur H, Wolf J, Meyer NH, Madl T, Saidowsky J, Hambruch E, Lazam A, Jung M, Sattler M, Schliebs W, Erdmann R (2014) A novel Pex14 protein-interacting site of human Pex5 is critical for matrix protein import into peroxisomes. J Biol Chem 289(1):437–448
Ni HM, Woolbright BL, Williams J, Copple B, Cui W, Luyendyk JP, Jaeschke H, Ding WX (2014) Nrf2 promotes the development of fibrosis and tumorigenesis in mice with defective hepatic autophagy. J Hepatol 61(3):617–625
Nordgren M, Wang B, Apanasets O, Fransen M (2013) Peroxisome degradation in mammals: mechanisms of action, recent advances, and perspectives. Front Physiol 4:145
Okamoto K (2014) Organellophagy: eliminating cellular building blocks via selective autophagy. J Cell Biol 205(4):435–445
O'Rourke EJ, Ruvkun G (2013) MXL-3 and HLH-30 transcriptionally link lipolysis and autophagy to nutrient availability. Nat Cell Biol 15(6):668–676
Pastore N, Ballabio A, Brunetti-Pierri N (2013) Autophagy master regulator TFEB induces clearance of toxic SERPINA1/alpha-1-antitrypsin polymers. Autophagy 9(7):1094–1096
Puls F, Goldschmidt I, Bantel H, Agne C, Brocker V, Dammrich M, Lehmann U, Berrang J, Pfister ED, Kreipe HH, Baumann U (2013) Autophagy-enhancing drug carbamazepine diminishes hepatocellular death in fibrinogen storage disease. J Hepatol 59(3):626–630
Puri P, Chandra A (2014) Autophagy modulation as a potential therapeutic target for liver diseases. J Clin Exp Hepatol 4(1):51–59
Rogov V, Dotsch V, Johansen T, Kirkin V (2014) Interactions between autophagy receptors and ubiquitin-like proteins form the molecular basis for selective autophagy. Mol Cell 53(2):167–178
Salih DA, Brunet A (2008) FoxO transcription factors in the maintenance of cellular homeostasis during aging. Curr Opin Cell Biol 20(2):126–136
Schneider JL, Suh Y, Cuervo AM (2014) Deficient chaperone-mediated autophagy in liver leads to metabolic dysregulation. Cell Metab 20(3):417–432
Schroeder B, Schulze RJ, Weller SG, Sletten AC, Casey CA, McNiven MA (2015) The small GTPase Rab7 as a central regulator of hepatocellular lipophagy. Hepatology 61(6):1896–1907
Schuck S, Gallagher CM, Walter P (2014) ER-phagy mediates selective degradation of endoplasmic reticulum independently of the core autophagy machinery. J Cell Sci 127(Pt 18):4078–4088

Schulze RJ, Weller SG, Schroeder B, Krueger EW, Chi S, Casey CA, McNiven MA (2013) Lipid droplet breakdown requires dynamin 2 for vesiculation of autolysosomal tubules in hepatocytes. J Cell Biol 203(2):315–326

Schwartz AL, Brandt RA, Geuze H, Ciechanover A (1992) Stress-induced alterations in autophagic pathway: relationship to ubiquitin system. Am J Phys 262(4 Pt 1):C1031–C1038

Sengupta A, Molkentin JD, Paik JH, DePinho RA, Yutzey KE (2011) FoxO transcription factors promote cardiomyocyte survival upon induction of oxidative stress. J Biol Chem 286(9):7468–7478

Seok S, Fu T, Choi SE, Li Y, Zhu R, Kumar S, Sun X, Yoon G, Kang Y, Zhong W, Ma J, Kemper B, Kemper JK (2014) Transcriptional regulation of autophagy by an FXR-CREB axis. Nature 516(7529):108–111

Settembre C, Di Malta C, Polito VA, Garcia Arencibia M, Vetrini F, Erdin S, Erdin SU, Huynh T, Medina D, Colella P, Sardiello M, Rubinsztein DC, Ballabio A (2011) TFEB links autophagy to lysosomal biogenesis. Science 332(6036):1429–1433

Settembre C, De Cegli R, Mansueto G, Saha PK, Vetrini F, Visvikis O, Huynh T, Carissimo A, Palmer D, Klisch TJ, Wollenberg AC, Di Bernardo D, Chan L, Irazoqui JE, Ballabio A (2013a) TFEB controls cellular lipid metabolism through a starvation-induced autoregulatory loop. Nat Cell Biol 15(6):647–658

Settembre C, Fraldi A, Medina DL, Ballabio A (2013b) Signals from the lysosome: a control centre for cellular clearance and energy metabolism. Nat Rev Mol Cell Biol 14(5):283–296

Shibata M, Yoshimura K, Furuya N, Koike M, Ueno T, Komatsu M, Arai H, Tanaka K, Kominami E, Uchiyama Y (2009) The MAP1-LC3 conjugation system is involved in lipid droplet formation. Biochem Biophys Res Commun 382(2):419–423

Shin HJ, Kim H, Oh S, Lee JG, Kee M, Ko HJ, Kweon MN, Won KJ, Baek SH (2016) AMPK-SKP2-CARM1 signalling cascade in transcriptional regulation of autophagy. Nature 534(7608):553–557

Singh R, Kaushik S, Wang Y, Xiang Y, Novak I, Komatsu M, Tanaka K, Cuervo AM, Czaja MJ (2009) Autophagy regulates lipid metabolism. Nature 458(7242):1131–1135

Spandl J, Lohmann D, Kuerschner L, Moessinger C, Thiele C (2011) Ancient ubiquitous protein 1 (AUP1) localizes to lipid droplets and binds the E2 ubiquitin conjugase G2 (Ube2g2) via its G2 binding region. J Biol Chem 286(7):5599–5606

Stolz A, Ernst A, Dikic I (2014) Cargo recognition and trafficking in selective autophagy. Nat Cell Biol 16(6):495–501

Strnad R, Zatloukal K, Stumptner C, Kulaksiz H, Denk H (2008) Mallory-Denk-bodies: lessons from keratin-containing hepatic inclusion bodies. Biochim Biophys Acta 1782(12):764–774

Subramani S (2015) A mammalian pexophagy target. Nat Cell Biol 17(11):1371–1373

Takagi A, Kume S, Kondo M, Nakazawa J, Chin-Kanasaki M, Araki H, Araki S, Koya D, Haneda M, Chano T, Matsusaka T, Nagao K, Adachi Y, Chan L, Maegawa H, Uzu T (2016) Mammalian autophagy is essential for hepatic and renal ketogenesis during starvation. Sci Rep 6:18944

Takamura A, Komatsu M, Hara T, Sakamoto A, Kishi C, Waguri S, Eishi Y, Hino O, Tanaka K, Mizushima N (2011) Autophagy-deficient mice develop multiple liver tumors. Genes Dev 25(8):795–800

Walter KM, Schonenberger MJ, Trotzmuller M, Horn M, Elsasser HP, Moser AB, Lucas MS, Schwarz T, Gerber PA, Faust PL, Moch H, Kofeler HC, Krek W, Kovacs WJ (2014) Hif-2alpha promotes degradation of mammalian peroxisomes by selective autophagy. Cell Metab 20(5):882–897

Wang HJ, Park JY, Kwon O, Choe EY, Kim CH, Hur KY, Lee MS, Yun M, Cha BS, Kim YB, Lee H, Kang ES (2015) Chronic HMGCR/HMG-CoA reductase inhibitor treatment contributes to dysglycemia by upregulating hepatic gluconeogenesis through autophagy induction. Autophagy 11(11):2089–2101

Wei H, Liu L, Chen Q (2015) Selective removal of mitochondria via mitophagy: distinct pathways for different mitochondrial stresses. Biochim Biophys Acta 1853(10 Pt B):2784–2790

Yamamura T, Ohsaki Y, Suzuki M, Shinohara Y, Tatematsu T, Cheng J, Okada M, Ohmiya N, Hirooka Y, Goto H, Fujimoto T (2014) Inhibition of Niemann-Pick-type C1-like1 by ezetimibe activates autophagy in human hepatocytes and reduces mutant alpha 1-antitrypsin Z deposition. Hepatology 59(4):1591–1599

Yang H, Ni HM, Guo F, Ding Y, Shi YH, Lahiri P, Frohlich LF, Rulicke T, Smole C, Schmidt VC, Zatloukal K, Cui Y, Komatsu M, Fan J, Ding WX (2016) Sequestosome-1/p62 is associated with autophagic removal of excess hepatic endoplasmic reticulum in mice. J Biol Chem 291(36):18663–18674

Yin XM, Ding WX, Gao W (2008) Autophagy in the liver. Hepatology 47(5):1773–1785

Youle RJ, Narendra DP (2011) Mechanisms of mitophagy. Nat Rev Mol Cell Biol 12(1):9–14

Yu L, McPhee CK, Zheng L, Mardones GA, Rong Y, Peng J, Mi N, Zhao Y, Liu Z, Wan F, Hailey DW, Oorschot V, Klumperman J, Baehrecke EH, Lenardo MJ (2010) Termination of autophagy and reformation of lysosomes regulated by mTOR. Nature 465(7300):942–946

Zatloukal K, French SW, Stumptner C, Strnad P, Harada M, Toivola DM, Cadrin M, Omary MB (2007) From Mallory to Mallory-Denk bodies: what, how and why? Exp Cell Res 313(10):2033–2049

Zhang J (2013) Autophagy and mitophagy in cellular damage control. Redox Biol 1(1):19–23

Zhao J, Brault JJ, Schild A, Cao P, Sandri M, Schiaffino S, Lecker SH, Goldberg AL (2007) FoxO3 coordinately activates protein degradation by the autophagic/lysosomal and proteasomal pathways in atrophying muscle cells. Cell Metab 6(6):472–483

Zirin J, Nieuwenhuis J, Perrimon N (2013) Role of autophagy in glycogen breakdown and its relevance to chloroquine myopathy. PLoS Biol 11(11):e1001708

Chapter 10
Hepatocyte Death in Liver Inflammation, Fibrosis, and Tumorigenesis

I-Fang Hsin and Ekihiro Seki

Abbreviations

ALT	Aminotransferase
APAP	Acetaminophen
AST	Aspartate aminotransferase
Bcl-xL	B-cell lymphoma-extra large
cIAP	Cellular inhibitor of apoptosis
CK	Cytokeratin
DAMP	Damage-associated molecular patterns
DEN	Diethylnitrosamine
FADD	Fas-associated protein with death domain
HCC	Hepatocellular carcinoma
HMGB	High-mobility group box
HSC	Hepatic stellate cell
IKK	IκB kinase
ILC	Innate lymphoid cell
JNK	c-Jun-N-terminal kinase
MAPK	Mitogen-activated protein kinase
Mcl-1	Myeloid cell leukemia 1
miRNA	microRNA
MKK	MAP kinase kinase
NF-κB	Nuclear factor κB
NEMO	NF-κB essential modulator
NK	Natural killer
RIPK	Receptor-interacting serine/threonine-protein kinase

I-F. Hsin • E. Seki, M.D., Ph.D. (✉)
Division of Gastroenterology, Department of Medicine, Cedars-Sinai Medical Center, 8700 Beverly Blvd, Davis Bldg. D2099, Los Angeles, CA 90048, USA
e-mail: Ekihiro.Seki@cshs.org

W.-X. Ding, X.-M. Yin (eds.), *Cellular Injury in Liver Diseases*, Cell Death in Biology and Diseases, DOI 10.1007/978-3-319-53774-0_10

X.-M. Yin and Z. Dong (Series eds.), *Cell Death in Biology and Diseases*

ROS	Reactive oxygen species
TAK	TGF-β activated kinase
TAB	TAK1-binding proteins
TLR	Toll-like receptor
TNF	Tumor necrosis factor
TNFR1	TNF receptor type I
TRADD	TNFR1-associated death domain protein

10.1 Introduction

Hepatocyte death is generally regarded as the initial event in almost all liver diseases, including viral hepatitis, alcoholic and non-alcoholic fatty liver diseases, drug and toxin-induced liver injury, biliary disorders, and autoimmune hepatitis. A number of biologically active mediators, receptors, intracellular signaling pathways, and different cell types contribute to the initiation and progression of liver diseases through hepatocyte death. Some known examples include: direct damage to hepatocytes by alcohol and its metabolites; excessive accumulation of bile acids and saturated fatty acids cause hepatocyte death; in Hepatitis B and C, the virus infection mediates cytotoxic T lymphocyte-mediated hepatocyte death. Hepatocyte death in these diseases will lead to liver inflammation, fibrosis, and ultimately induction of hepatocarcinogenesis. In a clinical setting, serum alanine aminotransferase (ALT) and aspartate aminotransferase (AST) are often measured to determine the extent of hepatocyte injury. Elevated levels reflect the degree of hepatocyte injury as hepatocytes release these enzymes into the blood stream when damaged. Since previous clinical observations determined a strong correlation between ALT levels and liver cancer development, persistent hepatocyte death is suggested to be associated with tumor development (Chen et al. 2011; Lee et al. 2010). These clinical observations are supported by recent studies using animal models that demonstrate hepatocellular death contributes to the triggering of hepatic inflammatory and fibrogenic reaction, and the induction of carcinogenesis. Notably, in contrast to increased cell death that drives tumorigenesis, loss of regular programmed cell death is also associated with malignant transformation and cancer development by escaping the death of malignant cells. This chapter will highlight the underlying molecular mechanisms of hepatocyte death and its contribution to liver inflammation, fibrosis, and tumorigenesis.

10.2 Death Receptor-Mediated Cell Death Signaling

Tumor necrosis factor α (TNFα) is a pleiotropic cytokine, and its most well-characterized receptor is TNF receptor type I (TNFR1). The signaling initiated by the TNFα-TNFR1 interaction activates both nuclear factor κB (NF-κB)-dependent cell

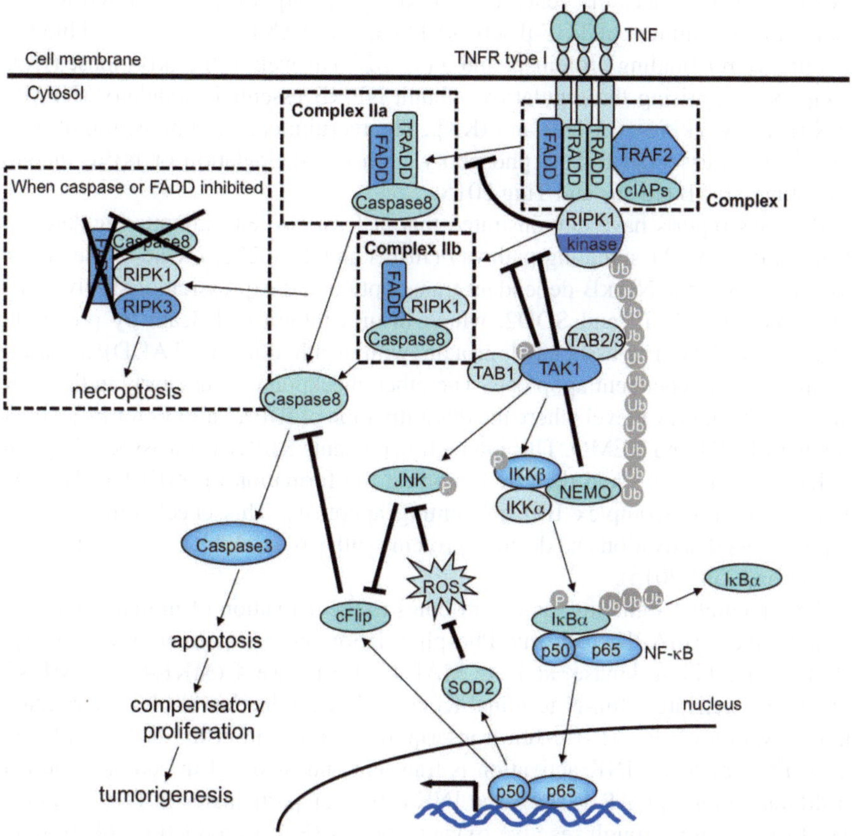

Fig. 10.1 IKK/NF-κB, JNK, and apoptosis in TNF receptor signaling. Binding of TNF to the TNF receptor type I (TNFR1) leads to the formation of "complex I" comprising TRADD, TRAF2, and cIAP. K63-ubiquitination of RIP1 recruits and activates TAK1 and IKK complex (IKKα/IKKβ/NEMO). TAK1 also activate JNK and its mediated AP-1/c-Jun transcription factor. Simultaneously, JNK1 promotes caspase-8-dependent apoptosis through inhibition of c-FLIP. TAK1 phosphorylates and activates IKK complex that leads to phosphorylation, unbiquitination, and degradation of IκBα, resulting in nuclear translocation and activation of NF-κB comprising p50 and p65 subunits. NF-κB inhibits ROS and caspase-8 through SOD2 and c-FLIP transcription, respectively. NEMO physiologically regulates NF-κB activation and simultaneously prevents the formation of the complex IIb (RIPK1, FADD, caspase-8). Complex IIb-mediated hepatocyte apoptosis requires RIPK1 kinase activity. NEMO and RIPK1 protein physiologically inhibit the formation of Complex IIa (TRADD, FADD, caspase-8). Alternatively, when caspase-8 and/or FADD are inhibited, RIP1 and RIP3 are phosphorylated and induce necroptosis

survival and death-inducing pathways, known as the best-characterized death receptor signaling that contributes to hepatocyte survival and death (Justus and Ting 2015). Binding of TNFα to TNFR1 leads to Complex I being formed at TNFR1 that consists of TNFR1-associated death domain protein (TRADD), receptor-interacting serine/threonine-protein kinase 1 (RIPK1), TNF receptor-associated factor 2 (TRAF2), and cellular inhibitor of apoptosis 1 and 2 (cIAP1/2) (Fig. 10.1). Complex

I subsequently induces the generation of K63 polyubiquitin chains on RIPK1 that enable the recruitment of TGF-β activated kinase 1 (TAK1) and associated binding proteins TAK1-binding proteins 2 and 3 (TAB2/3) as well as the IκB kinase (IKK) complex, comprising the regulatory subunit NF-κB essential modulator (NEMO/ IKKγ) and two kinases, IKKα and IKKβ. The recruitment and interaction of these complexes is essential for the phosphorylation and degradation of IκBα, thereby activating NF-κB (Justus and Ting 2015).

Previous reports have demonstrated that at least two checkpoints regulate cell death in the TNFR1 signaling pathway (Justus and Ting 2015). One of the checkpoints involves the NF-κB-dependent transcription and expression of survival proteins, such as c-FLIP and SOD2, which in turn inhibit cell death by protecting against TRADD/ Fas-associated protein with death domain (FADD)/caspase-8 (Complex IIa)-dependent apoptosis. The other checkpoint occurs early in the pathway at the Complex I level where the ubiquitination of RIPK1 allows for its association with TAK1 and NEMO. This interaction prevents RIPK1 from associating with FADD and caspase-8 and thereby inhibiting the formation of RIKP1/FADD/caspase-8 complex (Complex IIb), preventing apoptosis. This checkpoint does not require NF-κB activation and de novo protein synthesis to inhibit cell death pathway (Justus and Ting 2015).

TNF-mediated signaling also contributes to the activation of mitogen-activated protein kinase (MAPK) pathway. Phosphorylation and ubiquitination of TAK1, a MAP kinase kinase kinase, activate MAP kinase kinase 4 (MKK4) and MKK7, which then activate c-Jun-N-terminal Kinase (JNK), induction of phosphorylation, and activation of the AP-1/c-Jun transcription factor (Liu et al. 2002). Although early TNF-mediated JNK activation is transient and involved in cell survival and proliferation through AP-1, sustained JNK activation participates in TNF-mediated hepatocyte death through reactive oxygen species (ROS) production; likely independent of AP-1 (Wagner and Nebreda 2009; Micheau and Tschopp 2003; Ventura et al. 2004; Karin and Lin 2002; Kamata et al. 2005). TNF-induced JNK activation is negatively regulated by NF-κB-mediated survival factors that block both caspase-8-dependent cell death and sustained JNK activation (Liu et al. 2002).

10.3 IKK/NF-κB Negatively Regulates Hepatocyte Apoptosis

It has been reported that NEMO-deficient cells are susceptible to TNF-mediated apoptosis (Legarda-Addison et al. 2009). Since NEMO binding to RIPK1 polyubiquitin chains inhibits the interaction of RIPK1 with caspase-8, in the absence of NEMO Complex IIb formation is promoted, thereby inducing apoptosis. IKKα/ IKKβ kinases directly phosphorylate RIPK1 at the level of Complex I, thereby inhibiting the formation of Complex IIb and its mediated hepatocyte apoptosis, which is independently of NF-κB (Dondelinger et al. 2015; Koppe et al. 2016). In mice with deletion of NEMO in hepatocytes ($NEMO^{LPC-KO}$ mice), their livers become more sensitive to LPS- and TNFα-induced liver injury and spontaneously develops inflammation, steatosis, fibrosis, and hepatocellular carcinoma (Luedde et al. 2007; Ehlken et al. 2014). NEMO prevents spontaneous hepatocyte death by two distinct mechanisms: One is the function through NF-κB-mediated transcription of anti-apoptotic survival proteins, and the other is to inhibit RIPK1-mediated hepatocyte apoptosis by the NF-κB-independent manner (Kondylis et al. 2015). Since NEMO$^{LPC\text{-}KO}$ mice with bearing mutated RIPK1D138N of which kinase activity is absent showed less hepatocyte apoptosis, NEMO negatively regulates RIPK1 kinase activity to prevent hepatocyte apoptosis (Kondylis et al. 2015) (Fig. 10.1). Notably, RIPK1 also has a scaffold function that normally inhibits RIPK1 kinase-independent pro-apoptotic pathway through TRADD/FADD/caspase-8 complex (Complex IIa) (Fig. 10.1). Indeed, when whole RIPK1 protein is deleted, NEMO$^{LPC\text{-}KO}$ mice showed severe liver injury through formation of Complex IIa (Kondylis et al. 2015). Both FADD and Caspase-8 are required for forming the complex associated with either RIPK1 or TRADD. Importantly, TRADD-mediated alternative hepatocyte apoptosis pathway will not be activated until both RIPK1 and NEMO are inhibited.

10.4 JNK Activation and Hepatocyte Death

In hepatocytes, TNF signaling also activates JNK-dependent pathway, induction of phosphorylation, and activation of the AP-1/c-Jun transcription factor (Liu et al. 2002). TNF-induced JNK activation is negatively regulated by NF-κB-mediated survival factors that block both caspase-8-dependent cell death and sustained JNK activation (Liu et al. 2002). Sustained JNK activation promotes TNF-mediated hepatocyte death through ROS production but independent of AP-1 (Kamata et al. 2005) (Fig. 10.1). The liver expresses two JNK isoforms, JNK1 and JNK2. Compared to JNK1, JNK2 plays crucial roles in TNF-mediated hepatocyte apoptosis through caspase-8 activation, Bid cleavage, and mitochondrial cytochrome c release (Wang et al. 2006). In concanavalin A (ConA)-induced T cell-mediated liver injury, both JNK1 and JNK2 play a role (Maeda et al. 2003).

Mitochondrial Bcl-X_L, Mcl-1, and Sab (SH3BP5) are substrates for JNK. JNK translocation to the outer membrane of mitochondria plays a role in TNF-mediated hepatocyte death (Win et al. 2011). JNK phosphorylates mitochondrial Sab, which promotes translocation of MKK4, JNK, and Bax to mitochondrial outer membrane, thereby inducing mitochondrial ROS formation, sustained JNK activation, and hepatocyte death (Win et al. 2011). JNK also contributes to the cleavage of mitochondrial Bid, a BH3 only protein, through caspase-8 (Wang et al. 2006; Ni et al. 2009), which induces Bid translocation to mitochondria, cytochrome-c release, and caspase-9/caspase-3 activation, thereby promoting hepatocyte death (Wang et al. 2006; Takamura et al. 2007).

JNK is also involved in saturated fatty acids- and endoplasmic reticulum stress-induced hepatocyte death (Seki et al. 2012).

A variety of JNK-dependent biological functions are mediated through AP-1/c-Jun activation. In TNF-mediated hepatocyte apoptosis, c-Jun- and transcription-independent JNK-mediated signaling pathways are more important (Schwabe et al. 2004; Czaja 2003). AP-1/c-Jun may have protective effects against hepatocyte death (Eferl et al. 2003).

10.5 RIP3 and Regulation of Necroptosis and Hepatocyte Death

Necroptosis is defined as a programmed necrotic death, which differs from necrosis characterized by swelling of cells and cell organelles along with plasma membrane rupture. Necroptosis requires activation of both RIPK1 and RIPK3, which are normally inhibited through caspase-8 and/or FADD by cleaving RIPK1 and RIPK3. When caspase-8 and/or FADD are inhibited (this may occur in some diseases, such as viral infection), necroptosis (as a back-up cell death form) is induced through activating RIPK1 and RIPK3 (Fig. 10.1). RIPK3-mediated necroptosis contributes to hepatocyte death in acetaminophen (APAP)-induced liver injury, ethanol-mediated liver injury, and the development of methionine choline-deficient diet-induced non-alcoholic steatohepatitis (NASH) (Deutsch et al. 2015; Roychowdhury et al. 2013; Gautheron et al. 2014). However, recent studies demonstrated contradictory results that high-fat diet-induced non-alcoholic fatty liver disease (NAFLD) and insulin resistance were exacerbated in RIPK3 null mice (Roychowdhury et al. 2016; Gautheron et al. 2016). Further studies are needed for determining the role of RIP3K and necroptosis in liver disease.

10.6 Inflammasome Activation and Pyroptosis

Inflammasome and caspase-1 activation is required for IL-1β and IL-18 processing in macrophages. Notably, inflammasome and caspase-1 activation also contributes to "Pyroptosis," a type of programmed cell death, which is distinct from apoptosis and necrosis (Fink and Cookson 2005). Pyroptosis is characterized by pore formation in cell membrane, membrane disruption, and cell swelling, and shows positivity for PI and TUNEL staining. NLRP3D303N mutation is responsible for human cryopyrin-associated periodic syndrome, and causes excessive activation of inflammasome through NLRP3 (Wree et al. 2014a). A recent study showed mice with mutated NLRP3 developing spontaneous liver injury, inflammation, and fibrosis along with hepatocyte pyroptosis (Wree et al. 2014a). These mice are also sensitive to high nutrient stress and overt NASH development (Wree et al. 2014b). These findings indicate that hepatocyte pyroptosis is associated with liver inflammation and fibrosis.

10.7 Hepatocyte Death-Induced Inflammation through DAMPs

Hepatic inflammation is believed to be the first response after liver injury, which is associated with driving fibrogenesis and hepatocarcinogenesis. Dead or dying hepatocytes produce or release so-called damage-associated molecular patterns (DAMPs), which induce inflammation by recruiting and/or stimulating immune cells, production of inflammatory cytokines, and further perpetuating inflammation as well as hepatocyte injury. Hepatocyte-derived DAMPs include nucleotides, nuclear proteins, mitochondrial proteins and DNAs, lipids, and cytokines (Matzinger 2002; Chen and Nunez 2010; Rock et al. 2011; Kono and Rock 2008). DAMPs are generally released after hepatocyte necrosis and necroptosis due to cell membrane rupture, while other DAMPs are also known to leak out from apoptotic hepatocytes. Hepatocyte-derived DAMPs reported previously are listed in Table 10.1. A nuclear protein high-mobility group box 1 (HMGB1) is an early mediator released from dead hepatocytes, which can trigger inflammation in the setting of ischemia-reperfusion injury and APAP-induced liver injury, but not LPS, TNF, and Fas-mediated liver injury (Tsung et al. 2005; Huebener et al. 2015) (Fig. 10.2). HMGB1 mediates ischemia-reperfusion liver injury through toll-like receptor 4 (TLR4) (Tsung et al. 2005). In APAP-induced liver injury, RAGE is more important than TLR4 through recruiting neutrophils (Huebener et al. 2015). Hepatocytes contain a large amount of

Table 10.1 Hepatocyte-derived DAMPs and their receptors and actions

DAMPs	Receptors	Actions	References
HMGB1	TLR4, RAGE	Macrophage activation, neutrophil recruitment	(Huebener et al. 2015; Tsung et al. 2005; Wang et al. 2013)
		HSC proliferation, migration	
Mitochondrial DNA/nuclear DNA	TLR9	Macrophage activation, HSC activation, limit HSC chemotaxis	(Garcia-Martinez et al. 2016; Imaeda et al. 2009; Watanabe et al. 2007)
Formyl-peptides	FPR1	Neutrophil recruitment	(McDonald et al. 2010)
ATP	P2X7	Inflammasome activation	(Hoque et al. 2012)
Uric acids	N/A	Inflammasome activation	(Iracheta-Vellve et al. 2015; Petrasek et al. 2015; Wan et al. 2016)
Oxidized mitoDNA	NLRP3(?)	Inflammasome activation	(Shimada et al. 2012)
Adenosine	A2a receptor	HSC activation, limit HSC chemotaxis	(Hashmi et al. 2007)
IL-1α	IL-1R	Liver inflammation, HCC development	(Sakurai et al. 2008)
IL-33	ST2	IL-13 production in ILC2	(McHedlidze et al. 2013)
Apoptotic body	Phagocytosis(?)	Macrophage and HSC activation	(Canbay et al. 2003a, b)

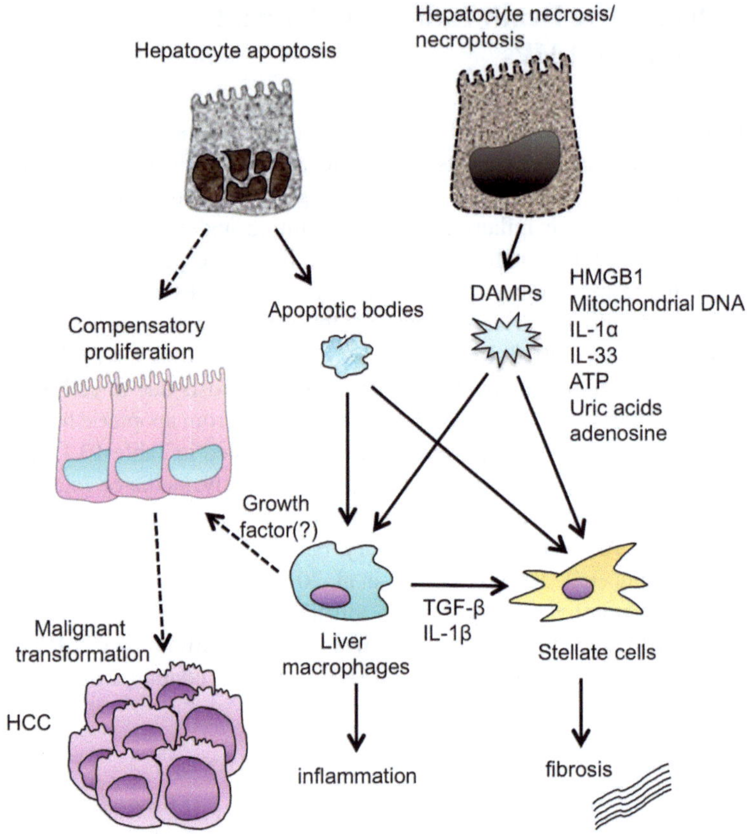

Fig. 10.2 Hepatocyte death and cellular crosstalk in liver inflammation, fibrosis, and carcinogenesis. In acute and chronic liver disease, hepatocyte apoptosis, necroptosis, and necrosis occur. Apoptotic hepatocytes release apoptotic bodies that are phagocytosed by Kupffer cells and hepatic stellate cells (HSCs). Necrotic cells release DAMPs, such as HMGB1, mitochondrial DNA, ATP, IL-1α, IL-33, uric acids, and adenosine. The DAMPs regulates Kupffer cells and HSC activation for liver inflammation and fibrosis. Hepatocyte apoptosis may participate in tumorigenesis through compensatory proliferation

ATP, a key factor for energy generation, which is released from dying hepatocytes. As a DAMP, extracellular ATP promotes liver inflammation through binding to its receptor P2X7 in APAP-induced liver injury (Hoque et al. 2012). During hepatocyte death, mitochondria damage also occurs. Since mitochondria contain ATP, formyl-peptides, and mitochondrial DNA, extracellularly leaked mitochondrial components recruit neutrophils, promotion of sterile inflammation and injury in the liver (McDonald et al. 2010). Notably, denatured host DNA and mitochondrial DNA activate TLR9, promoting liver inflammation in APAP-induced liver injury and in NAFLD (Imaeda et al. 2009; Garcia-Martinez et al. 2016) (Fig. 10.2). Kupffer cells are the major source of IL-1β and IL-18 that require processing to be active forms for secreting extracellularly. Inflammasome activation is a main mechanism for the activation of IL-1β and IL-18. Hepatocyte-derived DAMPs, such as ATP, uric acids, and oxidized mitochondrial DNA, are involved in NLRP3 inflammasome activation in APAP-induced liver injury, alcoholic and non-alcoholic steatohepatistis (Petrasek et al. 2015; Wan et al. 2016; Iracheta-Vellve et al. 2015) (Hoque et al. 2012; Shimada et al. 2012). Inflammatory cytokines, such as IL-1α and IL-33, are also released after hepatocyte death (Sakurai et al. 2008; McHedlidze et al. 2013). IL-1α released after carcinogen- and hepatotoxin-induced death mediates IL-6 production, promoting liver inflammation and hepatocyte proliferation (Sakurai et al. 2008). Hepatocyte-derived IL-33 promotes type 2 immune response by producing IL-13 from innate lymphoid cell type 2 (ILC2) in parasite infection (McHedlidze et al. 2013).

Viral hepatitis caused by hepatitis B or C virus induces hepatocyte apoptosis. Since hepatitis B and C virus themselves are minimal cytopathic, apoptosis is mainly caused by viral specific cytotoxic T cells and natural killer (NK) cells that eliminate infected hepatocytes (Malhi et al. 2010). T and NK cell-mediated hepatocyte killing is mediated by Fas and TRAIL-mediated signaling (Malhi et al. 2010). Anti-viral immunity does not always successfully eradicate all viral particles. In this setting, viral infection will be sustained and becomes chronic, which is characterized by persistent hepatocyte death and inflammation, ultimately induction of HSC activation and liver fibrosis/cirrhosis.

10.8 Hepatocyte Death as a Trigger for Liver Fibrosis

Liver fibrosis is an alternative repair response against liver injury and results in liver dysfunction, portal hypertension, and hepatocarcinogenesis. Activated hepatic stellate cells (HSCs) are the major contributors of the development of liver fibrosis by producing extracellular matrix, such as collagen fibers. Dead hepatocyte-released DAMPs directly stimulate HSC activation and/or indirectly activate HSC through activating other non-parenchymal cells, such as Kupffer cells, NK cells, T cells, B cells, and dendritic cells. Early studies showed that hepatocyte-derived apoptotic bodies engulfed by Kupffer cells and HSCs, induce HSC activation (Canbay et al. 2003a, b). Previous human cohorts demonstrated that fibrosis after viral hepatic injury is well correlated with elevated blood ALT levels, suggesting a possible link between hepatocyte death and liver fibrosis in human liver fibrosis (Fattovich et al. 2008; Wiese et al. 2014). Hepatocyte apoptosis has been proven to be an early event in a rapidly fibrosis progression in patients transplanted for hepatitis C (Meriden et al. 2010). In addition, there is a positive correlation of active caspase-3 and 7, and plasma-fragmented cytokeratin-18 (CK-18) levels (which is cleaved by caspase-3 from full-length CK-18) with the stage of hepatic fibrosis in NASH patients (Feldstein et al. 2003, 2009). A number of animal studies have also demonstrated that mice-specific deletion of anti-apoptotic molecules in hepatocytes enhances TNF-mediated hepatocyte death or spontaneously develops hepatocyte death. These anti-apoptotic molecules include NEMO, TAK1, Bcl-xL, and Mcl-1 (Luedde et al.

2007; Inokuchi et al. 2010; Takehara et al. 2004; Vick et al. 2009). In these mice, continuous hepatocyte apoptosis along with elevated ALT leads to liver fibrotic response.

As mentioned above, DAMPs released from dying cells trigger inflammation, linking to fibrosis. HMGB1 promotes HSC proliferation and migration through TLR4 (Wang et al. 2013). Interestingly, TLR9 activation by denatured host nucleic DNA and mitochondrial DNA is capable of stimulating HSCs in upregulation of fibrogenic genes (TGF-β and collagen) but limiting chemotaxis (Watanabe et al. 2007; Garcia-Martinez et al. 2016) (Fig. 10.2). Adenosine is released from dying hepatocytes, which makes higher adenosine levels in local liver injured sites. Adenosine is derived from the dephosphorylation of adenosine tri-, di, and mono-phosphates, and also from the degradation of nucleic acids through the uric acid pathway. Adenosine upregulates TGF-β and collagen α1(I) levels in HSCs but inhibits PDGF plus ATP-induced HSC chemotaxis via the A2a receptor (Hashmi et al. 2007). This system can hold HSCs when they migrate to an injured site with upregulation of fibrogenic molecules. Phagocytosis of dead hepatocyte-derived apoptotic bodies by HSCs induces fibrogenic response, such as upregulation of α-smooth muscle actin, TGF-β, and collagen α1(I) expression (Canbay et al. 2003b) (Fig. 10.2).

TGF-β is a potent fibrogenic cytokine that directly activates HSCs and stimulates collagen production. TGF-β also participates in hepatocyte apoptosis that indirectly promotes HSC activation and fibrosis in NASH and $TAK1^{LPC\text{-}KO}$ mice (Yang et al. 2013, 2014).

Parasite infection-induced hepatocyte death releases IL-33 that stimulates ILC2, producing IL-13 (McHedlidze et al. 2013). IL-13 then stimulates HSCs through IL-4Rα and STAT6 and contributes to parasite-mediated HSC activation and liver fibrosis.

In contrast to hepatocyte death that generally promotes fibrosis, death in HSC is implicated in fibrosis regression and resolution. When continuous fibrotic stimuli are removed, fully activated HSCs undergo apoptosis and fibrosis regression or resolution may occur (Kisseleva and Brenner 2006). NK cells play an important role in HSC killing through IFN-γ (Jeong et al. 2011). After full activation, some HSCs become senescent, which halt collagen production. Senescent HSCs become more susceptible to death through NK cells, and exogenous IL-22 treatment promotes HSC senescent and accelerates HSC apoptosis (Kong et al. 2012).

10.9 Hepatocyte Death as an Initiator of Hepatocarcinogenesis

Unlike other organs, the liver has the most powerful self-repairing capacity after tissue injury. After more than 50% of hepatectomy or acute liver damage by hepatotoxin exposure, liver may regenerate to original mass and recover its normal function Unfortunately, this event does not always occur in the setting of chronic liver disease. The complications of prolonged hepatocellular injury include fibrosis, which is abnormal wound-healing response, and become suitable "soil" for carcinogenesis. Hepatitis, fibrosis, and cirrhosis are well-established risk factors for hepatocellular carcinoma (HCC). Clinically, more than 80% of HCC patients have liver fibrosis/cirrhosis before being diagnosed with HCC (Fattovich et al. 2004). It suggests that long-term hepatocyte damage promotes malignant transformation of hepatocytes.

Through animal models, with pure cell death type and known mechanism, it is suggested that hepatocyte death itself and compensatory proliferation after massive liver injury are both the key factors for hepatocarcinogenesis. Mice with hepatocytes deficient in IKKβ, a crucial factor for NF-κB activation, cause massive hepatocyte death after exposure to chemical carcinogen diethylnitrosamine (DEN) (Maeda et al. 2005). These mice also showed more compensatory hepatocyte proliferation and hepatocarcinogenesis than wild-type counterparts. Mice with hepatocytes deficient in NEMO or TAK1, another regulator for NF-κB activation, showed spontaneous hepatocarcinogenesis in conjunction with hepatocyte injury and compensatory liver cell proliferation (Luedde et al. 2007; Inokuchi et al. 2010). Since death receptor-mediated hepatocyte apoptosis is associated with liver phenotype of aforementioned mutant mice, the deletion of FADD prevents the development of HCC in these mice. Hepatocarcinogenesis in $NEMO^{LPC\text{-}KO}$ mice is mediated by inhibiting NF-κB activation and promoting RIPK1 kinase activity-mediated hepatocyte apoptosis (Kondylis et al. 2015). However, since RIPK1 molecule also has a scaffold function for NF-κB activation, whole RIPK1 deletion does not reduce hepatocarcinogenesis in $NEMO^{LPC\text{-}KO}$ mice (Kondylis et al. 2015). In $NEMO^{LPC\text{-}KO}$ mice and $TAK1^{LPC\text{-}KO}$ mice, additional deletion of FADD or caspase-8 prevents hepatocyte apoptosis and hepatocarcinogenesis but additional RIPK3 deletion did not suppress hepatocyte injury and carcinogenesis, suggesting that hepatocyte apoptosis, but not necrosis/necroptosis, is an essential mechanism for hepatocarcinogenesis (Luedde et al. 2007; Vucur et al. 2013; Kondylis et al. 2015) (Fig. 10.2). One human study demonstrated that the downregulation of NEMO expression is correlated with the prognosis and survival of HCC patients (Aigelsreiter et al. 2012).

Deficient in anti-apoptotic proteins Mcl-1 and Bcl-xL in hepatocytes causes spontaneous hepatocyte apoptosis that results in the development of HCC (Takehara et al. 2004; Weber et al. 2010). In $Mcl\text{-}1^{LPC\text{-}KO}$ mice, additional deletion of Bak (a downstream molecule of Mcl-1 towards apoptosis) reduces HCC development (Hikita et al. 2012). These findings also support the mechanism that hepatocyte apoptosis is associated with HCC development. Notably, these genetic animal models might not recapitulate human HCC conditions since in most clinical settings external injuries caused by virus, toxin, alcohol, and lipids are associated with HCC formation.

microRNAs (miRNAs) are small, noncoding RNA molecules that negatively regulate the expression of corresponding mRNA. Hepatocytes predominantly express miR-122, which accounts for 70% of all miRNAs in hepatocyte. When hepatocytes are damaged, miR-122 is released to blood circulation then blood miR-122 levels increase significantly. miR-122 plays dominant roles as a tumor suppressor and a negative regulator for lipid metabolism as mice deficient in miR-122 spontaneously develop hepatic steatosis, fibrosis, and carcinogenesis (Tsai et al. 2012). Although evidence that released miR-122 plays a role in the development of liver inflammation and tumorigenesis is still lacking, released miR-122 may influence liver pathophysiology.

It is strongly suggested that apoptosis in healthy hepatocyte is associated with promoting tumorogenesis, while impaired death or programmed death of cancer or dysplastic cells may lead or enhance tumor formation and progression. Targeted ablation of Fas gene causes liver hyperplasia in mice (Adachi et al. 1995). In HCC, Fas expression is lower in human HCC tissue than non-cancerous tissue (Higaki et al. 1996). Another well-known example is NF-κB, which acts as both a promoter and a suppressor of HCC. In *Mdr2* knockout mice model, overexpression of mutated IκBα in hepatocytes inhibits inducible NF-κB activation and reduces liver tumor development compared with mice with normal NF-κB signaling. In this setting, NF-κB functions as a tumor promoter (Pikarsky et al. 2004).

10.10 Conclusions

A number of animal models have suggested a strong link between cell death and carcinogenesis. Several human studies also suggest this link but the direct molecular mechanism is still unclear. Mutations, such as p53, β-catenin, and TERT promoter, are often observed in human HCC. However, some mouse models enable developing HCC spontaneously without treatment with carcinogen. It is hypothesized that chronic hepatocyte damage induces extensive compensatory proliferation that increases the incidence of mutations on tumor suppressor gene(s) and eventually develops dysplastic cells or transformed cancer cells. Another hypothesis is that in some hepatocytes that escaped from cell death stimuli apoptotic signals directly affect mutations in tumor suppressor gene(s). Notably, cancer cells usually escape from the natural selection guard system by specific mutations, such as p53 mutation (Guichard et al. 2012). In summary, hepatocyte death is clearly associated with promoting liver inflammation and fibrosis whereas in hepatocarcinogenesis hepatocyte death has dual roles: On one hand, hepatocyte death signal functions as a tumor promoter. On the other hand, hepatocyte death acts as a tumor suppressor by eliminating cells with gene mutations.

Acknowledgment This work is supported by NIH grants R01DK085252 and P42 ES010337.

References

Adachi M, Suematsu S, Kondo T, Ogasawara J, Tanaka T, Yoshida N et al (1995) Targeted mutation in the Fas gene causes hyperplasia in peripheral lymphoid organs and liver. Nat Genet 11(3):294–300. doi:10.1038/ng1195-294

Aigelsreiter A, Haybaeck J, Schauer S, Kiesslich T, Bettermann K, Griessbacher A et al (2012) NEMO expression in human hepatocellular carcinoma and its association with clinical outcome. Hum Pathol 43(7):1012–1019. doi:10.1016/j.humpath.2011.08.009

Canbay A, Feldstein AE, Higuchi H, Werneburg N, Grambihler A, Bronk SF et al (2003a) Kupffer cell engulfment of apoptotic bodies stimulates death ligand and cytokine expression. Hepatology 38(5):1188–1198. doi:10.1053/jhep.2003.50472

Canbay A, Taimr P, Torok N, Higuchi H, Friedman S, Gores GJ (2003b) Apoptotic body engulfment by a human stellate cell line is profibrogenic. Lab Invest 83(5):655–663

Chen GY, Nunez G (2010) Sterile inflammation: sensing and reacting to damage. Nat Rev Immunol 10(12):826–837. doi:10.1038/nri2873

Chen CF, Lee WC, Yang HI, Chang HC, Jen CL, Iloeje UH et al (2011) Changes in serum levels of HBV DNA and alanine aminotransferase determine risk for hepatocellular carcinoma. Gastroenterology 141(4):1240–1248 . doi:10.1053/j.gastro.2011.06.0361248.e1-2

Czaja MJ (2003) The future of GI and liver research: editorial perspectives. III. JNK/AP-1 regulation of hepatocyte death. Am J Physiol Gastrointest Liver Physiol 284(6):G875–G879. doi:10.1152/ajpgi.00549.2002

Deutsch M, Graffeo CS, Rokosh R, Pansari M, Ochi A, Levie EM et al (2015) Divergent effects of RIP1 or RIP3 blockade in murine models of acute liver injury. Cell Death Dis 6:e1759. doi:10.1038/cddis.2015.126

Dondelinger Y, Jouan-Lanhouet S, Divert T, Theatre E, Bertin J, Gough PJ et al (2015) NF-kappaB-independent role of IKKalpha/IKKbeta in preventing RIPK1 kinase-dependent apoptotic and necroptotic cell death during TNF signaling. Mol Cell 60(1):63–76. doi:10.1016/j.molcel.2015.07.032

Eferl R, Ricci R, Kenner L, Zenz R, David JP, Rath M et al (2003) Liver tumor development. C-Jun antagonizes the proapoptotic activity of p53. Cell 112(2):181–192

Ehlken H, Krishna-Subramanian S, Ochoa-Callejero L, Kondylis V, Nadi NE, Straub BK et al (2014) Death receptor-independent FADD signalling triggers hepatitis and hepatocellular carcinoma in mice with liver parenchymal cell-specific NEMO knockout. Cell Death Differ 21(11):1721–1732. doi:10.1038/cdd.2014.83

Fattovich G, Stroffolini T, Zagni I, Donato F (2004) Hepatocellular carcinoma in cirrhosis: incidence and risk factors. Gastroenterology 127(5 Suppl 1):S35–S50

Fattovich G, Olivari N, Pasino M, D'Onofrio M, Martone E, Donato F (2008) Long-term outcome of chronic hepatitis B in Caucasian patients: mortality after 25 years. Gut 57(1):84–90. doi:10.1136/gut.2007.128496

Feldstein AE, Canbay A, Angulo P, Taniai M, Burgart LJ, Lindor KD et al (2003) Hepatocyte apoptosis and fas expression are prominent features of human nonalcoholic steatohepatitis. Gastroenterology 125(2):437–443

Feldstein AE, Wieckowska A, Lopez AR, Liu YC, Zein NN, McCullough AJ (2009) Cytokeratin-18 fragment levels as noninvasive biomarkers for nonalcoholic steatohepatitis: a multicenter validation study. Hepatology 50(4):1072–1078. doi:10.1002/hep.23050

Fink SL, Cookson BT (2005) Apoptosis, pyroptosis, and necrosis: mechanistic description of dead and dying eukaryotic cells. Infect Immun 73(4):1907–1916. doi:10.1128/IAI.73.4.1907-1916.2005

Garcia-Martinez I, Santoro N, Chen Y, Hoque R, Ouyang X, Caprio S et al (2016) Hepatocyte mitochondrial DNA drives nonalcoholic steatohepatitis by activation of TLR9. J Clin Invest 126(3):859–864. doi:10.1172/JCI83885

Gautheron J, Vucur M, Reisinger F, Cardenas DV, Roderburg C, Koppe C et al (2014) A positive feedback loop between RIP3 and JNK controls non-alcoholic steatohepatitis. EMBO Mol Med 6(8):1062–1074. doi:10.15252/emmm.201403856

Gautheron J, Vucur M, Schneider AT, Severi I, Roderburg C, Roy S et al (2016) The necroptosis-inducing kinase RIPK3 dampens adipose tissue inflammation and glucose intolerance. Nat Commun 7:11869. doi:10.1038/ncomms11869

Guichard C, Amaddeo G, Imbeaud S, Ladeiro Y, Pelletier L, Maad IB et al (2012) Integrated analysis of somatic mutations and focal copy-number changes identifies key genes and pathways in hepatocellular carcinoma. Nat Genet 44(6):694–698. doi:10.1038/ng.2256

Hashmi AZ, Hakim W, Kruglov EA, Watanabe A, Watkins W, Dranoff JA et al (2007) Adenosine inhibits cytosolic calcium signals and chemotaxis in hepatic stellate cells. Am J Physiol Gastrointest Liver Physiol 292(1):G395–G401. doi:10.1152/ajpgi.00208.2006

Higaki K, Yano H, Kojiro M (1996) Fas antigen expression and its relationship with apoptosis in human hepatocellular carcinoma and noncancerous tissues. Am J Pathol 149(2):429–437

Hikita H, Kodama T, Shimizu S, Li W, Shigekawa M, Tanaka S et al (2012) Bak deficiency inhibits liver carcinogenesis: a causal link between apoptosis and carcinogenesis. J Hepatol 57(1):92–100. doi:10.1016/j.jhep.2012.01.027

Hoque R, Sohail MA, Salhanick S, Malik AF, Ghani A, Robson SC et al (2012) P2X7 receptor-mediated purinergic signaling promotes liver injury in acetaminophen hepatotoxicity in mice. Am J Physiol Gastrointest Liver Physiol 302(10):G1171–G1179. doi:10.1152/ajpgi.00352.2011

Huebener P, Pradere JP, Hernandez C, Gwak GY, Caviglia JM, Mu X et al (2015) The HMGB1/RAGE axis triggers neutrophil-mediated injury amplification following necrosis. J Clin Invest 125(2):539–550. doi:10.1172/JCI76887

Imaeda AB, Watanabe A, Sohail MA, Mahmood S, Mohamadnejad M, Sutterwala FS et al (2009) Acetaminophen-induced hepatotoxicity in mice is dependent on Tlr9 and the Nalp3 inflammasome. J Clin Invest 119(2):305–314. doi:10.1172/JCI35958

Inokuchi S, Aoyama T, Miura K, Osterreicher CH, Kodama Y, Miyai K et al (2010) Disruption of TAK1 in hepatocytes causes hepatic injury, inflammation, fibrosis, and carcinogenesis. Proc Natl Acad Sci U S A 107(2):844–849. doi:10.1073/pnas.0909781107

Iracheta-Vellve A, Petrasek J, Satishchandran A, Gyongyosi B, Saha B, Kodys K et al (2015) Inhibition of sterile danger signals, uric acid and ATP, prevents inflammasome activation and protects from alcoholic steatohepatitis in mice. J Hepatol 63(5):1147–1155. doi:10.1016/j.jhep.2015.06.013

Jeong WI, Park O, Suh YG, Byun JS, Park SY, Choi E et al (2011) Suppression of innate immunity (natural killer cell/interferon-gamma) in the advanced stages of liver fibrosis in mice. Hepatology 53(4):1342–1351. doi:10.1002/hep.24190

Justus SJ, Ting AT (2015) Cloaked in ubiquitin, a killer hides in plain sight: the molecular regulation of RIPK1. Immunol Rev 266(1):145–160. doi:10.1111/imr.12304

Kamata H, Honda S, Maeda S, Chang L, Hirata H, Karin M (2005) Reactive oxygen species promote TNFalpha-induced death and sustained JNK activation by inhibiting MAP kinase phosphatases. Cell 120(5):649–661. doi:10.1016/j.cell.2004.12.041

Karin M, Lin A (2002) NF-kappaB at the crossroads of life and death. Nat Immunol 3(3):221–227. doi:10.1038/ni0302-221

Kisseleva T, Brenner DA (2006) Hepatic stellate cells and the reversal of fibrosis. J Gastroenterol Hepatol 21(Suppl 3):S84–S87. doi:10.1111/j.1440-1746.2006.04584.x

Kondylis V, Polykratis A, Ehlken H, Ochoa-Callejero L, Straub BK, Krishna-Subramanian S et al (2015) NEMO prevents steatohepatitis and hepatocellular carcinoma by inhibiting RIPK1 kinase activity-mediated hepatocyte apoptosis. Cancer Cell 28(5):582–598. doi:10.1016/j.ccell.2015.10.001

Kong X, Feng D, Wang H, Hong F, Bertola A, Wang FS et al (2012) Interleukin-22 induces hepatic stellate cell senescence and restricts liver fibrosis in mice. Hepatology 56(3):1150–1159. doi:10.1002/hep.25744

Kono H, Rock KL (2008) How dying cells alert the immune system to danger. Nat Rev Immunol 8(4):279–289. doi:10.1038/nri2215

Koppe C, Verheugd P, Gautheron J, Reisinger F, Kreggenwinkel K, Roderburg C et al (2016) IkappaB kinasealpha/beta control biliary homeostasis and hepatocarcinogenesis in mice by phosphorylating the cell-death mediator receptor-interacting protein kinase 1. Hepatology 64(4):1217–1231. doi:10.1002/hep.28723

Lee MH, Yang HI, Lu SN, Jen CL, Yeh SH, Liu CJ et al (2010) Hepatitis C virus seromarkers and subsequent risk of hepatocellular carcinoma: long-term predictors from a community-based cohort study. J Clin Oncol 28(30):4587–4593. doi:10.1200/JCO.2010.29.1500

Legarda-Addison D, Hase H, O'Donnell MA, Ting AT (2009) NEMO/IKKgamma regulates an early NF-kappaB-independent cell-death checkpoint during TNF signaling. Cell Death Differ 16(9):1279–1288. doi:10.1038/cdd.2009.41

Liu H, Lo CR, Czaja MJ (2002) NF-kappaB inhibition sensitizes hepatocytes to TNF-induced apoptosis through a sustained activation of JNK and c-Jun. Hepatology 35(4):772–778. doi:10.1053/jhep.2002.32534

Luedde T, Beraza N, Kotsikoris V, van Loo G, Nenci A, De Vos R et al (2007) Deletion of NEMO/IKKgamma in liver parenchymal cells causes steatohepatitis and hepatocellular carcinoma. Cancer Cell 11(2):119–132. doi:10.1016/j.ccr.2006.12.016

Maeda S, Chang L, Li ZW, Luo JL, Leffert H, Karin M (2003) IKKbeta is required for prevention of apoptosis mediated by cell-bound but not by circulating TNFalpha. Immunity 19(5):725–737

Maeda S, Kamata H, Luo JL, Leffert H, Karin M (2005) IKKbeta couples hepatocyte death to cytokine-driven compensatory proliferation that promotes chemical hepatocarcinogenesis. Cell 121(7):977–990. doi:10.1016/j.cell.2005.04.014

Malhi H, Guicciardi ME, Gores GJ (2010) Hepatocyte death: a clear and present danger. Physiol Rev 90(3):1165–1194. doi:10.1152/physrev.00061.2009

Matzinger P (2002) The danger model: a renewed sense of self. Science 296(5566):301–305. doi:10.1126/science.1071059

McDonald B, Pittman K, Menezes GB, Hirota SA, Slaba I, Waterhouse CC et al (2010) Intravascular danger signals guide neutrophils to sites of sterile inflammation. Science 330(6002):362–366. doi:10.1126/science.1195491

McHedlidze T, Waldner M, Zopf S, Walker J, Rankin AL, Schuchmann M et al (2013) Interleukin-33-dependent innate lymphoid cells mediate hepatic fibrosis. Immunity 39(2):357–371. doi:10.1016/j.immuni.2013.07.018

Meriden Z, Forde KA, Pasha TL, Hui JJ, Reddy KR, Furth EE et al (2010) Histologic predictors of fibrosis progression in liver allografts in patients with hepatitis C virus infection. Clin Gastroenterol Hepatol 8(3):289–296 . doi:10.1016/j.cgh.2009.10.034296.e1-8

Micheau O, Tschopp J (2003) Induction of TNF receptor I-mediated apoptosis via two sequential signaling complexes. Cell 114(2):181–190

Ni HM, Chen X, Shi YH, Liao Y, Beg AA, Fan J et al (2009) Genetic delineation of the pathways mediated by bid and JNK in tumor necrosis factor-alpha-induced liver injury in adult and embryonic mice. J Biol Chem 284(7):4373–4382. doi:10.1074/jbc.M807259200

Petrasek J, Iracheta-Vellve A, Saha B, Satishchandran A, Kodys K, Fitzgerald KA et al (2015) Metabolic danger signals, uric acid and ATP, mediate inflammatory cross-talk between hepatocytes and immune cells in alcoholic liver disease. J Leukoc Biol 98(2):249–256. doi:10.1189/jlb.3AB1214-590R

Pikarsky E, Porat RM, Stein I, Abramovitch R, Amit S, Kasem S et al (2004) NF-kappaB functions as a tumour promoter in inflammation-associated cancer. Nature 431(7007):461–466. doi:10.1038/nature02924

Rock KL, Lai JJ, Kono H (2011) Innate and adaptive immune responses to cell death. Immunol Rev 243(1):191–205. doi:10.1111/j.1600-065X.2011.01040.x

Roychowdhury S, McMullen MR, Pisano SG, Liu X, Nagy LE (2013) Absence of receptor interacting protein kinase 3 prevents ethanol-induced liver injury. Hepatology 57(5):1773–1783. doi:10.1002/hep.26200

Roychowdhury S, McCullough RL, Sanz-Garcia C, Saikia P, Alkhouri N, Matloob A et al (2016) Receptor interacting protein 3 protects mice from high-fat diet-induced liver injury. Hepatology 64(5):1518–1533. doi:10.1002/hep.28676

Sakurai T, He G, Matsuzawa A, Yu GY, Maeda S, Hardiman G et al (2008) Hepatocyte necrosis induced by oxidative stress and IL-1 alpha release mediate carcinogen-induced compensatory proliferation and liver tumorigenesis. Cancer Cell 14(2):156–165. doi:10.1016/j.ccr.2008.06.016

Schwabe RF, Uchinami H, Qian T, Bennett BL, Lemasters JJ, Brenner DA (2004) Differential requirement for c-Jun NH2-terminal kinase in TNFalpha- and Fas-mediated apoptosis in hepatocytes. FASEB J 18(6):720–722. doi:10.1096/fj.03-0771fje

Seki E, Brenner DA, Karin M (2012) A liver full of JNK: signaling in regulation of cell function and disease pathogenesis, and clinical approaches. Gastroenterology 143(2):307–320. doi:10.1053/j.gastro.2012.06.004

Shimada K, Crother TR, Karlin J, Dagvadorj J, Chiba N, Chen S et al (2012) Oxidized mitochondrial DNA activates the NLRP3 inflammasome during apoptosis. Immunity 36(3):401–414. doi:10.1016/j.immuni.2012.01.009

Takamura M, Matsuda Y, Yamagiwa S, Tamura Y, Honda Y, Suzuki K et al (2007) An inhibitor of c-Jun NH2-terminal kinase, SP600125, protects mice from D-galactosamine/lipopolysaccharide-induced hepatic failure by modulating BH3-only proteins. Life Sci 80(14):1335–1344. doi:10.1016/j.lfs.2006.12.034

Takehara T, Tatsumi T, Suzuki T, Rucker EB 3rd, Hennighausen L, Jinushi M et al (2004) Hepatocyte-specific disruption of Bcl-xL leads to continuous hepatocyte apoptosis and liver fibrotic responses. Gastroenterology 127(4):1189–1197

Tsai WC, Hsu SD, Hsu CS, Lai TC, Chen SJ, Shen R et al (2012) MicroRNA-122 plays a critical role in liver homeostasis and hepatocarcinogenesis. J Clin Invest 122(8):2884–2897. doi:10.1172/JCI63455

Tsung A, Sahai R, Tanaka H, Nakao A, Fink MP, Lotze MT et al (2005) The nuclear factor HMGB1 mediates hepatic injury after murine liver ischemia-reperfusion. J Exp Med 201(7):1135–1143. doi:10.1084/jem.20042614

Ventura JJ, Cogswell P, Flavell RA, Baldwin AS Jr, Davis RJ (2004) JNK potentiates TNF-stimulated necrosis by increasing the production of cytotoxic reactive oxygen species. Genes Dev 18(23):2905–2915. doi:10.1101/gad.1223004

Vick B, Weber A, Urbanik T, Maass T, Teufel A, Krammer PH et al (2009) Knockout of myeloid cell leukemia-1 induces liver damage and increases apoptosis susceptibility of murine hepatocytes. Hepatology 49(2):627–636. doi:10.1002/hep.22664

Vucur M, Reisinger F, Gautheron J, Janssen J, Roderburg C, Cardenas DV et al (2013) RIP3 inhibits inflammatory hepatocarcinogenesis but promotes cholestasis by controlling caspase-8- and JNK-dependent compensatory cell proliferation. Cell Rep 4(4):776–790. doi:10.1016/j.celrep.2013.07.035

Wagner EF, Nebreda AR (2009) Signal integration by JNK and p38 MAPK pathways in cancer development. Nat Rev Cancer 9(8):537–549. doi:10.1038/nrc2694

Wan X, Xu CY, Lu C, Li D, Sang J et al (2016) Uric acid regulates hepatic steatosis and insulin resistance through the NLRP3 inflammasome-dependent mechanism. J Hepatol 64(4):925–932. doi:10.1016/j.jhep.2015.11.022

Wang Y, Singh R, Lefkowitch JH, Rigoli RM, Czaja MJ (2006) Tumor necrosis factor-induced toxic liver injury results from JNK2-dependent activation of caspase-8 and the mitochondrial death pathway. J Biol Chem 281(22):15258–15267. doi:10.1074/jbc.M512953200

Wang FP, Li L, Li J, Wang JY, Wang LY, Jiang W (2013) High mobility group box-1 promotes the proliferation and migration of hepatic stellate cells via TLR4-dependent signal pathways of PI3K/Akt and JNK. PLoS One 8(5):e64373. doi:10.1371/journal.pone.0064373

Watanabe A, Hashmi A, Gomes DA, Town T, Badou A, Flavell RA et al (2007) Apoptotic hepatocyte DNA inhibits hepatic stellate cell chemotaxis via toll-like receptor 9. Hepatology 46(5):1509–1518. doi:10.1002/hep.21867

Weber A, Boger R, Vick B, Urbanik T, Haybaeck J, Zoller S et al (2010) Hepatocyte-specific deletion of the antiapoptotic protein myeloid cell leukemia-1 triggers proliferation and hepatocarcinogenesis in mice. Hepatology 51(4):1226–1236. doi:10.1002/hep.23479

Wiese M, Fischer J, Lobermann M, Gobel U, Grungreiff K, Guthoff W et al (2014) Evaluation of liver disease progression in the German hepatitis C virus (1b)-contaminated anti-D cohort at 35 years after infection. Hepatology 59(1):49–57. doi:10.1002/hep.26644

Win S, Than TA, Han D, Petrovic LM, Kaplowitz N (2011) c-Jun N-terminal kinase (JNK)-dependent acute liver injury from acetaminophen or tumor necrosis factor (TNF) requires mitochondrial Sab protein expression in mice. J Biol Chem 286(40):35071–35078. doi:10.1074/jbc.M111.276089

Wree A, Eguchi A, McGeough MD, Pena CA, Johnson CD, Canbay A et al (2014a) NLRP3 inflammasome activation results in hepatocyte pyroptosis, liver inflammation, and fibrosis in mice. Hepatology 59(3):898–910. doi:10.1002/hep.26592

Wree A, McGeough MD, Pena CA, Schlattjan M, Li H, Inzaugarat ME et al (2014b) NLRP3 inflammasome activation is required for fibrosis development in NAFLD. J Mol Med (Berl) 92(10):1069–1082. doi:10.1007/s00109-014-1170-1

Yang L, Inokuchi S, Roh YS, Song J, Loomba R, Park EJ et al (2013) Transforming growth factor-beta signaling in hepatocytes promotes hepatic fibrosis and carcinogenesis in mice with hepatocyte-specific deletion of TAK1. Gastroenterology 144(5):1042–1054 . doi:10.1053/j.gastro.2013.01.056e4

Yang L, Roh YS, Song J, Zhang B, Liu C, Loomba R et al (2014) Transforming growth factor beta signaling in hepatocytes participates in steatohepatitis through regulation of cell death and lipid metabolism in mice. Hepatology 59(2):483–495. doi:10.1002/hep.26698

Chapter 11
Cell Death and Autophagy in Liver Tumorigenesis and Liver Cancer

Meng-Xin Tian, Yuan-Fei Peng, Han Wang, Jia Fan, and Ying-Hong Shi

Abbreviations

3-MA	3-methyladenine
ATD	Alpha1-antitrypsin deficiency
ATG	Autophagy-related gene
CBZ	Carbamazepine
CQ	Chloroquine
HBV	Hepatitis B virus
HCC	Hepatocellular carcinoma
HCQ	Hydroxychloroquine
HCV	Hepatitis C virus
ICC	Intrahepatic cholangiocarcinoma
MDBs	Mallory-denk bodies
NAFLD	Nonalcoholic fatty liver disease

In recent years, autophagy has been a research focus. Cellular autophagy is particularly well studied in liver cancer, among the different tumor types. Confusingly, autophagy shows complex and dual roles in both suppressing and promoting tumors in liver cancer.

M.-X. Tian, M.D • Y.-F. Peng, M.D., Ph.D. • H. Wang, M.D
J. Fan, M.D., Ph.D. • Y.-H. Shi, M.D., Ph.D. (✉)
Department of Liver Surgery, Liver Cancer Institute, Zhongshan Hospital,
Fudan University, 180 FengLin Road, Shanghai 200032, China

Key Laboratory of Carcinogenesis and Cancer Invasion of Ministry of Education,
Shanghai, China
e-mail: shi.yinghong@zs-hospital.sh.cn

W.-X. Ding, X.-M. Yin (eds.), *Cellular Injury in Liver Diseases*, Cell Death in Biology and Diseases, DOI 10.1007/978-3-319-53774-0_11

X.-M. Yin and Z. Dong (Series eds.), *Cell Death in Biology and Diseases*

11.1 Autophagy in Tumor Suppression

Autophagy was proposed to be a tumor suppressor due to the direct evidence, showing that liver tumors were developed in autophagy-related gene (ATG)-deficient mouse models. Currently, some critical autophagy genes, including beclin1, ATG5 and ATG7, are proven tumor suppressor genes. In an ATG5 knockout mouse model, multiple benign tumors were developed only in the liver, but not in other tissues, indicating that continuous autophagy is important for the suppression of tumorigenesis. Moreover, swollen mitochondria, oxidative stress, and genomic damage responses were detected in the hepatic tumor cells. Similar results were also observed in liver-specific ATG7-deficient mice (Takamura et al. 2011). Beclin1 is a mammalian autophagy gene and a haploinsufficient tumor suppressor and negatively regulates tumorigenesis (Liang et al. 1999). Beclin1$^{-/-}$ mutant mice not only died early in embryogenesis but also suffered from a high incidence of spontaneous tumors (Qu et al. 2003; Yue et al. 2003). Heterozygous disruption of beclin1 could increase the frequency of spontaneous malignancies and accelerate the development of hepatitis B virus (HBV)-induced premalignant lesions. In addition, beclin1 heterozygous disruption also leads to an increase in cellular proliferation and reduces autophagy in vivo (Qu et al. 2003). Recently, studies revealed that the expression of beclin1 and bax in hepatocellular carcinoma (HCC) tissues might provide a synergistic effect that inhibits tumor proliferation, infiltration, metastasis, and angiogenesis (Qiu et al. 2014). All of these data from animal models provide clear evidence that autophagy is an important suppressor mechanism to prevent tumorigenesis. p62 is a selective autophagy substrate. Autophagy defects could result in sustained expression of p62 and promote tumor progression (Mathew et al. 2009). In human cancers, p62 accumulation has been detected (Inami et al. 2011), it is involved in heterozygous mutations of the beclin1 gene (Komatsu et al. 2007; Mathew et al. 2009; Qu et al. 2003; Yue et al. 2003). In a mouse model, liver-specific ATG7 knockout mice developed hepatocellular adenoma accompanied by excessive p62 accumulation and subsequent Nrf2 activation, and persistent Nrf2 activation contributes to the HCC formation (Inami et al. 2011). After a concomitant knockout of the p62 gene, tumor size in this mouse model significantly decreased (Takamura et al. 2011). Furthermore, the tumor suppressor gene p53 plays a dual role in cellular autophagy. In the cytoplasm, autophagy is suppressed by p53, and p53 degradation is also required for the induction of autophagy (Levine and Abrams 2008). Additional evidence indicates that p62 is increased in HCC tissues compared with non-tumorous liver tissues, suggesting that human HCC is autophagy defective (Bao et al. 2014). The suppression of liver cancer is also reflected in another aspect: the autophagy level is negatively correlated with the degree of HCC malignancy. Under starvation conditions, decreased basal expression of autophagic genes and their corresponding autophagic activity in HCC cell lines, and autophagy defects are correlated well with the highly malignant phenotype (Ding et al. 2008). These results suggest that autophagy defects in combination with altered apoptotic activity might facilitate tumor progression and poor prognosis because autophagy might interact with apoptosis in the regulation of HCC (Shi et al. 2009). Due to its suppression effect on liver tumor, autophagy might be of therapeutic value. Tissue injury, inflammation, and genomic instability are risk factors for carcinogenesis. In precancerous cells, these risk factors could be inhibited by autophagy, and autophagy induction may be beneficial for the prevention of liver cancer. For example, in patients with α-1 antitrypsin deficiency, inflammation and carcinogenesis could be inhibited by the upregulation of basal autophagy level in liver cells (Perlmutter 2009). It has been demonstrated that carbamazepine (CBZ) could increase the autophagy level, reduce the α-1 antitrypsin load, and lower the risks of liver fibrosis and cancer in an α-1 antitrypsin deficiency (ATD) mouse model (Hidvegi et al. 2010), indicating that increased autophagy might provide effective and preventive effects against liver cancer.

11.2 Autophagy in Cancer Promotion

Growing evidences indicate that autophagy is needed by tumor cells under disadvantageous conditions, including starvation, deficiency of growth factor, hypoxia, injury, and drug medication. Generally, up-regulation of basal autophagy is required for the survival of tumor cells, implying that cancer cells use the catabolic function of autophagy to tolerate stress. The autophagosome is common in ischemic areas, and the autophagy inhibition induced by decreased levels of beclin1 causes cellular death (Degenhardt et al. 2006). In RAS-driven tumors, the basal autophagy level is increased. Moreover, RAS-expressing $ATG5^{-/-}$ or $ATG7^{-/-}$ cells displayed reduced tumor growth in vivo (Guo et al. 2011).

Recently, increasing evidences showed that HCC metastasis could be suppressed by autophagy, and the intrinsic mechanism underlying metastasis might involve facilitating anoikis resistance and lung colonization (Peng et al. 2013b, c). Moreover, studies demonstrated that hypoxia-induced autophagy contributed to HCC chemoresistance. In vitro, by enhancing cellular survival and decreasing the apoptotic potential, autophagy may contribute to the resistance of HCC cells to chemotherapeutic agents under hypoxia; thus, the cellular survivability is changed by autophagy suppression (Song et al. 2009). Because autophagy is essential for tumor survival pathways, autophagy modulation is a novel approach for enhancing the efficacy of existing cancer therapy. Current (angiogenesis inhibitors and growth factor receptors) and conventional cancer treatments (radiotherapy and chemotherapy) could lead to metabolic stress and induce autophagy. To improve the clinical efficacy, autophagy inhibitors could be used to eliminate the protective effect of autophagy in combination therapy (Wu et al. 2012). Antimalarial drugs (chloroquine and hydroxychloroquine) have been evaluated and shown promising results in preclinical and clinical trials, and research on specific small molecule autophagy inhibitors is currently in progress (details as follows).

Autophagy could exert a strong influence on carcinogenesis and tumor progression by altering the risk factors of liver cancer, including HBV, hepatitis C virus

(HCV), alcohol intake, and others. HBV infection is the most important leading cause of liver cancer, and it induces autophagy by enhancing viral DNA replication with only a slight effect on viral RNA transcription (Sir et al. 2010). During a natural HBV infection, autophagy facilitates HBV replication, which has primarily been demonstrated in cell cultures and in a mouse model (Li et al. 2011; Tian et al. 2011). HCV is also correlated with liver cancer, and studies have shown that autophagy contributes to the self-replication of HCV particles. HCV RNA replication could promote the expression of UPR and CHOP, thus leading to the accumulation of autophagosomes and activation of autophagy (Ait-Goughoulte et al. 2008; Sir et al. 2008). Some ATGs (BECN1, ATG4B, ATG5, and ATG12) are involved in the translation of viral mRNA and initiation of replication (Dreux et al. 2009). HCV RNA replication may block the maturation of the autophagosome and autolysosome (Sir et al. 2008). Furthermore, HCV seems to avoid recognition by the autophagic machinery (Alavian et al. 2011; Rautou et al. 2010). Chronic alcohol consumption has been widely recognized as one of the most common causes, and it may lead to several types of liver status: fatty liver, hepatic fibrosis, cirrhosis, and alcoholic hepatitis. The mechanism underlying the liver injury may include the induction of oxidative stress. Hepatic steatosis and oxidative stress are regulated by autophagy in liver cells, indicating the importance of autophagy in ethanol-induced liver diseases (Czaja 2011; Wang et al. 2010). Needless to say, the exact role of autophagy in alcohol-related liver cancer remains to be further studied. Nonalcoholic fatty liver disease (NAFLD) is a common chronic liver disease. The loss of lipid droplets is promoted by autophagy and is called fat endocytosis (Singh et al. 2009). In hepatic stellate cells, blocking autophagy could result in the accumulation of lipid droplets and attenuation of liver fibrosis (Hernandez-Gea et al. 2012), implying that autophagy may prevent the formation of NAFLD-related tumors.

It should be noted that the majority of current research is focused on HCC, but autophagy also plays a similar role in two other types of liver cancer, including intrahepatic cholangiocarcinoma (ICC) and hepatoblastoma. In nutrient starvation and xenograft cholangiocarcinoma cells, autophagy is activated (Hou et al. 2011). Inhibiting autophagy (by an autophagy inhibitor or beclin1 siRNA) could induce apoptosis during starvation and increase the sensitivity of cholangiocarcinoma cells to chemotherapy (Hou et al. 2011). Similarly, blocking autophagy by regulating autophagic genes could suppress the formation of hepatoblastoma. However, the exact mechanism of autophagy in ICC and hepatoblastoma requires in-depth studies.

11.3 Autophagy as a New Therapeutic Target

Due to its critical role in liver cancer, autophagy has been considered a new therapeutic target, but there is currently no consensus about autophagy therapy in clinics. Theoretically, autophagy inhibitors could eliminate or weaken the protective roles of chemotherapy induced by autophagy, which could enhance the tumor-killing effect and promote the therapeutic effect. In addition, autophagy inducers could lead to cellular autophagic death when combined with chemotherapy. By activating the PI3K-Akt and Akt signaling pathways, sorafenib or bortezomib treatment could activate autophagy, and combining these treatments with chloroquine (CQ) or 3-methyladenine (3-MA) could improve their cytotoxicity (Hui et al. 2012; Shimizu et al. 2012; Yu et al. 2013). Cisplatin and oxaliplatin share similar effects (Ding et al. 2011; Xu et al. 2012). However, the molecule machinery of certain drugs is not similar to those above. Anticancer drugs (nilotinib, cannabinoids tetrahydrocannabinol, JWH-015, and SAHA) can upregulate autophagy by activating the AMPK signaling pathway, but their treatment efficacy was decreased when they were combined with inhibitors. Autophagy may protect cancer cells, but the induced autophagy mediated by the AMPK signaling pathway seems to induce cytotoxicity. The above results indicate the complexity of autophagy treatments in liver cancer.

11.4 Autophagy Inducers and Inhibitors

Owing to the multi-faceted role of autophagy, inhibitor applications depend on the specific circumstances (Amaravadi et al. 2011, 2007). Autophagy is considered a cellular death mechanism that promotes apoptosis or autophagic cell death, leading to better treatment (Edinger and Thompson 2004; Eisenberg-Lerner et al. 2009; Shen and Codogno 2011). In apoptosis-deficient tumor cells, induced autophagy activity and increased autophagic cell death are effective ways to facilitate cellular death (Maycotte and Thorburn 2011). The autophagy inducers and inhibitors that have been developed using in vitro or in vivo systems and have been used in clinical trials are summarized in Table 11.1.

Autophagy inducers: important cellular autophagy inducers for liver cancer (based on the concept that autophagy could improve the treatment effect by enhancing cellular death) include the following: (1) rapamycin and its analogs: in HCC, the PI3K/Akt/mTOR pathway is the key signaling cascade because it modulates cellular growth, proliferation, angiogenesis, and apoptosis (Sabatini 2006), and this pathway is activated in 15–41% of HCC tumors. mTOR inhibitors have anti-HCC activity (Sieghart et al. 2007). Rapamycin (sirolimus), an mTOR inhibitor, is widely used as an autophagy inducer, and it has been applied for its anti-proliferation and anti-angiogenesis effects. In clinical research, rapamycin and its analogs (everolimus) have proven anti-tumor activity (Huynh et al. 2009). (2) Tyrosine kinase inhibitors: tyrosine kinase plays an important role in tumor progression, and it has been utilized in cancer treatment. In combination with the HDAC inhibitor SAHA, sorafenib can enhance liver cancer death by inducing autophagy (Martin et al. 2009; Park et al. 2008). However, adriamycin (DOX)-induced autophagic cell death can be inhibited by sorafenib, thus promoting the cell cycle progress, improving the survival rate, and decreasing cancer autophagy (Manov et al. 2011). Therefore, the potent antagonistic effect of sorafenib and adriamycin treatment needs to be further considered. (3) Others: a new naphthalimide polyamine conjugate

Table 11.1 Autophagy inducers and inhibitors used in clinical trials and developed in in vivo and in vitro systems

	Drug	Experimental system
Autophagy inducer	Rapamycin	C57BL/6 mice, Hep G2; mice
	Everolimus	Xenograft mouse model
	Everolimus + BEZ235	Hepatocyte cell line, mice, Stage I clinical trial
	Bortezomib	Huh7 cells/FVB and transgenic mice
	SAHA/OSU-HDAC42	Hep 3B, Hep G2, Huh7
	Panobinostat	Hep 3B, Hep G2, Huh7, human, xenograft mouse model
	Sorafenib	Hep 3B, Hep G2, Huh7
	Nilotinib	Hepatocyte cell line, mice
	NPC-16	Hep G2, Bel-7402
	Berberine	MHCC97-L, HepG2
	MLN4924	Huh7, Hep G2, xenograft mouse model
	Δ9-THC/JWH-015	Hep G2, Huh7, xenograft mouse model
	THC	Hepatocyte cell line, mice
	Fangchinoline	Hep G2, PLC/PRF/5
	SB203580	Hep G2, Hep3B, PLC/PRF/5, Huh7
	Melatonin	H22, H22 mouse models
	Cisplatin	Hepatocyte cell line, mice
	Oxaliplatin	Hepatocyte cell line, mice
	miR-100	Hepatocyte cell line, mice
Autophagy inhibitor	3-MA	H22, Hep G2; xenograft mouse model, PLC/PRF/5; SMMC7721
	Wortmannin	HCCC9810
	Bafilomycin A1	Hep 3B, Hep G2, Huh7; HA22T/VGH
	CQ/HCQ	HA22T/VGH; Mahlavu
	siRNA (siATG5, siBeclin1, siATG7, shBeclin1, shATG5)	Hep G2, H22; PLC/PRF/5; HA22T/VGH; Huh7, HCCLM3, MHCC97H, SMMC7721
	MicroRNAs (mir-375, miR-101, miR-199a-5p)	Huh7 and Hep3B, xenograft mice

can induce autophagy and apoptosis, and the involvement of the mTOR pathway in the autophagy mediated by NPC-16 has been confirmed (Xie et al. 2011). Berberine, which inhibits the mTOR pathway by up-regulating the AMPK signaling pathway, can also induce autophagic cellular death in liver cancer (Wang et al. 2010a; Yu et al. 2014). The cannabinoid Δ9-THC and its receptor agonist (JWH-015) could inhibit autophagy induction and decrease the growth of subcutaneous xenografts through the Akt-mTORC1 axis and stimulation of the AMPK pathway. Evidence demonstrated that Δ9-THC and the agonist JWH-015 led to the death of liver cancer. As a new type of antitumor agent, fangchinoline can induce autophagic death in liver cancer through the p53/sestrin2/AMPK pathway (Wang et al. 2011).

Furthermore, MLN4924 (a small molecule inhibitor of NEDD8-activating enzyme) can suppress the growth of liver cancer by inducing autophagy in vivo and in vitro. The intrinsic mechanism might be the suppression of mTOR activity by Deptor (mTOR-binding protein) (Luo et al. 2012).

Autophagyinhibitors: autophagy is an essential survival mechanism, through which cancer cells can survive under various stress conditions, including drug treatment. Therefore, inhibitors can increase the treatment effect by abolishing the protective effect of autophagy, increasing cytotoxicity, and limiting treatment resistance. (1) CQ and hydroxychloroquine (HCQ): CQ and HCQ are widely used as autophagy inhibitors. Many reports have demonstrated that drug treatments, including CQ and HCQ treatments, have a sensitizing effect on cell apoptosis both in vivo and in vitro. For example, CQ-induced autophagy increases the cellular death caused by oxaliplatin, improving the chemosensitivity in liver cancer, and a similar result was shown by ding et al. (Ding et al. 2011). In HCC cell lines and a nude mouse model, the sorafenib group combined with CQ significantly suppressed the tumor growth compared with that for sorafenib alone (Shi et al. 2011; Shimizu et al. 2012). Currently, 25 autophagy-related clinical trials are in progress that use CQ/HCQ alone or in combination with other drugs to treat tumors (http://www.clinicaltrials.gov). However, no related clinical trials are being performed for liver cancer. (2) siRNA and shRNA: due to the high specificity and selectiveness of siRNA or shRNA, side effects could be avoided to a large extent by silencing certain autophagy genes. Many reports have reported cancer treatments that knock out specific ATG genes, but there are no clinical trials targeting liver cancer. Chen demonstrated that suppressed autophagy, which was mediated by knocking down beclin1, promoted increased cellular death in liver cancer (Chen et al. 2011). Beclin1 silencing also facilitates the H22 cellular death induced by melatonin and improves the antitumor effects of melatonin (Liu et al. 2012). Overall, silencing-specific ATG genes to inhibit autophagy could improve the clinical effect of chemotherapy. However, it is still unclear whether the siRNA-induced treatment effect would occur in liver cancer. Via the construction of viral vectors carrying shRNA or miRNA targeting critical autophagy genes, autophagy inhibition in vivo might be a new field in the future. An autophagy-related and specific shRNA designed by Peng et al. was demonstrated to be of great value in suppressing autophagy and lung metastasis both in vivo and in vitro (Peng et al. 2013a, c). (3) microRNAs (miRNAs) can regulate autophagy through multiple genes and pathways (Kim and Kim 2014). miRNAs could regulate autophagy in liver cancer by acting on ATG genes. Many miRNAs (such as miR-101, miR-30a, miR-34a, miR-204, and miR-375) could serve as inhibitors. In liver cancer, hypoxia-induced autophagy could be suppressed effectively by the direct actions of miR-375 on ATG7 (Kim and Kim 2014). Transfection with a miR-375 mimic attenuated the protective effect of autophagy and impaired the vitality of tumor cells (Pastore et al. 2013). By down regulating STMN, RAB5A, and ATG4D, miR-101 could suppress HCC autophagy. When combined with adriamycin and fluorouracil, miR-101 could increase their chemosensitivity. In liver cancer patients treated with cisplatin, the miR-199a-5p level is decreased significantly; this miRNA acts on ATG7 and abolishes the drug resistance induced by autophagy.

Autophagy inhibition by miRNA is becoming a new focus. However, some miRNAs will induce autophagy. miR-100 facilitates HCC autophagy by preventing mTOR and IGF-1R expression. Overexpression of miR-100 can lead to cellular death, and this effect can be inhibited by ATG7 silencing and CQ, providing a potent target for HCC treatment. (4) Others: 20(S)-Ginsenoside Rg3 is a new type of autophagy inhibitor that can improve the adriamycin treatment effect (Kim et al. 2014). Adriamycin-induced autophagy has a protective effect in liver cancer. Ginsenoside Rg3 inhibits late-period autophagy. Combined with doxorubicin, Rg3 is able to kill HCC cells, which exhibit good tolerance to Rg3, and able to inhibit liver cancer xenografts in mice. The combination of inhibitors and traditional chemotherapy drugs is becoming an effective strategy in liver cancer treatment.

11.5 Conclusion

In liver cancer, autophagy has a Janus-face in that being a tumor survival mechanism, it can also lead to autophagic cell death. Accordingly, taking into account the essential role of autophagy in liver cancer and the existence of therapies targeting this process, may improve clinical efficacy of liver cancer. Exploring the potential clinical value of autophagy in liver cancer may lead to a new avenue of cancer therapeutics.

References

Ait-Goughoulte M, Kanda T, Meyer K, Ryerse JS, Ray RB, Ray R (2008) Hepatitis C virus genotype 1a growth and induction of autophagy. J Virol 82(5):2241–2249. doi:10.1128/JVI.02093-07

Alavian SM, Ande SR, Coombs KM, Yeganeh B, Davoodpour P, Hashemi M, Ghavami S et al (2011) Virus-triggered autophagy in viral hepatitis - possible novel strategies for drug development. J Viral Hepat 18(12):821–830. doi:10.1111/j.1365-2893.2011.01530.x

Amaravadi RK, Yu D, Lum JJ, Bui T, Christophorou MA, Evan GI et al (2007) Autophagy inhibition enhances therapy-induced apoptosis in a Myc-induced model of lymphoma. J Clin Invest 117(2):326–336. doi:10.1172/JCI28833

Amaravadi RK, Lippincott-Schwartz J, Yin XM, Weiss WA, Takebe N, Timmer W et al (2011) Principles and current strategies for targeting autophagy for cancer treatment. Clin Cancer Res 17(4):654–666. doi:10.1158/1078-0432.CCR-10-2634

Bao L, Chandra PK, Moroz K, Zhang X, Thung SN, Wu T, Dash S (2014) Impaired autophagy response in human hepatocellular carcinoma. Exp Mol Pathol 96(2):149–154. doi:10.1016/j.yexmp.2013.12.002

Chen LH, Loong CC, Su TL, Lee YJ, Chu PM, Tsai ML et al (2011) Autophagy inhibition enhances apoptosis triggered by BO-1051, an N-mustard derivative, and involves the ATM signaling pathway. Biochem Pharmacol 81(5):594–605. doi:10.1016/j.bcp.2010.12.011

Czaja MJ (2011) Functions of autophagy in hepatic and pancreatic physiology and disease. Gastroenterology 140(7):1895–1908. doi:10.1053/j.gastro.2011.04.038

Degenhardt K, Mathew R, Beaudoin B, Bray K, Anderson D, Chen G et al (2006) Autophagy promotes tumor cell survival and restricts necrosis, inflammation, and tumorigenesis. Cancer Cell 10(1):51–64. doi:10.1016/j.ccr.2006.06.001

Ding ZB, Shi YH, Zhou J, Qiu SJ, Xu Y, Dai Z et al (2008) Association of autophagy defect with a malignant phenotype and poor prognosis of hepatocellular carcinoma. Cancer Res 68(22):9167–9175. doi:10.1158/0008-5472.CAN-08-1573

Ding ZB, Hui B, Shi YH, Zhou J, Peng YF, Gu CY et al (2011) Autophagy activation in hepatocellular carcinoma contributes to the tolerance of oxaliplatin via reactive oxygen species modulation. Clin Cancer Res 17(19):6229–6238. doi:10.1158/1078-0432.CCR-11-0816

Dreux M, Gastaminza P, Wieland SF, Chisari FV (2009) The autophagy machinery is required to initiate hepatitis C virus replication. Proc Natl Acad Sci U S A 106(33):14046–14051. doi:10.1073/pnas.0907344106

Edinger AL, Thompson CB (2004) Death by design: apoptosis, necrosis and autophagy. Curr Opin Cell Biol 16(6):663–669. doi:10.1016/j.ceb.2004.09.011

Eisenberg-Lerner A, Bialik S, Simon HU, Kimchi A (2009) Life and death partners: apoptosis, autophagy and the cross-talk between them. Cell Death Differ 16(7):966–975. doi:10.1038/cdd.2009.33

Guo JY, Chen HY, Mathew R, Fan J, Strohecker AM, Karsli-Uzunbas G et al (2011) Activated Ras requires autophagy to maintain oxidative metabolism and tumorigenesis. Genes Dev 25(5):460–470. doi:10.1101/gad.2016311

Hernandez-Gea V, Ghiassi-Nejad Z, Rozenfeld R, Gordon R, Fiel MI, Yue Z et al (2012) Autophagy releases lipid that promotes fibrogenesis by activated hepatic stellate cells in mice and in human tissues. Gastroenterology 142(4):938–946. doi:10.1053/j.gastro.2011.12.044

Hidvegi T, Ewing M, Hale P, Dippold C, Beckett C, Kemp C et al (2010) An autophagy-enhancing drug promotes degradation of mutant alpha1-antitrypsin Z and reduces hepatic fibrosis. Science 329(5988):229–232. doi:10.1126/science.1190354

Hou YJ, Dong LW, Tan YX, Yang GZ, Pan YF, Li Z et al (2011) Inhibition of active autophagy induces apoptosis and increases chemosensitivity in cholangiocarcinoma. Lab Investig 91(8):1146–1157. doi:10.1038/labinvest.2011.97

Hui B, Shi YH, Ding ZB, Zhou J, Gu CY, Peng YF et al (2012) Proteasome inhibitor interacts synergistically with autophagy inhibitor to suppress proliferation and induce apoptosis in hepatocellular carcinoma. Cancer 118(22):5560–5571. doi:10.1002/cncr.27586

Huynh H, Chow KH, Soo KC, Toh HC, Choo SP, Foo KF et al (2009) RAD001 (everolimus) inhibits tumour growth in xenograft models of human hepatocellular carcinoma. J Cell Mol Med 13(7):1371–1380. doi:10.1111/j.1582-4934.2008.00364.x

Inami Y, Waguri S, Sakamoto A, Kouno T, Nakada K, Hino O et al (2011) Persistent activation of Nrf2 through p62 in hepatocellular carcinoma cells. J Cell Biol 193(2):275–284. doi:10.1083/jcb.201102031

Kim KM, Kim SG (2014) Autophagy and microRNA dysregulation in liver diseases. Arch Pharm Res 37(9):1097–1116. doi:10.1007/s12272-014-0439-9

Kim DG, Jung KH, Lee DG, Yoon JH, Choi KS, Kwon SW et al 20(S)-Ginsenoside Rg3 is a novel inhibitor of autophagy and sensitizes hepatocellular carcinoma to doxorubicin. Oncotarget 5(12):4438–4451. doi:10.18632/oncotarget.2034

Komatsu M, Waguri S, Koike M, Sou YS, Ueno T, Hara T et al (2007) Homeostatic levels of p62 control cytoplasmic inclusion body formation in autophagy-deficient mice. Cell 131(6):1149–1163. doi:10.1016/j.cell.2007.10.035

Levine B, Abrams J (2008) p53: the Janus of autophagy? Nat Cell Biol 10(6):637–639. doi:10.1038/ncb0608-637

Li J, Liu Y, Wang Z, Liu K, Wang Y, Liu J et al (2011) Subversion of cellular autophagy machinery by hepatitis B virus for viral envelopment. J Virol 85(13):6319–6333. doi:10.1128/JVI.02627-10

Liang XH, Jackson S, Seaman M, Brown K, Kempkes B, Hibshoosh H, Levine B (1999) Induction of autophagy and inhibition of tumorigenesis by beclin 1. Nature 402(6762):672–676. doi:10.1038/45257

Liu C, Jia Z, Zhang X, Hou J, Wang L, Hao S et al (2012) Involvement of melatonin in autophagy-mediated mouse hepatoma H22 cell survival. Int Immunopharmacol 12(2):394–401. doi:10.1016/j.intimp.2011.12.012

Luo Z, Yu G, Lee HW, Li L, Wang L, Yang D et al (2012) The Nedd8-activating enzyme inhibitor MLN4924 induces autophagy and apoptosis to suppress liver cancer cell growth. Cancer Res 72(13):3360–3371. doi:10.1158/0008-5472.CAN-12-0388

Manov I, Pollak Y, Broneshter R, Iancu TC (2011) Inhibition of doxorubicin-induced autophagy in hepatocellular carcinoma Hep3B cells by sorafenib—the role of extracellular signal-regulated kinase counteraction. FEBS J 278(18):3494–3507. doi:10.1111/j.1742-4658.2011.08271.x

Martin AP, Park MA, Mitchell C, Walker T, Rahmani M, Thorburn A et al (2009) BCL-2 family inhibitors enhance histone deacetylase inhibitor and sorafenib lethality via autophagy and overcome blockade of the extrinsic pathway to facilitate killing. Mol Pharmacol 76(2):327–341. doi:10.1124/mol.109.056309

Mathew R, Karp CM, Beaudoin B, Vuong N, Chen G, Chen HY et al (2009) Autophagy suppresses tumorigenesis through elimination of p62. Cell 137(6):1062–1075. doi:10.1016/j.cell.2009.03.048

Maycotte P, Thorburn A (2011) Autophagy and cancer therapy. Cancer Biol Ther 11(2):127–137

Park MA, Zhang G, Martin AP, Hamed H, Mitchell C, Hylemon PB et al (2008) Vorinostat and sorafenib increase ER stress, autophagy and apoptosis via ceramide-dependent CD95 and PERK activation. Cancer Biol Ther 7(10):1648–1662

Pastore N, Blomenkamp K, Annunziata F, Piccolo P, Mithbaokar P, Maria Sepe R et al (2013) Gene transfer of master autophagy regulator TFEB results in clearance of toxic protein and correction of hepatic disease in alpha-1-anti-trypsin deficiency. EMBO Mol Med 5(3):397–412. doi:10.1002/emmm.201202046

Peng YF, Shi YH, Ding ZB, Zhou J, Qiu SJ, Hui B et al (2013a) alpha-Fetoprotein promoter-driven Cre/LoxP-switched RNA interference for hepatocellular carcinoma tissue-specific target therapy. PLoS One 8(2):e53072. doi:10.1371/journal.pone.0053072

Peng YF, Shi YH, Shen YH, Ding ZB, Ke AW, Zhou J et al (2013b) Promoting colonization in metastatic HCC cells by modulation of autophagy. PLoS One 8(9):e74407. doi:10.1371/journal.pone.0074407

Peng YF, Shi YH, Ding ZB, Ke AW, Gu CY, Hui B et al (2013c) Autophagy inhibition suppresses pulmonary metastasis of HCC in mice via impairing anoikis resistance and colonization of HCC cells. Autophagy 9(12):2056–2068. doi:10.4161/auto.26398

Perlmutter DH (2009) Autophagic disposal of the aggregation-prone protein that causes liver inflammation and carcinogenesis in alpha-1-antitrypsin deficiency. Cell Death Differ 16(1):39–45. doi:10.1038/cdd.2008.103

Qiu DM, Wang GL, Chen L, Xu YY, He S, Cao XL et al (2014) The expression of beclin-1, an autophagic gene, in hepatocellular carcinoma associated with clinical pathological and prognostic significance. BMC Cancer 14:327. doi:10.1186/1471-2407-14-327

Qu X, Yu J, Bhagat G, Furuya N, Hibshoosh H, Troxel A et al (2003) Promotion of tumorigenesis by heterozygous disruption of the beclin 1 autophagy gene. J Clin Invest 112(12):1809–1820. doi:10.1172/JCI20039

Rautou PE, Mansouri A, Lebrec D, Durand F, Valla D, Moreau R (2010) Autophagy in liver diseases. J Hepatol 53(6):1123–1134. doi:10.1016/j.jhep.2010.07.006

Sabatini DM (2006) mTOR and cancer: insights into a complex relationship. Nat Rev Cancer 6(9):729–734. doi:10.1038/nrc1974

Shen HM, Codogno P (2011) Autophagic cell death: Loch Ness monster or endangered species? Autophagy 7(5):457–465

Shi YH, Ding ZB, Zhou J, Qiu SJ, Fan J (2009) Prognostic significance of Beclin 1-dependent apoptotic activity in hepatocellular carcinoma. Autophagy 5(3):380–382

Shi YH, Ding ZB, Zhou J, Hui B, Shi GM, Ke AW et al (2011) Targeting autophagy enhances sorafenib lethality for hepatocellular carcinoma via ER stress-related apoptosis. Autophagy 7(10):1159–1172. doi:10.4161/auto.7.10.16818

Shimizu S, Takehara T, Hikita H, Kodama T, Tsunematsu H, Miyagi T et al (2012) Inhibition of autophagy potentiates the antitumor effect of the multikinase inhibitor sorafenib in hepatocellular carcinoma. Int J Cancer 131(3):548–557. doi:10.1002/ijc.26374

Sieghart W, Fuereder T, Schmid K, Cejka D, Werzowa J, Wrba F et al (2007) Mammalian target of rapamycin pathway activity in hepatocellular carcinomas of patients undergoing liver transplantation. Transplantation 83(4):425–432. doi:10.1097/01.tp.0000252780.42104.95

Singh R, Kaushik S, Wang Y, Xiang Y, Novak I, Komatsu M et al (2009) Autophagy regulates lipid metabolism. Nature 458(7242):1131–1135. doi:10.1038/nature07976

Sir D, Chen WL, Choi J, Wakita T, Yen TS, Ou JH (2008) Induction of incomplete autophagic response by hepatitis C virus via the unfolded protein response. Hepatology 48(4):1054–1061. doi:10.1002/hep.22464

Sir D, Tian Y, Chen WL, Ann DK, Yen TS, Ou JH (2010) The early autophagic pathway is activated by hepatitis B virus and required for viral DNA replication. Proc Natl Acad Sci U S A 107(9):4383–4388. doi:10.1073/pnas.0911373107

Song J, Qu Z, Guo X, Zhao Q, Zhao X, Gao L et al (2009) Hypoxia-induced autophagy contributes to the chemoresistance of hepatocellular carcinoma cells. Autophagy 5(8):1131–1144

Takamura A, Komatsu M, Hara T, Sakamoto A, Kishi C, Waguri S et al (2011) Autophagy-deficient mice develop multiple liver tumors. Genes Dev 25(8):795–800. doi:10.1101/gad.2016211

Tian Y, Sir D, Kuo CF, Ann DK, Ou JH (2011) Autophagy required for hepatitis B virus replication in transgenic mice. J Virol 85(24):13453–13456. doi:10.1128/JVI.06064-11

Wang N, Feng Y, Zhu M, Tsang CM, Man K, Tong Y, Tsao SW (2010a) Berberine induces autophagic cell death and mitochondrial apoptosis in liver cancer cells: the cellular mechanism. J Cell Biochem 111(6):1426–1436. doi:10.1002/jcb.22869

Wang Y, Singh R, Xiang Y, Czaja MJ (2010b) Macroautophagy and chaperone-mediated autophagy are required for hepatocyte resistance to oxidant stress. Hepatology 52(1):266–277. doi:10.1002/hep.23645

Wang N, Pan W, Zhu M, Zhang M, Hao X, Liang G, Feng Y (2011) Fangchinoline induces autophagic cell death via p53/sestrin2/AMPK signalling in human hepatocellular carcinoma cells. Br J Pharmacol 164(2b):731–742. doi:10.1111/j.1476-5381.2011.01349.x

Wu WK, Coffelt SB, Cho CH, Wang XJ, Lee CW, Chan FK et al (2012) The autophagic paradox in cancer therapy. Oncogene 31(8):939–953. doi:10.1038/onc.2011.295

Xie SQ, Li Q, Zhang YH, Wang JH, Mei ZH, Zhao J, Wang CJ (2011) NPC-16, a novel naphthalimide-polyamine conjugate, induced apoptosis and autophagy in human hepatoma HepG2 cells and Bel-7402 cells. Apoptosis 16(1):27–34. doi:10.1007/s10495-010-0537-1

Xu N, Zhang J, Shen C, Luo Y, Xia L, Xue F, Xia Q (2012) Cisplatin-induced downregulation of miR-199a-5p increases drug resistance by activating autophagy in HCC cell. Biochem Biophys Res Commun 423(4):826–831. doi:10.1016/j.bbrc.2012.06.048

Yu HC, Hou DR, Liu CY, Lin CS, Shiau CW, Cheng AL, Chen KF (2013) Cancerous inhibitor of protein phosphatase 2A mediates bortezomib-induced autophagy in hepatocellular carcinoma independent of proteasome. PLoS One 8(2):e55705. doi:10.1371/journal.pone.0055705

Yu R, Zhang ZQ, Wang B, Jiang HX, Cheng L, Shen LM (2014) Berberine-induced apoptotic and autophagic death of HepG2 cells requires AMPK activation. Cancer Cell Int 14:49. doi:10.1186/1475-2867-14-49

Yue Z, Jin S, Yang C, Levine AJ, Heintz N (2003) Beclin 1, an autophagy gene essential for early embryonic development, is a haploinsufficient tumor suppressor. Proc Natl Acad Sci U S A 100(25):15077–15082. doi:10.1073/pnas.2436255100

Index

W.-X. Ding, X.-M. Yin (eds.), *Cellular Injury in Liver Diseases*, Cell Death in Biology and Diseases, DOI 10.1007/978-3-319-53774-0

X.-M. Yin and Z. Dong (Series eds.), *Cell Death in Biology and Diseases*

B

I

O

P

MIX
Papier aus verantwortungsvollen Quellen
Paper from responsible sources
FSC® C105338

If you have any concerns about our products,
you can contact us on
ProductSafety@springernature.com

In case Publisher is established outside the EU,
the EU authorized representative is:
Springer Nature Customer Service Center GmbH
Europaplatz 3, 69115 Heidelberg, Germany

Printed by Libri Plureos GmbH
in Hamburg, Germany